P9-EKS-699

DETERGENCY

THEORY AND TEST METHODS

SURFACTANT SCIENCE SERIES

Volume 1: NONIONIC SURFACTANTS, edited by Martin J. Schick

Volume 2: SOLVENT PROPERTIES OF SURFACTANT SOLUTIONS, edited by Kozo Shinoda *(out of print)*

Volume 3: SURFACTANT BIODEGRADATION, by Robert D. Swisher *(out of print)*

Volume 4: CATIONIC SURFACTANTS, edited by Eric Jungermann

Volume 5: DETERGENCY: THEORY AND TEST METHODS *(in three parts)*, edited by W. G. Cutler and R. C. Davis

Volume 6: EMULSIONS AND EMULSION TECHNOLOGY *(in two parts)*, edited by Kenneth J. Lissant

Volume 7: ANIONIC SURFACTANTS *(in two parts)*, edited by Warner M. Linfield

Volume 8: ANIONIC SURFACTANTS—CHEMICAL ANALYSIS, edited by John Cross

Volume 9: STABILIZATION OF COLLOIDAL DISPERSIONS BY POLYMER ADSORPTION, by Tatsuo Sato and Richard Ruch

Volume 10: ANIONIC SURFACTANTS—BIOCHEMISTRY, TOXICOLOGY, DERMATOLOGY, edited by Christian Gloxhuber

Volume 11: ANIONIC SURFACTANTS—PHYSICAL CHEMISTRY OF SURFACTANT ACTION, edited by E. H. Lucassen-Reynders

OTHER VOLUMES IN PREPARATION

DETERGENCY

THEORY AND TEST METHODS

(in three parts)

PART III

edited by W. G. CUTLER and R. C. Davis

Whirlpool Corporation
Research and Engineering Center
Benton Harbor, Michigan

MARCEL DEKKER, INC. New York and Basel

Library of Congress Cataloging in Publication Data
Main entry under title:

Detergency; theory and test methods.

(Surfactant science series, v. 5)
Includes bibliographical references.
1. Cleaning compounds--Testing. I. Cutler, W. Gale, [date] ed. II. Davis, Richard C., ed.
TP990.D39 668'14'0287 79-163921
ISBN 0-8247-6982-1 (v. 5)

MARCEL DEKKER, INC.
270 Madison Avenue, New York, New York 10016

Current printing (last digit):
10 9 8 7 6 5 4 3 2 1

PRINTED IN THE UNITED STATES OF AMERICA

PREFACE

The detergent process is very widely used today. Despite its acceptance and wide usage, it is a process not completely defined and understood. There is no general agreement as to test methods for measuring the effectiveness of the detergent process. This volume of the Surfactant Science Series is intended to provide to both industrial and academic research workers a compilation of test methods in use today and background information as to the theoretical basis of these tests. Perhaps such a volume will stimulate new interest in further investigation in the field of detergency.

The author list has been chosen to provide representatives from detergent and detergent-component producers, those specialized in applications of detergents, and universities. There has been no attempt to conform to a group presentation or to render a group opinion. Authors have been left free, within the constraint of a general outline, to present and interpret data from their viewpoint and experience.

For the convenience of the reader and to expedite publication, this volume of the Surfactant Science Series is being presented in three parts. Part I treats the fundamentals of the soil removal process, soil redeposition, and test methods for soil removal and soil redeposition. Part I is concerned with the soil removal process primarily as it applies to fibers.

Part II consists of topics related to a more complete understanding of the detergency process. The emphasis on the detergency process as related to fibers continues with such topics as the rinsing process, bleaching, physical damage, the action of enzymes, and ancillary tests. Topics of considerable current interest are sequestration and test methods in toxicology and dermatology. Departing from the fiber emphasis, a chapter summarizing the cleaning of metals is provided.

The current part, Part III, completes this volume of the Surfactant Science Series. It contains a detailed explanation of fluorescent whitening agents with appropriate test methods, an examination of the relation of detergents to the environment, and a treatment of the topic of dishwashing as an example of the role of detergents in hard surface cleaning. The preparation and publication of this volume has spanned several years. These years have seen major changes in detergent formulation and performance. A chapter discussing these changes is provided as the final chapter of Part

III. Part III also contains the indexes to the entire volume. These indexes permit the reader to examine contributions to the theory of detergency in the scientific literature and serve to identify those researchers who have made contributions to this theory.

A number of people have contributed to Part III. The editors wish to thank the contributing authors and their companies. The support of the Whirlpool Corporation in permitting the editors to undertake the task of compiling this volume is also acknowledged. Special thanks are also due to Dr. Martin Schick for his many suggestions and critical review of the book outline and to Mrs. Carol Hauch for the secretarial assistance so necessary to this undertaking. The editors also acknowledge the help of Mrs. Barbara Johns and Mrs. Jean Prillwitz in manuscript preparation, of Mrs. Lucille Kulin in compilation of reference lists and of Associate Professor Emeritus Hugh Ackert, Notre Dame University, in preparation of illustrations.

Benton Harbor, Michigan

W. G. Cutler
R. C. Davis

CONTRIBUTORS TO PART III

JUDITH B. CARBERRY, Department of Civil Engineering, University of Delaware, Newark, Delaware

WILLIAM G. MIZUNO, Economics Laboratory, Inc., St. Paul, Minnesota

BEVERLY J. RUTKOWSKI, Whirlpool Corporation, Benton Harbor, Michigan

PER S. STENSBY, Ciba-Geigy Corporation, Greensboro, North Carolina

MARK W. TENNEY, TenEch Environmental Consultants, Inc., South Bend, Indiana

CONTENTS

Preface		iii
Contributors to Part III		v
Contents of Parts I and II		ix
20.	FLUORESCENT WHITENING AGENTS Per S. Stensby	729
I.	Introduction	730
II.	Chemistry, Properties and Analysis	730
III.	Role as Detergent Additive	749
IV.	Evaluation of Whiteness	786
V.	Product Safety	805
	References	806
21.	DISHWASHING William G. Mizuno	815
I.	Introduction and History	816
II.	Theory of Dishwashing	825
III.	Detergent Evaluation	838
IV.	Functions and Properties of Detergent Components	848
V.	Rinsing and Rinse Additives	885
VI.	Summary	888
	References	889
22.	DETERGENTS AND OUR ENVIRONMENT Mark W. Tenney and Judith B. Carberry	899
I.	Introduction	900
II.	Environmental Aspects	903
III.	Pollutionary Characteristics of Synthetic Detergents	923
IV.	Wastewater Treatment	940
V.	Summary	980
	Glossary	986
	References	990

23. RECENT CHANGES IN LAUNDRY DETERGENTS 993
Beverly J. Rutkowski

I. Introduction 994
II. Nonphosphate Detergent Builders 998
III. Surfactant Changes in Laundry Detergents 1006
IV. Performance Characteristics of Phosphate-Restricted Detergents 1009
V. Detergents of the Future 1017
References 1018

AUTHOR INDEX (Parts I, II, III) 1021

SUBJECT INDEX (Parts I, II, III) 1051

CONTENTS OF PART I

INTRODUCTION, W. G. Cutler

DEFINITION OF TERMS, Oscar W. Neiditch

LAUNDRY SOILS, William C. Powe

THEORIES OF PARTICULATE SOIL ADHERENCE AND REMOVAL, W. G. Cutler, R. C. Davis, and H. Lange

REMOVAL OF ORGANIC SOIL FROM FIBROUS SUBSTRATES, Hans Schott

REMOVAL OF PARTICULATE SOIL, Hans Schott

ROLE OF MECHANICAL ACTION IN THE SOIL REMOVAL PROCESS, B. A. Short

SOIL REDEPOSITION, Richard C. Davis

EVALUATION METHODS FOR SOIL REMOVAL AND SOIL REDEPOSITION, J. J. Cramer

TEST EQUIPMENT, W. G. Spangler

CONTENTS OF PART II

SEQUESTRATION, William W. Morgenthaler

EVALUATION OF THE RINSING PROCESS, Walter J. Marple

BLEACHING AND STAIN REMOVAL, Charles P. McClain

ASSESSMENT OF PHYSICAL DAMAGE TO A TEXTILE SUBSTRATE DURING LAUNDERING AND DRY CLEANING, J. A. Dayvault

ANCILLARY TESTS, Paul Sosis

ENZYMES, Theodore Cayle

TEST METHODS IN TOXICOLOGY, Leonard J. Vinson

TEST METHODS IN DERMATOLOGY, Christian Gloxhuber and K. H. Schulz

CLEANING OF METALS, E. B. Saubestre

Chapter 20

FLUORESCENT WHITENING AGENTS

Per S. Stensby

Ciba-Geigy Corporation
Greensboro, North Carolina

I. INTRODUCTION 730

II. CHEMISTRY, PROPERTIES, AND ANALYSIS 730

A. Classification 730
B. Purity and Physical Form 738
C. Optical Properties 738
D. Identification 743
E. Activity 747

III. ROLE AS DETERGENT ADDITIVE 749

A. Reason for Use 749
B. Selection 750
C. Incorporation and Detergent Whitening 753
D. Whiteners in the Wash Liquor 753
E. Whitener on Fabric 784

IV. EVALUATION OF WHITENESS 786

A. Physical Factors 787
B. Physiological/Psychological Factors 792
C. Visual Evaluation 796
D. Instrumental Evaluation 797

V. PRODUCT SAFETY 805

REFERENCES 806

I. INTRODUCTION

Fluorescent whitening agents* (FWAs)—also named fluorescent brightening agents, optical brighteners, optical brightening agents, optical bleaches, optical dyes, etc.—have been used by the detergent industry since 1940 [1]. About 1929 [2], it was discovered that certain organic chemicals with a conjugated double-bond system could whiten highly reflectant organic textile and paper substrates. By 1970, FWAs were counted among the most important components of modern laundry products. Laundering with whitener-free detergents generally restores, but does not improve, the initial whiteness of soiled white fabrics. Fluorescent whiteners are needed to increase the total spectral radiance, which normally results in better whiteness and an impression of superior cleanliness [3]. Practically all of the 4.5 billion pounds of home laundry products formulated during 1970 in the United States contained FWAs. In 1970, FWA consumption in the United States (see Fig. 1) was approximately 21 million pounds of full strength materials at a value of $36.6MM. This poundage was equal to around half of the worldwide usage by the laundry industry. However, in the 1970s, more effective FWAs became available and, for economic reasons and also because of the decline of presoaks, the consumption in the United States became somewhat less, declining to approximately 18 million pounds per year and subsequently to approximately 15 million pounds per year.

II. CHEMISTRY, PROPERTIES, AND ANALYSIS

A. Classification

Fewer than 10 chemically different FWAs have been used to any significant extent in soaps, detergents, and other laundry products in the 1970s. The whiteners used are selected from among more than 200 chemically different compounds commercially available around the world under more than 1000 different names and designations. These commercially offered whiteners represent only a minute portion of the tens of thousands of different FWA structures covered by the patent literature. Most FWA patents are published in Holland, Britain, France, Belgium, Germany, Switzerland, Japan, Canada, and the U.S. Representative of these are Refs. 4-19.

The general chemistry, related properties, and field of application of the important FWA types have been discussed in the chemical and trade

*Name established by ASTM and generally adopted.

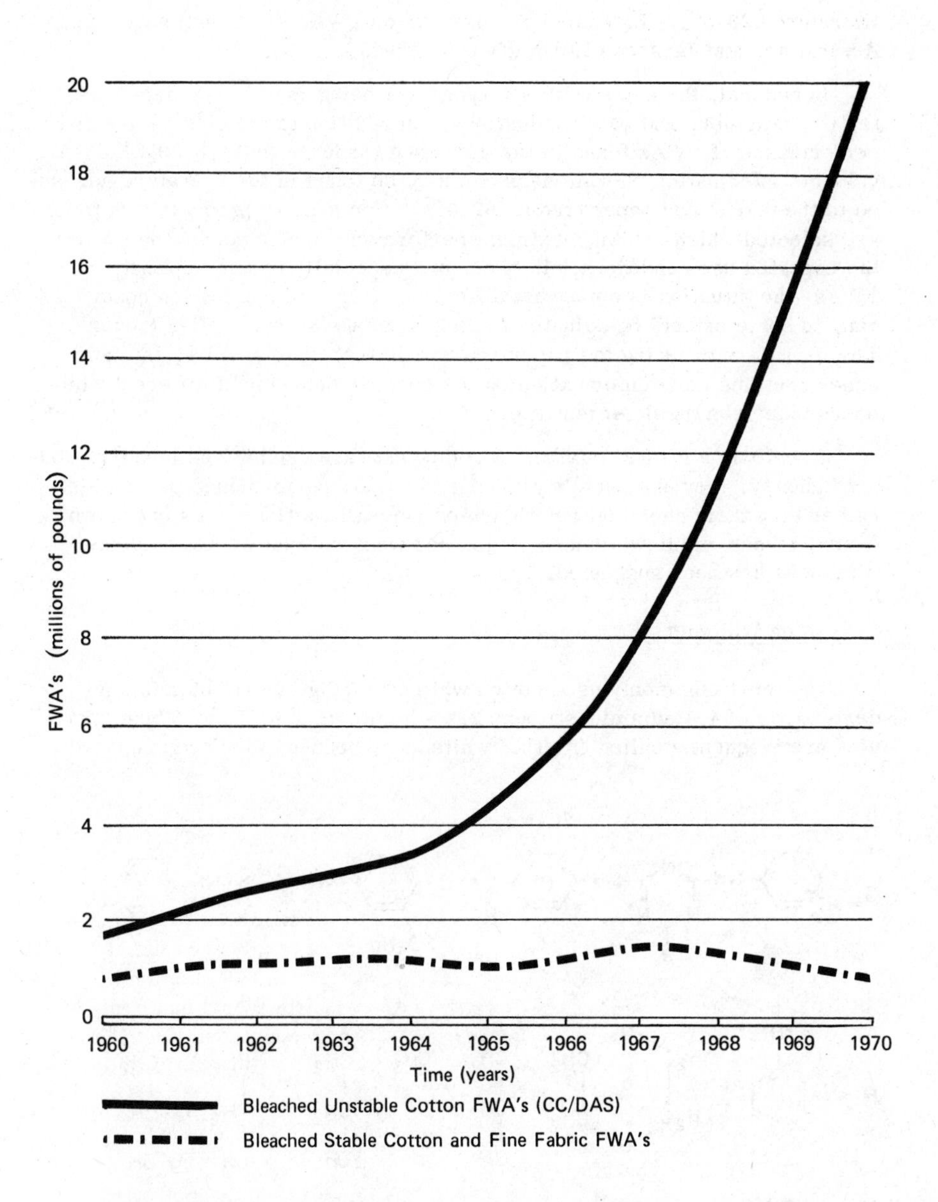

FIG. 1. FWA consumption in U. S. home laundry detergents from 1960 to 1970.

literature [20-32]. Extensive discussions on FWAs are found in European Australian, and Japanese literature [20, 33-45].

In general, the same whitener types are being used by the detergent, textile, polymer, and paper industries. In addition to articles on use and performance of FWAs found in the detergent trade journals [3, 23, 24, 46-51], valuable information on whiteners can also be found in the literature adressed to the textile and paper trades [52-61]. For laundry products, whiteners are selected which exhibit maximum performance under conditions close to the expected use conditions. In paper and especially textile applications of FWAs, the situation is somewhat different. Here the application conditions can, to some extent, be adjusted to maximize a whitener's effectiveness. The literature covering the paper and textile trades, therefore, in many cases contains more information on the specific behavior of different whiteners than does the detergent literature.

Since only a limited number of compounds are used by the laundry product industry, they are usually classified by their performance properties rather than their chemical relationships [3]. Classification as cotton whiteners, bleach stable whiteners, nylon and wool whiteners, and polyester whiteners has been suggested.

1. Cotton Whiteners

The most commonly used cotton whiteners (Fig. 2) are bistriazinyl derivatives of 4,4'-diaminostilbene 2,2'-disulfonic acid [62]. These products are frequently called CC/DAS whiteners, because they are prepared

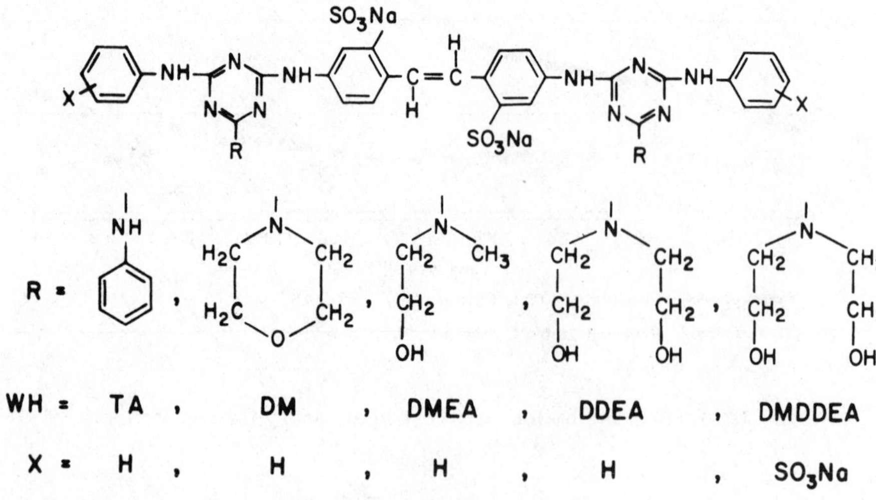

WH = TA , DM , DMEA , DDEA , DMDDEA

X = H , H , H , H , SO_3Na

FIG. 2. CC/DAS whiteners.

from the reaction product of 2 mol of cyanuric chloride (CC) with 1 mol of the disodium salt of diaminostilbene disulfonic acid (DAS). These cotton whiteners, with the exception of whitener DMDDEA, are reaction products of 1 mol of CC/DAS with 2 mol of aniline (whitener DMDDEA is the reaction product of 1 mol CC/DAS and 2 mol of sulfanilic or metanilic acid). The first four whiteners (TA, DM, DMEA, and DDEA) differ only in the amine used for reaction with the two remaining reactive chlorine groups on the triazine rings: for whitener TA, 2 mol of aniline; for whitener DM, 2 mol of morpholine; for whitener DMEA, 2 mol of N-methyl-ethanolamine; and for whitener DDEA and whitener DMDDEA, 2 mol of diethanolamine.

2. Bleach-Stable Whiteners

Typical structures of conventional whiteners stable in the presence of chlorine bleach in the wash liquor appear in Fig. 3. These include benzidine sulfone disulfonic acid (whitener BS), naphthotriazolyl-stilbene sulfonic acid (whitener NTS, R = H), and benzimidazolyl derivatives (whitener BBI). Whitener BS builds up only to a limited degree on cotton and does not have affinity for nylon. Both whiteners NTS and BBI whiten cotton and nylon. Only whitener NTS is used in the U. S. to any significant extent.

The classification of whiteners BS, NTS, and BBI as bleach-stable products and of whiteners TA, DM, DMEA, and DDEA as bleach unstable products is somewhat arbitrary. No commonly accepted definition or standard test procedure classifying whiteners in regard to their bleach stability exists at the present time. There is even a difference in bleach stability among the CC/DAS whiteners, as indicated in Table 1 and discussed in more detail by Villaume [23] and Stensby [3]. In general, whiteners with primary

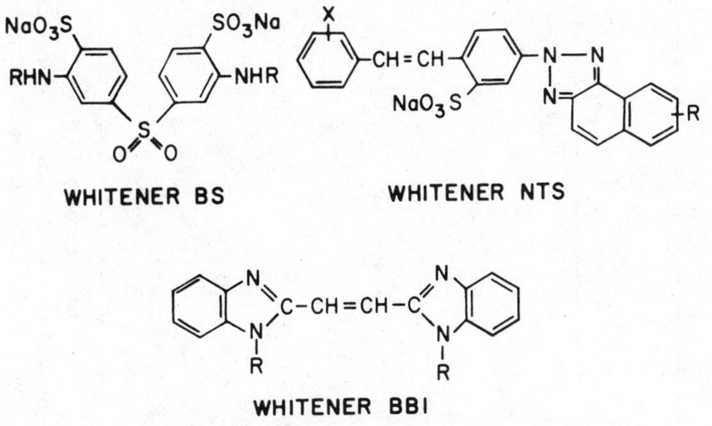

NaO3S SO3Na RHN NHR S O O

WHITENER BS

X CH=CH N N N NaO3S R

WHITENER NTS

N N C-CH=CH-C N N R R

WHITENER BBI

FIG. 3. Bleach-stable whiteners.

TABLE 1

General Properties of CC/DAS Whiteners

Whitener	Relative[a] solubility	Fiber affinity	Product[c] whitening	Relative[c] bleach stability	Exhaust at pH 4-5 (sour)	Recommended for: System	Recommended for: Temp. range
TA	Low	Cellulosics, nylon	Good	Medium	Not suited	Anionics	Above 110° F
DM	Low-medium	Cellulosics	Good	Good	Not suited	Anionics, nonionics	All temps., best above 110° F
DMEA	Medium-high	Cellulosics	Good - medium	Good	Too fast, unlevel	Anionics, nonionics, liquids	All temps.
DDEA	High[b]	Cellulosics	Good - medium	Good - medium	Too fast, unlevel	Anionics, nonionics, cationics, liquids	All temps.
DMDDEA	High	Cellulosics	—	—	Good	Acid - sour	All temps.

[a] In aqueous systems.
[b] As salt prepared from acid-form, free of inorganic electrolyte.
[c] Depends on purity, physical form, and detergent base.

amino substituents on both triazine rings (e.g., the tetraamino derivatives) are less stable than those with secondary substituents (e.g., the tetraanilino derivative). Whiteners of the CC/DAS class with tertiary amino substituents (e.g., the tetramorpholino derivative) exhibit still higher resistance to chlorine bleach. However, a change of a substituent on the triazine ring, which might improve bleach resistance, can lower affinity of the compound for cotton fabrics. Consequently, a more bleach-resistant whitener can exhibit inferior performance relative to less bleach-resistant FWA in the presence of hypochlorite bleach.

Still better bleach resistance than that of the CC/DAS type compounds with tertiary amino substituents on the triazine ring can be obtained by elimination of the secondary amine groups (between stilbene and the triazine ring). Whitener NTS is an example of such a compound. Whiteners like BBI, BBO, and NOS (see polyester whiteners for the latter two), which are practically water insoluble, also exhibit good bleach stability. Oxidation of these whiteners by sodium hypochlorite (NaOCl) is more difficult because the whiteners are not significantly dissolved in the aqueous phase and are thus not as susceptible to attack by the bleach. The latter three hydrophobic whiteners are, however, unsuited for whitening of cotton.

While whitener NTS is bleach stable, it does not yield as high whiteness levels in the presence or absence of hypochlorite in the washbath as does the cotton whitener DMEA or DM. A need consequently existed for a bleach-stable cotton whitener with improved buildup characteristics equal to the best CC/DAS type in the absence of bleach [50]. One such FWA (whitener DSBP) has been used for several years outside the U.S. and is now also being employed in domestic detergents. The structure of this new class of FWA is shown in Fig. 4. Products of the distyrylbisphenyl series may become the successors of the CC/DAS compounds in many instances.

3. Polyamide Whiteners

The general structures of typical nylon and wool whiteners for fine-fabric detergents are illustrated in Fig. 5. These are amino coumarin (whitener AC) and diphenyl pyrazoline (whitener DP) derivatives, neither of which are stable to chlorine bleach. The pyrazoline types are preferred to the amino coumarin types in Europe because they exhibit a more attractive

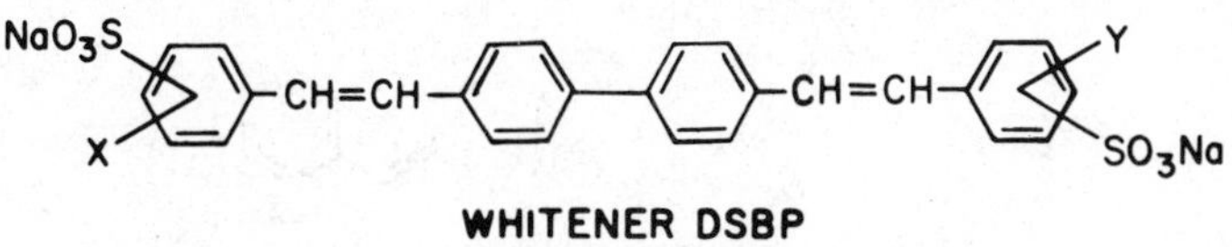

WHITENER DSBP

FIG. 4. Bleach-stable cotton whitener.

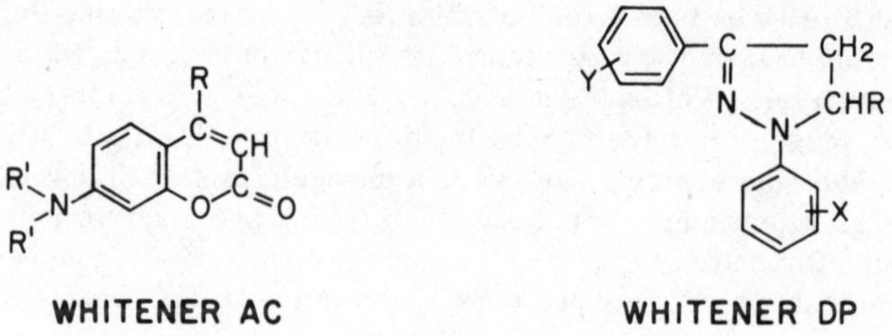

WHITENER AC

WHITENER DP

FIG. 5. Polyamide whiteners.

hue and better fastness to light. Some of the pyrazoline derivatives can discolor polyester, however, through staining.

4. Polyester Whiteners

The general structures of polyester shiteners which also have affinity for polyamide fibers and exhibit good bleach stability are shown in Fig. 6. In this group belong bis-benzoxazolyl (whitener BBO) and naphthoxazolyl (whitener NOS) derivatives. Under typical U. S. home laundry conditions, these whiteners exhibit only limited effectiveness on mass prewhitened polyester [47].

5. Summary of General Properties

The effect of the different amines (and of the two additional sulfonate groups in the case of whitener DMDDEA) on the performance properties of

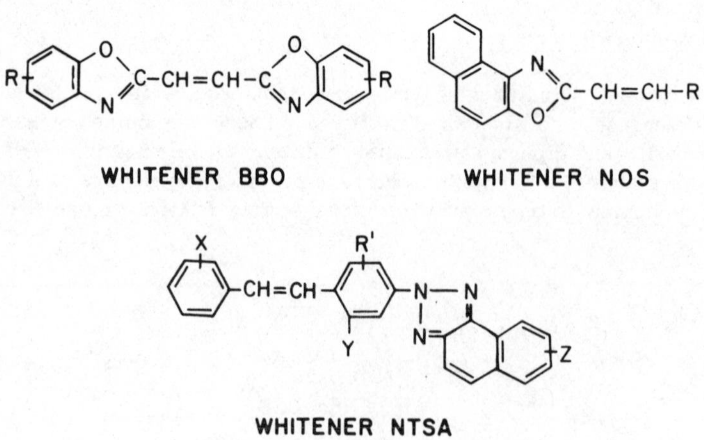

WHITENER BBO

WHITENER NOS

WHITENER NTSA

FIG. 6. Polyester and polyamide whiteners.

CC/DAS whiteners are presented in Table 1. These properties, including relative solubility, fiber affinity, laundry product whitening properties, bleach stability, performance in different surfactant systems, temperature dependency, and leveling properties, etc., together with other factors, determine which cotton whitener is recommended for use in a specific laundry product. The most important properties of bleach-stable, fine-fabric, and product whiteners are listed in Table 2. Additional information on the performance and properties of the whitener types discussed here is available in technical bulletins supplied by the whitener manufacturers and also in the trade literature.

TABLE 2

General Properties of Bleach-Stable, Fine-Fabric, and Product Whiteners

Whitener	Relative[a] solubility	Nylon affinity	Polyester affinity	Bleach stability	Used especially for
BS	Low	No	No	Good	Bleach stability, cotton
NTS (R = H)	Low	Good	Low	Good	Bleach stability, cotton, nylon
DSBP	High	No	No	Good	Bleach stability, cotton, liquids, cationics
NTS (R = SO_3Na)	High	No	No	Good	Product whitening (soap bars etc.)
BBI	No	Good	No	Good	Bleach stability, cotton, nylon
AC	No	Good	No	No	Nylon, wool
DP	Low	Good	No	No	Nylon, wool
BBO	No	Good	Medium	Good	Polyester, nylon
BOS	No	Good	Medium	Good	Polyester, nylon
NTSA	No	Good	Medium	Good	Nylon, polyester (nonionics)

[a]In aqueous systems.

B. Purity and Physical Form

The chemical nature of a whitener is only partly responsible for its performance characteristics. Other important factors are purity and physical form.

1. Purity

Purity of whiteners differs, depending on the manufacturer, though today's products are generally far superior to those produced a decade ago. The high whitener levels currently used would not have been feasible during the late 1950s; buildup by impurities then present in the products caused discoloration of both detergent and fabric.

2. Physical Form

The physical form (crystal form) of the whitener, as well as the incorporation of techniques employed by the detergent formulators, can affect (1) whiteness (or color) of the laundry product itself; (2) the time required for the whitener to go into complete solution in the wash liquor; and (3) the whitener's performance in the presence of hypochlorite bleach. Performance differences in the wash liquor caused by differences in crystal form between two whiteners can, if required, be eliminated by predissolving the whitener prior to or during incorporation into the finished product. On the other hand, change in crystal habit might result in detergent powder discoloration.

C. Optical Properties

1. Absorption Spectra

The ultraviolet absorption spectra of the mentioned CC/DAS whiteners peak in the 350-nm region. Spectra of the tetraanilino (whitener TA) derivatives, the dianilino di-diethanolamino (whitener DDEA) derivatives, as well as that of the tetradiethanolamino CC/DAS (whitener TDEA) derivatives appear in the left-hand part of Fig. 7. The 350-nm absorption is related to the conjugated [52] bonding system of the planar trans-stilbene molecule. At equimolar concentrations in solution, the three compounds absorb light to approximately the same extent in the 350-nm region. The absorption in the 270-nm region is characteristic of the phenyl-amino groups. Whitener TA, with four phenylamino groups, absorbs more in the 270-nm region than does whitener DDEA with two phenyl-amino groups. The trans-tetradiethanolamino derivative without phenylamino groups does not absorb significantly at 270 nm.

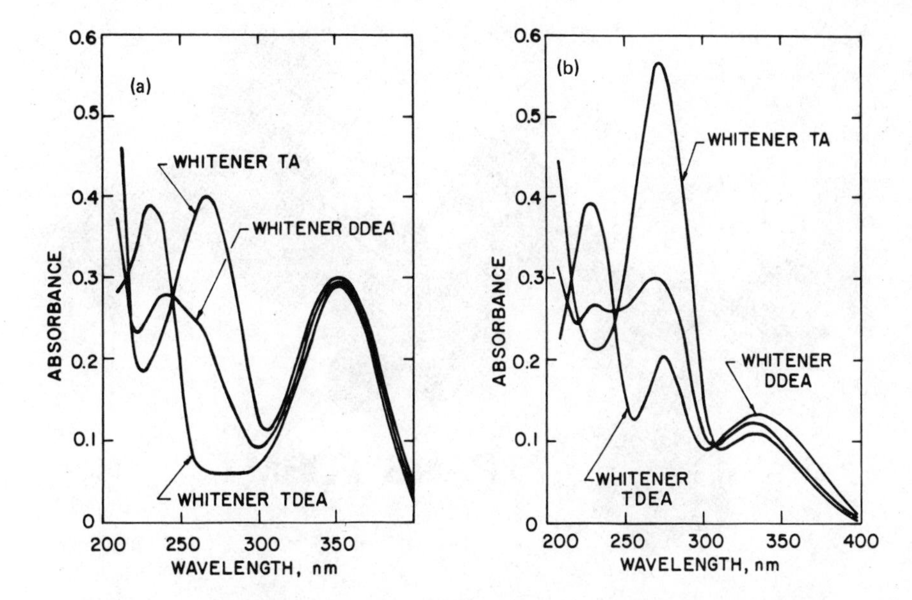

FIG. 7. Ultraviolet spectra of CC/DAS derivatives (a) before light exposure and (b) after exposure to 100 Langleys. From Ref. 3.

On exposure of the whitener solutions to ultraviolet light (100 Langleys of sunlight), isomerization [35, 63-65] occurs from the relatively stable trans form to the somewhat less stable cis form (Fig. 8). The trans-stilbene double bond is converted into radicals on exposure to ultraviolet light. In this state, rotation of the stilbene phenyl rings occurs, permitting formation of the cis form, which is not planar. Because of steric hindrance between the hydrogen atoms in the ortho position, the resonance is reduced. In the nonplanar cis form, the π-electron clouds cannot overlap and the conjugation is destroyed. In the cis form, therefore, the two phenylamino groups of the diamino stilbene system will absorb in the 270-nm region, resulting in stronger total absorption in this region by the cis form than by the trans form. The cis form without the conjugated stilbene system will not absorb at 350 nm and, consequently, lacks the fluorescence exhibited by the trans-stilbene form. Absorption of whitener TA, whitener DDEA, and whitener TDEA after exposure to 100 Langleys of sunlight exposure is presented in the right-hand part of Fig. 7. The figure demonstrates that the exposed solutions do indeed have a stronger absorption in the 270-nm range and a weaker absorption in the 350-nm range than do the unexposed whitener solutions. Ultraviolet exposure of aqueous solutions of the cis form will result in gradual extinction of the compound. On the other hand, the cis form, if isolated and dissolved in ethanol or if present on textile or paper substrates, will be converted back to the trans form on light exposure [64-65]. In other

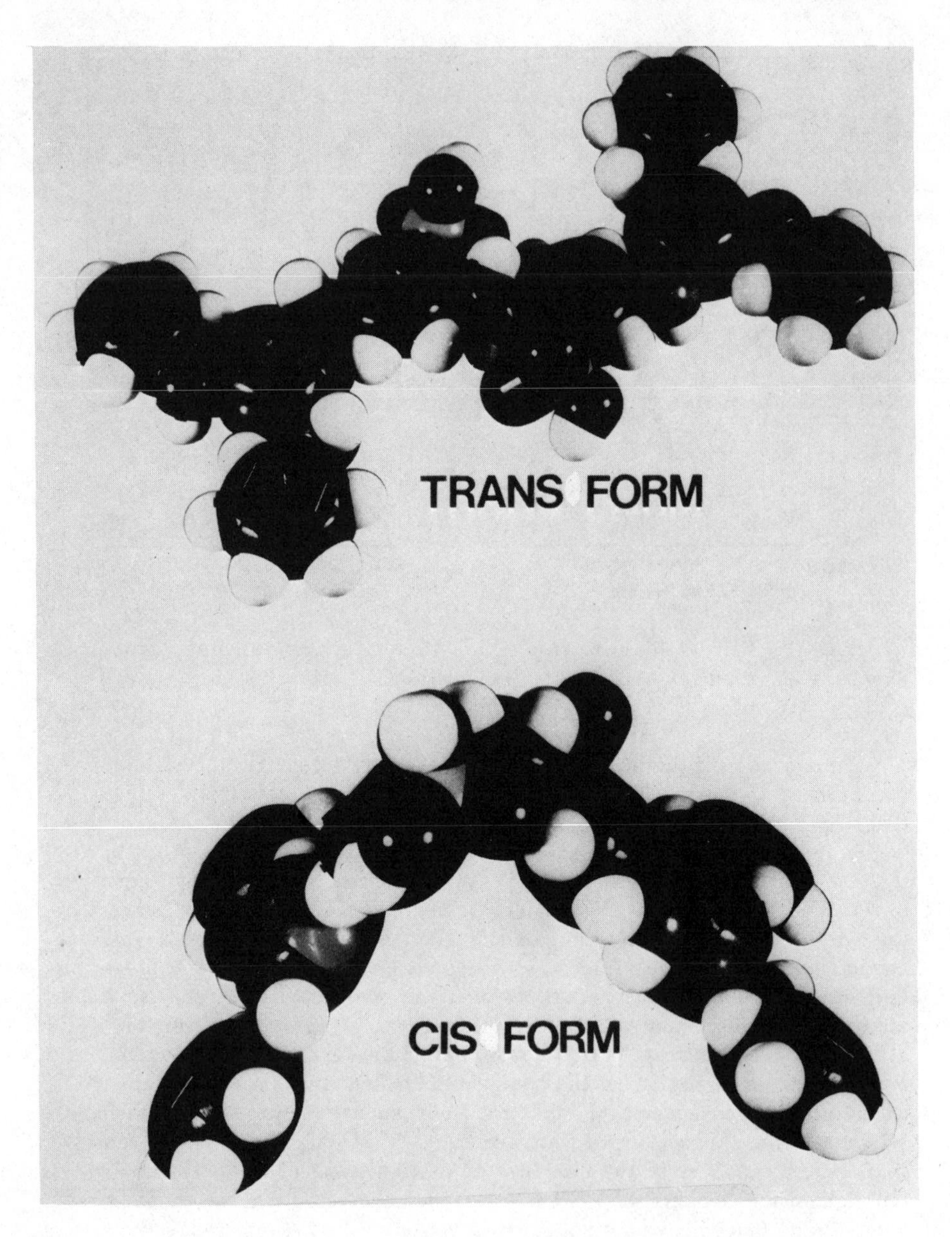

FIG. 8. Isomers of whitener TA. From Ref. 3.

words, the CC/DAS whitener, regardless of its isomeric form in the wash liquor, will be present as a whitener (trans form) on the fabric. The affinity for cellulosic fibers of the nonplanar cis form is very low compared to that of the planar trans form (see Sec. III. D. 2). Light exposure of whitener liquors, therefore, should be avoided.

2. Excitation and Emission Spectra

Absorption of light by a molecule results in conversion of radiant energy into energy of internal motion within the molecule. Such motions consist of rotation of the molecule as a whole, vibration of the atoms with respect to each other (bond stretching or bending), and change in motion of electrons. The energies of all these motions have their minimum values when the molecule is in its ground state. By absorption of energy, the molecule can be transferred to excited states. Whiteners, for example, can be excited by absorption of quanta of ultraviolet radiation.

A schematic diagram [66-68] of molecular energy levels appears in Fig. 9. The electronic excitation occurs from the lowest vibrational level of the ground state of a whitener. Not all the electrons are excited to the same energy level of the ground state. Radiation of different frequencies (different energies) are absorbed, thus explaining the absorption spectra measured for fluorescent whiteners. Part of the absorbed energy is lost by internal dissipation of heat, whereby the electrons fall back to the lowest energy level of the first excited state. From this first excited state the electrons return to the different energy levels of the ground state, producing broad-band emission of light. Figure 9 indicates that (1) light of different energies is emitted; (2) the light emitted is of lower energy (longer wavelength) than the radiation absorbed; and (3) the relative energy distribution of the emitted light is independent of the wavelength of excitation (within the region that the whitener absorbs).

Spectrofluorometers (such as those supplied by Aminco, Baird-Atomic, Farrand, Perkin-Elmer, and Turner, etc.) can be used to record excitation and emission spectra of whiteners in solution and in solid substrates. Figure 10 shows the excitation and emission spectra of whiteners NTS (R = H) and NTS (R = SO_3Na) on cotton. Such spectra (of solutions or fabrics) can be used, in some cases, to identify whiteners of different chemical classes. In addition, the emission spectrum gives an indication of the hue exhibited by the whitener on the substrate. An emission spectrum peaking at 435 nm, as a general rule, is related to a neutral blue hue, a maximum at longer wavelengths generally is related to a violet blue hue. It must be realized, however, that not only is location of the maximum important but so also is the relative energy distribution of the emitted light. One must keep in mind that emission spectra represent physical measurements and are not directly

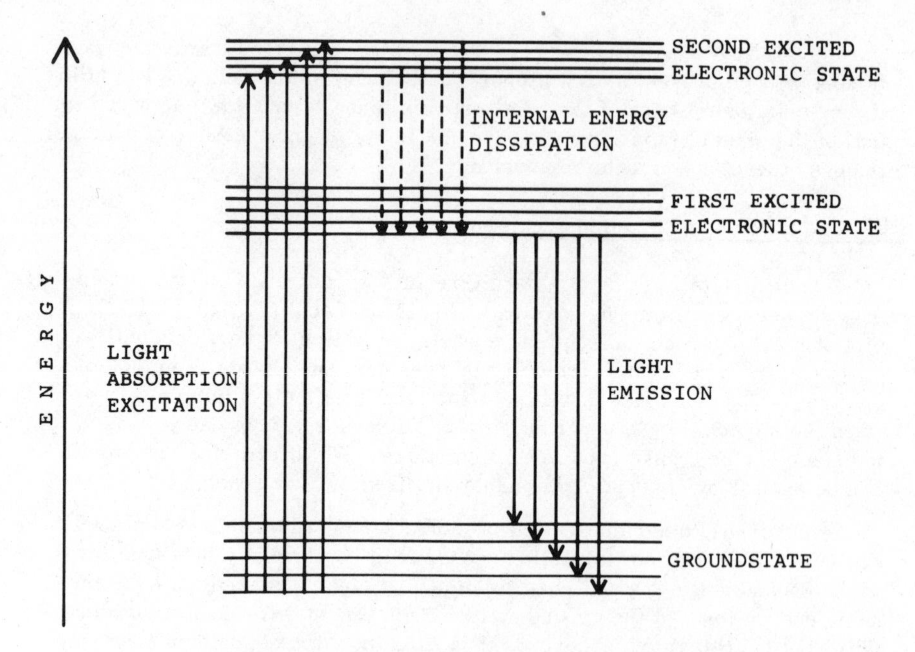

FIG. 9. Molecular energy levels. From Ref. 3.

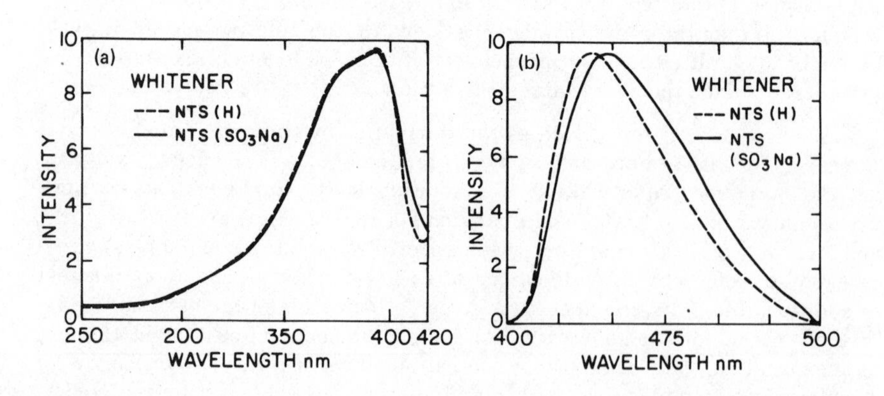

FIG. 10. (a) Excitation and (b) emission of spectra of whitener NTS (R = H) and NTS (R = SO_3Na). From Ref. 3.

related to psychophysical response and that these spectra depend on the substrate. In other words, the emission spectra of an FWA dissolved in water is different from that exhibited by the same whitener on a textile substrate.

Excitation spectra can, however, contribute important information regarding the properties of the whitener. The relative intensity of the most common light sources decreases rapidly between 400 nm and 300 nm. On the other hand, energy emitted by whiteners applied to fabrics at normal levels will increase with the amount of ultraviolet energy absorbed. Whiteners with strongest absorption close to (but not in) the visible region will consequently best utilize the available energy. The excitation spectra will furnish information on the absorbing properties of whiteners and give some indication as to their relative effectiveness.

D. Identification

Many of the whiteners currently offered under different trade names to the laundry products industry are chemically equivalent. These whiteners may contain different amounts of active ingredient and may also be offered in various forms (powder, paste, liquid, etc.). Before a laundry products manufacturer evaluates a new whitener, its chemical class and, if possible, exact chemical type must be determined. Although costly and time-consuming chemical analysis usually is the only way to identify precisely the structure of an unknown whitener, physical methods are available that in most cases will furnish sufficiently reliable information at a much lower cost. These physical identification methods can be used on the whitener itself, as well as to determine the whitener or whitener systems in a laundry product. The chemical type can be determined with a good degree of certainty by several of the following analytical tools: thin-layer chromatography, spot tests, wash tests, ultraviolet spectra, and fluorescence spectra.

1. Thin-Layer Chromatography

Thin-layer chromatography (TLC) is the most versatile and time-saving identification procedure currently available [35, 65, 69-74]. The separation by TLC of a mixture of whiteners TA, DM, and DDEA is shown in Fig. 11. Figure 12 shows how the different whiteners (BBO, NTS, and DM) in a typical heavy-duty detergent can be separated and identified. Reflectance (R_f) values of whiteners in a defined system are not necessarily absolute, repeatable numbers. Both known and unknown products, therefore, should be run simultaneously, separately, and in mixtures, as well as in different solvent systems, to increase the certainty of correct identification.

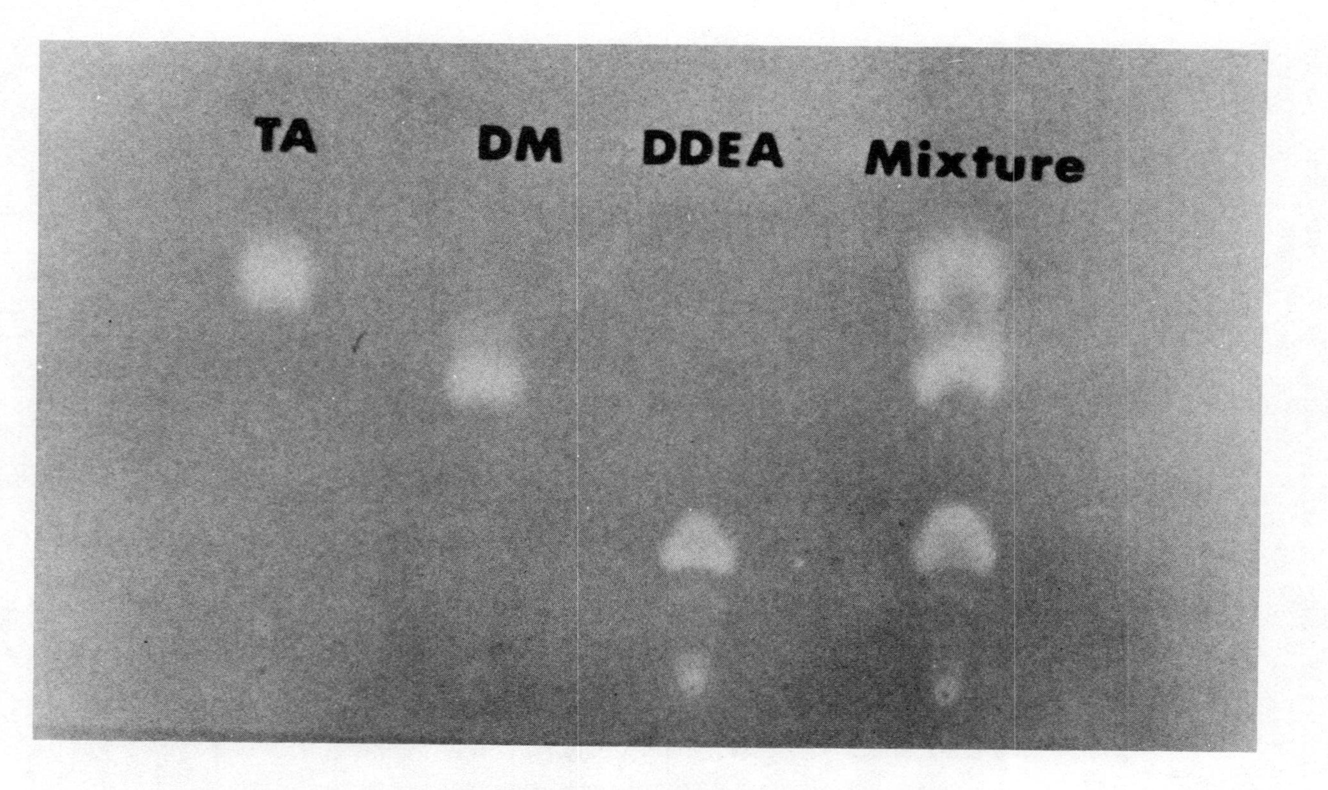

FIG. 11. Separation of whiteners TM, DM, and DDEA by TLC.

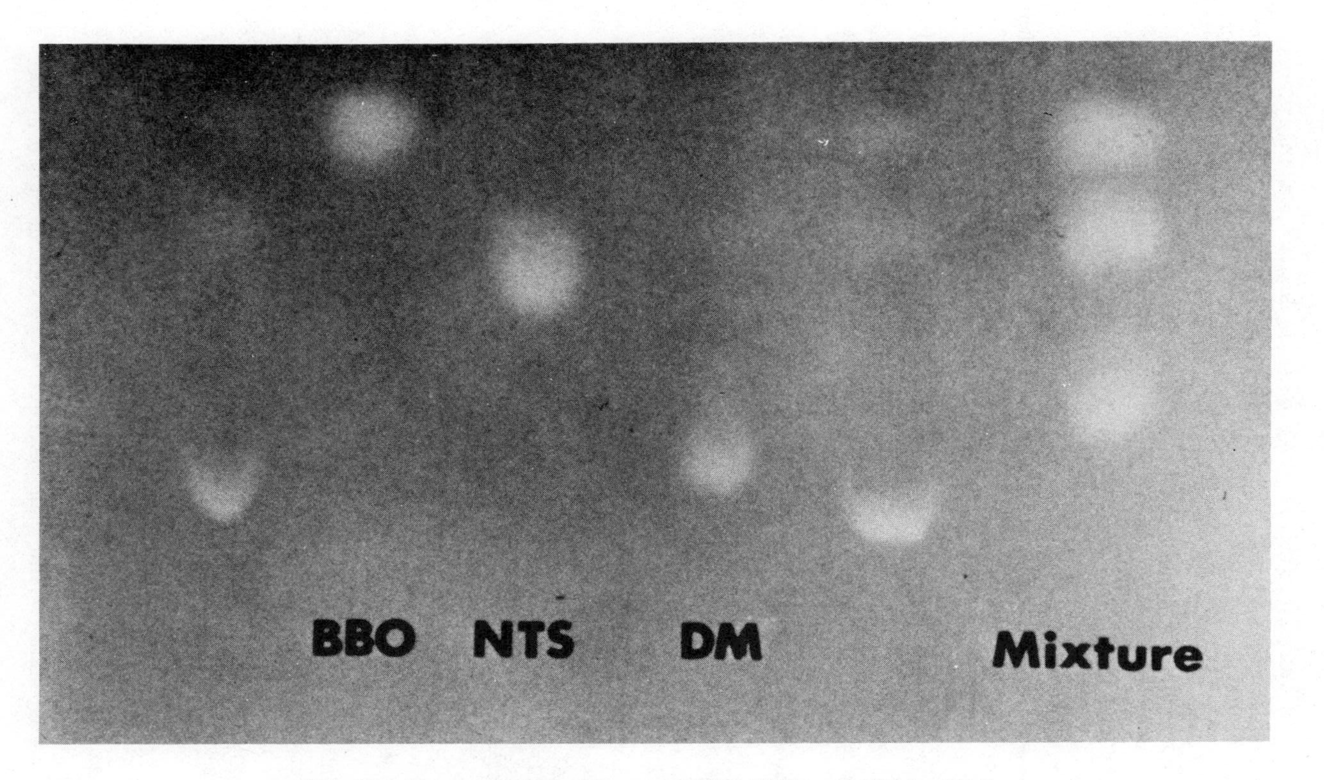

FIG. 12. Separation of whiteners BBO, NTS, and DM by TLC.

2. Spot Tests

Spotting of whitened swatches with, for example, 10% sodium hydroxide or 10% sulfuric acid solutions and observation of color reactions under black light can furnish information on whitener type. Such tests do not distinguish among different CC/DAS whiteners, but will differentiate between a CC/DAS derivative and most of the bleach-stable and fine-fabric whiteners currently in use. Cotton fabrics treated with CC/DAS-type whiteners produce a bright yellow when spotted with sodium hydroxide, but react only slightly with sulfuric acid. Cotton treated with whitener BBO turns pale orange in contact with the acid, while whitener BBI turns bright green on spotting with sodium hydroxide. Whitener NTS (R = H) produces a bright pink spot in contact with sulfuric acid; caustic soda turns whitener BS orange. The fluorescence of swatches treated with whitener AC will be destroyed by both sulfuric acid and caustic soda. No color reactions are produced by either acid or base when applied to cotton treated with whitener DSBP (X and Y = H). Because the color reaction changes with time, known and unknown whiteners should be compared side by side.

3. Wash Tests

An indication of whitener type can also be obtained by ovserving the whitener's performance in wash liquors or on substrates [75]. For example, whiteners can be identified by differences in: (1) bleach stability; (2) affinity to various fibers; (3) affinity to fibers as a function of bath pH; (4) affinity at various temperatures; (5) affinity as a function of the concentration of the inorganic electrolyte in the bath; and (6) fastness properties on substrates.

4. Ultraviolet Spectra

The relationship between the chemical structures and the ultraviolet absorption spectra of some CC/DAS whiteners has been discussed already. Ultraviolet spectra of whiteners in solution can provide identifying "fingerprints." In most cases, comparing the spectra between 200 nm and 400 nm of a known whitener with those of an unknown sample at approximately equal concentrations will give enough information, at least in combination with TLC, to characterize the sample. Comparing the spectra in different solvent systems and at various pH values will provide a double check if two whiteners appear to be equivalent.

5. Fluorescence Spectra

Excitation and emission spectra of whiteners in solution or on substrates will also furnish an indication of whitener-type [66, 76-77]. This technique is especially useful when a sample of the whitener itself is not available. For example, whiteners (contained in a detergent) with selective affinity for certain fibers can be applied to the respective substrates. Comparing excitation and emission spectra of known and unknown whiteners on fabrics helps to identify the unknown product. Similarly, whiteners applied to fibers in the mass can be identified in this manner. It is important, however, that the FWA concentrations in the known and unknown samples are comparable because the fluorescence spectra of whiteners change with concentration.

E. Activity

Strictly speaking, the active content of a whitener can be determined only by comparing a sample of unknown strength with one of identical chemical structure and known strength. A reference sample for each different chemical is frequently selected. The active-ingredient content of unknown materials is then compared with that of the reference.

Strength comparisons [64, 65, 78-80] usually are based on absorption measurements, on fluorescence measurements of highly dilute (1 ppm) solutions, and on fabric fluorescence. Determination of active-ingredient content by chemical analysis is less common.

1. Absorbance in Solution

The absorption of whitener solutions (2 to 10 ppm, depending on type and strength) in water, water/solvent mixtures, or solvent solutions prepared in the absence of ultraviolet light) is usually measured in the 340-nm to 370-nm wavelength range at the maximum. Measuring the absorption of whiteners at the wavelength of the isosbestic point has also been suggested [63]. The absorption of CC/DAS-type whiteners is independent of trans/cis concentration ratios at that wavelength. Absorption is only a measure of the concentration of material that absorbs radiation and does not indicate the concentration of fluorescent material present; nor does it indicate the concentration of absorbing, fluorescent material with substantivity for the substrate. Determination of the active content of a whitener by absorption, therefore, should be regarded only as a first approximation.

2. Fluorescence in Solution

The amount of fluorescent material in a whitener can be determined by comparing the fluorescence of a highly diluted solution (0.1 to 15 ppm, depending on the whitener) of a whitener of unknown strength with that of a reference sample. Instruments such as those supplied by Aminco, Coleman, Farrand, Photovolt, Turner, etc., can be used for this purpose. The intensity of the exciting light should be low because of the instability of dilute whitener solutions to ultraviolet light. The strength of two chemically different whiteners cannot be compared by this method, however, because of the exciting light source (frequently a mercury vapor lamp with a strong 365-nm band, combined with filter cutting off around 400 nm) and the photorecording system used may favor one whitener over another. The fluorescence values measured on a certain instrument depend on each whitener's excitation and emission characteristics in the solvent system used. Relative quantum yield of the whiteners in the selected solvent system is also a factor affecting relative fluorescence. The relationship between whitener concentration and relative fluorescence values is close to linear when highly diluted solutions are used. Standard concentration curves can be used to correct for nonlinearity if required. Fluorescence in solution gives an indication of the concentration of light-absorbing and fluorescing material but does not yield information on fabric substantivity of the whitener.

3. Fabric Fluorescence

To compare the relative substantivity of a sample of unknown strength with that of a reference sample of the same chemical structure, the whitener should be applied in a liquor to a fiber substrate for which the whitener has high substantivity. To assure complete solution and maximum performance, the whiteners should be added to the bath predissolved. Concentration of the laundry aid and whitener, temperature, liquor-to-fabric ratio, agitation, and length of wash cycle should approximate the expected conditions of use.

The reference sample is usually applied at standard strength; the sample of unknown activity is applied at concentrations somewhat lower, equal to, and slightly higher than that of the reference. The fabric fluorescence of washed, dried, and conditioned swatches are then compared to determine relative strength.

Comparisons of fabric fluorescence can be made instrumentally or visually. Several instruments (such as those supplied by Jarrell-Ash, Engineering Equipment, Photovolt, Beckman, Turner, and Galvanek-Morrison) are available for these measurements. If the fabric fluorescence values of several concentrations of the sample are plotted, the concentration required to match that of the reference sample can be determined; a strength determination can be made visually in a similar manner. Fabric fluorescence can be compared under black light and swatches of equal strength can be matched visually.

The relative strength of whiteners of different chemical structures cannot be compared properly by this method because of their different excitation and emission characteristics. The relative fluorescence recorded (gross fluorescence) represents only a physical measurement and is not related to a psychophysical expression of whiteness.

Naturally, the relative performance of two chemically different whiteners cannot be determined by absorption, fluorescence in solution, or fabric fluorescence obtained in wash tests using stock solutions. Likewise, the active content of chemically equivalent whiteners, as determined by the procedures discussed above, does not yield information on the relative performance properties of the two whiteners. As described in Sec. III, the performance value (or the technical value) of whiteners can only be derived from tests run under conditions closely simulating use conditions.

III. ROLE AS DETERGENT ADDITIVE

A. Reason for Use

Fabrics are laundered for reasons of health, hygiene, cleanliness, and appearance. The presence of light-absorbing soils is most easily observed on white and pastel-colored substrates because such surfaces have relatively high light reflection in the visible wavelength region. For this reason, the performance of a detergent is customarily evaluated on white fabrics. A detergent free of FWA can increase the reflectance of soiled substrate only to a level approximately equal to the presoiled level. Bleach in the laundry liquor can assist in soil removal and provide a higher-than-initial fabric reflectance.

Figure 13 shows the relative spectral radiance curves (W = whiteness), against $BaSO_4$, of soiled cotton (W = 54.6), soiled cotton washed once with a whitener-free detergent (W = 69.9), soiled cotton washed with a whitener-free detergent and hypochlorite bleach (W = 71.9), soiled cotton washed with a commercial detergent containing whitener (W = 93.0), and soiled cotton washed with a commercial detergent and bleach (W = 89.4). The whiteness values given were obtained from Hunter Colorimeter measurements by using the $W = L + 3a - 3b$ function (discussed below). The use of detergent and bleach results in a fabric with a higher reflectance than that of soiled cotton. The total light energy coming from the fabric is higher still when whitener is used. Repeated washings of soiled cotton with a whitener-free detergent and bleach would result eventually in a fabric reflectance similar to that of unsoiled cotton (W = 99.4).

Also shown in Fig. 13 are the contributions of whitener to the total spectral radiance of unsoiled cotton, after one wash with a commercial detergent alone (W = 134.6) and with a commercial detergent plus bleach (W = 119.6).

Theoretically, the maximum reflectance that can be obtained by removal of light-absorbing bodies is 100% (measured against MgO) at all wavelengths between 400 nm and 700 nm. In practice, however, even a reflectance of 92 to 95% at all wavelengths in the visible region is hard to

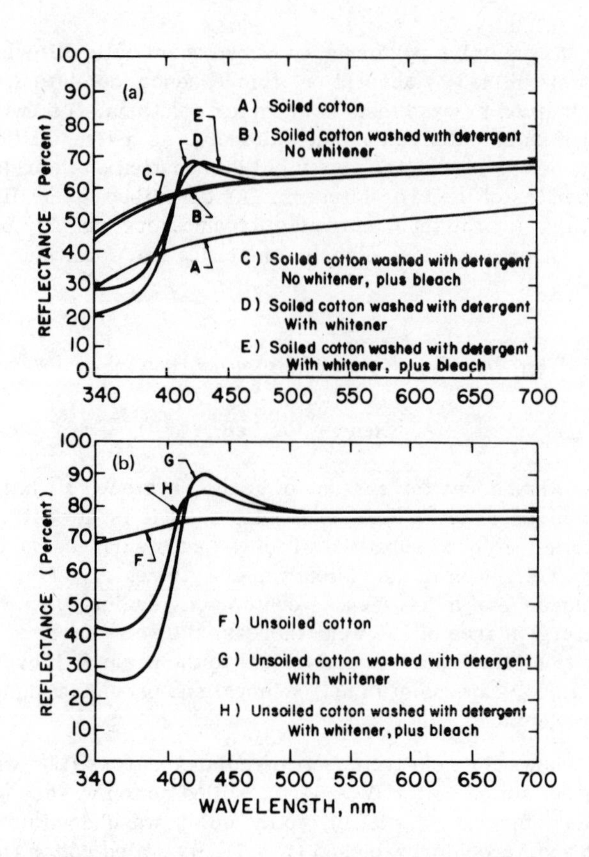

FIG. 13. Radiance of (a) soiled and (b) unsoiled cotton.

reach. Also, this type of "neutral white" is not necessarily the white preferred by the average homemaker. "Whites" are preferred with a higher radiance in the 420-nm to 450-nm region than in other ranges of the visible spectrum; this is the region in which naturally yellowed substrates show a deficiency in reflectance. Fluorescent whitening agents that exhibit their strongest emission in this region are able to compensate for the deficiency in blue relfectance of the substrate and to contribute additional radiance to produce the appearance of preferred "whites." This is precisely why whiteners are used in detergents and other laundry products.

B. Selection

The choice of a whitener is commonly based on its performance per unit cost. Before different whiteners are compared for performance, it is important to ascertain that representative samples are used for comparison, because the performance properties of a whitener are related not only to its chemical structure and activity but also to its physical form and purity. As shown in Table 3, a large number of performance properties are considered in selecting a whitener for a laundry product.

The laundry product type (see Table 4) in which the whitener is to be used, will influence the selection. For example, a heavy-duty detergent for a mixed fabric load requires whiteners different from those intended for a fine-fabric detergent. Similarly, acid sours used by commercial

TABLE 3

Performance Properties

A. Incorporation	B. Detergent
Ease of incorporation	Detergent whitening or coloring
Stability during incorporation	Storage stability
Compatibility	
Toxicity	

C. In Wash Liquor

Whitener	Conditions
Solubility/dispersibility	Surfactant
Rate of solution	pH
Affinity	Temperature
Exhaust rate	Inorganic electrolyte
Exhaust equilibrium	Agitation
Leveling	Detergent/fabric/liquor ratio
Buildup	Duration
	Hypochlorite stability

D. On Fabric

Whiteness	Fastness
Initially	Light
On buildup	Gas
Alone	Softeners
With other whiteners	Perspiration
	Ironing
	Humidity
	Washing
	Aging
	Metal ions

Source: From Ref. 62.

TABLE 4

Laundry Products Containing Whiteners

HOME LAUNDRY
Heavy Duty Detergents
Powdered and liquid products
Anionic, Nonionic
Surfactant mixtures
Cold-water products
Light Duty Detergents
Fine-fabric products
Dry Bleaches
Oxygen
Chlorine
Reducing
Miscellaneous
Enzyme presoaks
Fabric softeners
Starches
Borax formulations
Bluing products
Toilet soaps
Etc.
POWER LAUNDRY
Breaks
Soaps and detergents
Antichlors
Sours
Fabric softeners

laundries do not contain the same whiteners as those found in an oxygen-type dry bleach. Selection of FWA also depends on laundry product composition. The amount and type of surfactant [46, 81], the builder system, and other components used in the finished laundry product are also important for whitener choice. For example, whiteners requiring an inorganic electrolyte in the liquor to exhaust well onto cotton are not recommended for unbuilt detergent systems.

Highly bleach-stable FWAs are required for dichloroisocyanurate-type dry bleaches. Components such as sodium perborate, on the other hand, do not influence whitener selection directly but, since perborate will react with hypochlorite if added to the wash liquor, the performance of so-called bleach-unstable whiteners will be affected indirectly.

For a heavy-duty home laundry detergent, the selection will depend on the washing practices of the homemaker. "Average" home laundry conditions (such as temperature, fabric/detergent/liquor ratio, duration of cycle, and agitation, etc.) and washload composition, to a large extent, will determine the whitener systems selected. The relative importance assigned to the various performance properties depends largely on marketing considerations. The most important factors and their influence on whitener performance and selection are discussed in Sec. IV.

C. Incorporation and Detergent Whitening

Besides being easy to incorporate in a detergent formulation, an FWA should be stable to processing conditions. The most severe conditions are probably those in which a detergent whitener is preslurried in a strong alkaline medium, added to the feed slurry, and spray dried at high temperatures. Useful whiteners must not only be resistant to such conditions but also be stable on storage in the product, so that the latter will exhibit its full whitening power when used. Whiteners used in liquid laundry products must be compatible with the formulation, remain stable in the solution, and not cause separation or otherwise decrease the performance characteristics of the product. Proper particle size is necessary for satisfactory mixing and for avoiding separation in dry-blended products. At the whitener levels used (0.3 to 0.7%), the effect of whiteners on the appearance of the detergent powder has become quite important. Some FWAs improve the whiteness of a detergent powder, whereas others do not. Detergent whitening is generally related to the chemical structure, purity, and physical properties of the whitener, to FWA concentration, to other components in the formulation, and to the incorporation techniques used by the detergent producers. The whitening effect of some FWAs in anionic and nonionic detergent powders is shown in Table 5.

The whitener should not change the shade of colored products. If a shade change does occur, it should be reproducible. In special cases, FWAs are used as colorants.

D. Whiteners in the Wash Liquor

1. Steps in the Whitening Process

Detergent whiteners are commonly introduced into the wash liquor with the laundering products, usually in powder form (Fig. 14). To become effective, the whitener must dissolve or disperse in the liquor (step 1), and

TABLE 5

Detergent Whitening

Whitener (%)	Whiteness (W) of powder Anionic	Nonionic
0.4 TA (fast)	113.3	140.4
0.4 TA (slow)	100.8	140.3
0.4 DM	100.9	135.7
0.1 NTS (R = H)	111.0	125.2
0.1 BBO	112.9	104.7
No whitener	82.9	90.2

Source: From Ref. 62.

then exhaust from the liquor onto the surface of the fiber substrate (step 2). After deposition on the fiber surface, most FWAs will subsequently distribute themselves (migrate) throughout the fiber (step 3). Generally either step 1 or step 3 will govern the rate of whitening. The rate of step 1 is related primarily not only to the chemical structure of the FWA and to its physical form in the detergent but also to the wash temperature and to the degree of agitation. The adsorption of the whitener molecule onto the fiber surface (step 2) takes relatively little time. Step 3, the diffusion or migration of the FWA into the fiber, is slow because of mechanical obstructions and the restraining forces between the fiber and the whitener. The mechanism of whitener-fiber interaction depends both on the chemical structure of the whitener and on the fiber type.

2. Whitener-Fiber Interaction and Washload

A variety of theories has been proposed to explain how different dyestuffs act on various types of fibers [82-91]. These same theories also explain the affinity of whiteners for fibers [3, 92].

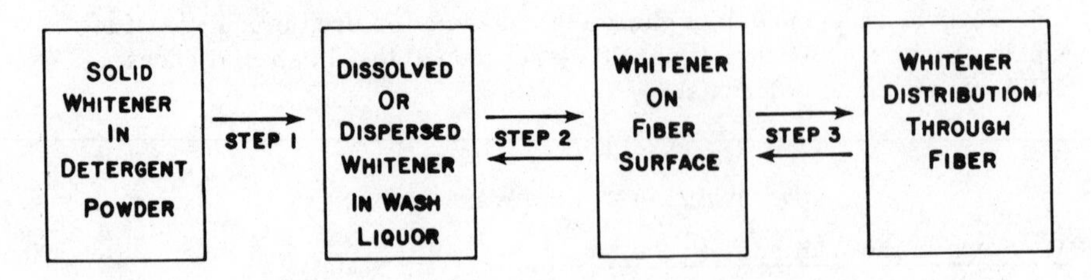

FIG. 14. Steps in the whitening process.

a. Cellulosic fibers

The CC/DAS and DSBP products behave like direct dyestuffs on cotton [3]. The cellulosic fibers, being hydrophilic, swell in water and the pores in the amorphous region grow to 15 to 30 Å in diameter, large enough to admit the whitener molecules. The molecules are not believed to penetrate the crystalline regions of the cellulosic fibers. The classic theories for dyeing or whitening cellulose are:

Hydrogen bonding theory. For high affinity to cellulosic fibers the dyestuff molecule should be linear, have aromatic nuclei capable of coplanar configuration, contain conjugated double bonds, and be capable of forming hydrogen bonds.

Van der Waals' forces theory. For high affinity to cellulosics, the dyestuff molecule should be planar to allow the closest possible contact with the fiber. The attraction is by interaction of dipoles between nonionic portions of the molecules. Substantivity of a dye increases with the number of conjugated double bonds. The attractive forces depend on the size and mobility of the electron bonds.

Aggregation theory. Dye molecules pass into the fiber singly, and then aggregate after losing their water of hydration. Coplanarity of the dye molecule facilitates penetration of the cellulose and enhances the forces of aggregation between dye molecules. Aggregation will also be enhanced by the number of double bonds: the more double bonds there are, the stronger the van der Walls' forces between the dye molecules.

Ion-pair theory. To exhibit substantivity to cellulosic fibers, a dye or whitener must be water soluble, neutral (no charge), contain conjugated double bonds, and be able to deaggregate into a monomolecular form. All other frequently mentioned factors, such as linear structure, coplanarity, and degree of conjugation, etc., might affect the substantivity but are not basic requirements. Wegmann [82] concludes that previously proposed dyeing theories (such as those mentioned above) are based on too few, and partly incorrectly run, experiments. The ion-pair concept is suggested by him as a more general replacement theory. According to Wegmann, and as indicated in Fig. 15, anionic whiteners can be present in solution in three different forms: as dissociated whiteners, as ion pairs, and as aggregates. Dissociated, negatively charged whitener anions cannot exhaust onto the negatively charged fiber. Neither can aggregates exhaust because they are too large to penetrate the narrow pores in the fiber. Only when the whitener is present in the wash liquor as a neutral ion pair may it diffuse into the cellulose.

DW ⟷ IW ⟷ (WW)

FIG. 15. Whitener in solution: D - dissociated; I - ion pair; W - whitener; o - aggregate.

It can be concluded from the above that, to achieve high affinity for cellulosic fibers, the whitener molecule should have several conjugated double bonds and aromatic nuclei of coplanar configuration. Of course, the latter is also a requisite for fluorescence of the molecules.

It can be seen from Fig. 8 that the trans form of whitener TA (and also of the other CC/DAS and DSBP derivatives) fulfill the above requirements, while the cis form does not. How a "direct" type of whitener (DTW) is attached to the cellulosic chains (CE) is shown schematically in the upper part of Fig. 16.

b. Resin-treated cellulosic fibers

The reason why regular CC/DAS products do not build up as well on cross-linked cotton as on untreated cotton is shown schematically in the lower part of Fig. 16. The degree of buildup is a function of resin type and concentration used for cross-linking. The cotton part in durable-press-treated polyester/cotton contains up to 20% resin. Extensive cross-linking hinders the buildup of the regular cotton whiteners. By cross-linking, cotton is rendered less hydrophilic. Dry-cured, resin-treated durable-press cotton does not swell in water to the same extent as untreated cotton. Therefore, FWAs with less "attraction" for water (lower relative solubility) can be expected to exhibit higher affinity for resin-treated cotton than do the more soluble types. Indeed, it has been reported [93], that whitener TA and whitener NTS (R = H) with relatively low solubility, and whitener BBO which is a dispersed water-insoluble product, achieve relatively better buildup on resin-treated cotton than do more soluble whiteners. Cotton, resin treated and wet cured in the swollen state, will allow a higher degree of FWA penetration.

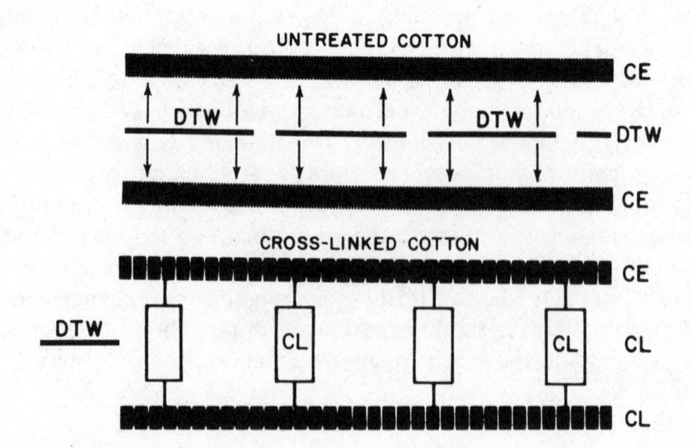

FIG. 16. Whitening of cotton: CE - cellulosic chain; DTW - direct-type whitener; CL - cross-linking agent; ↓↑ - forces.

c. Hydrophobic fibers

It is believed that nylon, polyester, acetate, and other hydrophobic fibers are whitened in the wash liquor by a procedure similar to that employed when dyeing these fibers with dispersed dyes [3]. Dispersed-type FWA molecules are relatively small in size, are nonionic, and exhibit low solubility. The mechanism of dyeing, however, is a controversial subject. Some investigators believe that solid dyes dissolve in the solid fiber, while others feel that the dyes penetrate into canals between fiber molecules. A third group theorizes that the so-called water-insoluble dispersed dyes are in reality very slightly soluble and deposit on the fiber from true solution. The small amount of dissolved dye enters into and is absorbed by the fiber, permitting more dye to dissolve, etc. [46]. Weisz and Zollinger [91] believe that exhaust of ionic dyes into hydrophilic fibers and exhaust of dispersed dyes into hydrophobic fibers are both based on the penetration of the aqueous phase of the dyebath into the textile substrate. Based on this concept, the rate of diffusion of an FWA into all kinds of fiber materials is related not only mainly to the degree of porosity of the substrate but also to the molecule size and form of the whitener.

Some synthetic fibers (for example, polyester) are extremely compact below the softening temperature. They do not swell to any significant extent in water, at least not at temperatures below 150°F employed in the U. S. home laundry. It is fortunate that the mass whitener in polyester fibers exhibits high lightfastness, because the effectiveness of currently available polyester whiteners is limited, especially on mass-whitened polyester [47]. With the advent of mass-whitened polyester and polyamide treated at the textile mill, the need to use fluorescent whitener for fine fabrics in the wash liquor declined, in spite of the growing percentage of these materials in the washload. The whiteners used during the processing of synthetic fibers and fabrics are stable to wash and wear, so that their effects are retained throughout a significant portion of the life of the garment. This is fortunate, especially in the case of polyester, because of the physical nature of the fiber. Only limited penetration of polyester fiber is possible by presently available whiteners at the low wash-water temperatures used in the United States. They are generally below the glass transition temperature of the fiber [51]. Although few, if any, new types of fiber are expected during the next decade, modifications of presently available ones are probable [94]. These modifications in fibers might permit additional enhancement of whiteness in the wash liquor.

3. Washload Composition

The fiber composition of the washload is another determining factor in the selection of an FWA for a laundry product. For commercial and institutional laundries, the problem is a fairly easy one. The washloads can be sorted into groups of all cotton, all synthetic, or all cellulosic/synthetic blends. Knowing the composition of the load permits the operator to select the wash conditions and whitener that give maximum results, much as in a dyehouse. Manufacturers of detergents for the home face a much more difficult task because the composition of the washload varies widely from household to household.

The selection of whitener systems for heavy-duty detergents is dependent upon the makeup of the "average" U. S. washload. Because of advances in synthetic-fiber and textile-finishing technologies, the composition of the average U. S. washload changed during the 1960s and this trend has continued during the 1970s.

Information on past, annual natural and synthetic fiber production in the U. S., import, export, consumption, etc. is readily available [95, 96]. Similarly, forecasts on fiber production can be found in the literature [97]. Although the fiber composition of the average U. S. washload is related to the amount of the various fibers produced, there is no direct relationship. The following factors operate to make the relationship indirect:

1. Only a fraction of the textile fibers produced will go into fabrics washed in home laundries. Also the distribution of fiber types used in washable articles differs from that of textile articles not generally washed at home.
2. Some articles are washed more frequently than others. For example, underwear is generally laundered more often than bed linens.
3. The lag factor accounts for the time lapse between the year X, during which a certain percentage of the various fibers were produced, and the year Y when such blends were found in the average U. S. home laundry washload (Fig. 17). After all, inspection of a linen closet might reveal that many articles are 5 to 10 years old.

As far as future whitener requirements are concerned, in addition to fiber type, it is important to know how much of the textile fabrics will be finished as white, pastel-colored or printed goods. Such information is available (Table 6) [98]. It is significant that over the last few years the ratio of white finished to colored cotton goods has changed very little. As a matter of fact, production of white goods has increased slightly despite the popularity of color. Colored business shirts have been promoted heavily and one could conclude that few white shirts are being worn. The results of small private surveys during business and trade meetings over the past several years have, with few exceptions, indicated a colored/white shirt ratio of around 1:1.

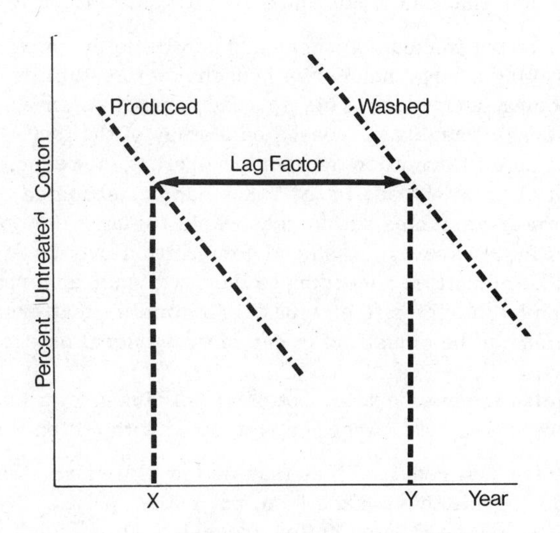

FIG. 17. Lag factor. From Ref. 50.

TABLE 6

Ratio of White to Colored Goods
(U.S. Cotton Fabric Production)

	Total Production (%)		
Year	White	Print/Dyed	Ratio
1966	45	55	0.82
1967	46	54	0.85
1968	47	53	0.89
1969	47	53	0.89
1970	47	53	0.89

Source: From Ref. 50.

For the future, predictions are that sales of white shirts will increase in the coming years. Fulfillment of this prediction, however, might be related to the detergent industry's capability under a restrictive situation to formulate products that can clean white articles satisfactorily.

Past and current makeup of washloads can be determined most accurately by surveying a large number of households to establish which types of fabric are being laundered. This approach cannot be employed to forecast future average washloads. Based on a study of the factors mentioned above and available information on fiber production, however, the future composition of U. S. washloads in the 1980s can be estimated by varying the involved factors or variables within reasonable limits. For this purpose, the anticipated future average washload composition may be forecast. Fibers should be classified according to their whitener absorption properties rather than by fabric or fiber types. To simplify matters, the washload composition can be classified in one of two general categories:

1. Articles (such as cotton, cotton cross-linked with low levels of resin, rayon, etc.) that can be treated with cotton whiteners.
2. Articles that require other than cotton whiteners. Such fabrics would include those made from polyester, polyamide, acrylic, durable-press-treated cotton cross-linked with high levels of resin in the dry stage, and blends of these fibers.

Estimates of the composition of the average washload in the United States during 1964-1980 were developed using one selected value for each variable; an example is shown in Table 7. From this table it can be seen that fabrics which can be whitened by a fluorescent agent designed for use on cotton remained a significant percentage of the washload throughout the 1970s. During this decade, therefore, the goal of the detergent formulator would appear to be to provide products which retain or improve the whiteness of both fine fabrics and cotton.

4. Whitener Properties in Solution

a. Solubility and solubility rate

Although the solubility of the commonly used FWAs in a detergent wash liquor is relatively low, it is still high enough to allow a sufficient amount to dissolve (step 1). The exhaust of whitener onto the fiber (step 2) and migration into the fiber (step 3) allow more whitener to dissolve in the liquor. Actually, the total concentration of whitener in a modern home laundry liquor in the United States would only be 3 to 10 mg/liter even if all the whitener were in solution at one time. However, differences in relative solubility of chemically different products are at least partly responsible for their degree of affinity for a substrate. The affinity depends also on the substrate type and wash liquor conditions (Table 3, part C).

TABLE 7

Estimated Average U. S. Washload Compositions (1964-1980)

Year	Washload Requiring Cotton FWA (%)
1964	92
1965	86
1966	83
1967	78
1968	75
1969	73
1970	70
1971	66
1972	66
1973	63
1974	60
1975	58
1976	56
1977	54
1978	52
1979	50
1980	47

Source: From Ref. 50.

The rate of solution (step 1) of a whitener added to the wash liquor in a detergent powder also depends on its physical form and on the procedure used to incorporate it into the detergent powder. The rate of solution is often more important for the relative performance of two chemically similar whiteners than are the equilibrium solubilities of the products. If a whitener requires more time to dissolve and exhaust onto the fabric than the duration of the wash cycle (approximately 15 min in the United States), a large portion of it will simply go down the drain. Similarly, for water insoluble products, both the degree and rate of dispersion are important for whitener performance.

The effect of rate of solution on results of performance testing of whiteners is discussed by Stensby [92]. It is shown in Table 8 that whitener TA (with slow rate of solution), for example, is much less effective on cotton when added with the detergent powder in solid form than when added predissolved from stock solution. Whitener TA (with fast rate of solution) and whitener DM, however, build up to the same fabric fluorescence regardless of the method of whitener addition. Whitener NTS, with a fast rate of solution, gives approximately the same effect independent of the method.

TABLE 8

Method of Whitener Addition

Whitener (%)	Substrate	With detergent powder	From stock solution
		F	F
0.4 TA (slow)	C	327	400
0.4 TA (fast)	C	403	401
0.4 DM	C	407	403
0.1 NTS	C	299	303
No whitener	C	20	20
0.1 BBO	P	122	179
No whitener	P	10	10

Source: From Ref. 62.

Whitener BBO, a water-insoluble product, exhibits poorer buildup on polyester from a detergent powder than if it is predissolved before addition.

b. Affinity and exhaust

The degree of exhaust (step 2) of a whitener onto a substrate is dependent on wash-liquor conditions. Under a selected set of these conditions and with sufficient time, an exhaust equilibrium will be reached. Ideally, at equilibrium, 90 to 95% of the whitener should be on the fabric and the rest in the bath to permit leveling. However, the short wash cycles in the United States do not always allow steps 2 and 3 to reach equilibrium. Sometimes less than 50% of the total whitener is found on the cloth. In practice, the distribution of whitener among detergent powder, wash liquor, fiber surface, and fiber interior is not a simple phenomenon. Usually the detergent contains a combination of whiteners, each with a different degree of affinity for the various fiber types in a mixed washload; further complications arise from the presence of whiteners already in the fibers—whiteners which may be, to a limited extent according to the reversible steps 2 and 3, transferred to the wash liquor.

c. Leveling

Fortunately, steps 2 and 3 (wash liquor/fiber surface/fiber interior), are two-way processes. Initially, the whitener exhausts onto the fiber, but unevenly. However, a properly chosen whitener establishes an equilibrium between fiber and liquor. Part of the whitener returns to the liquor and re-exhausts back onto the fiber, gradually reducing uneven FWA distribution until, at the end of the wash cycle, a perfectly uniform whitening is obtained. An FWA like DMEA which, in alkaline detergent liquors has the ability to

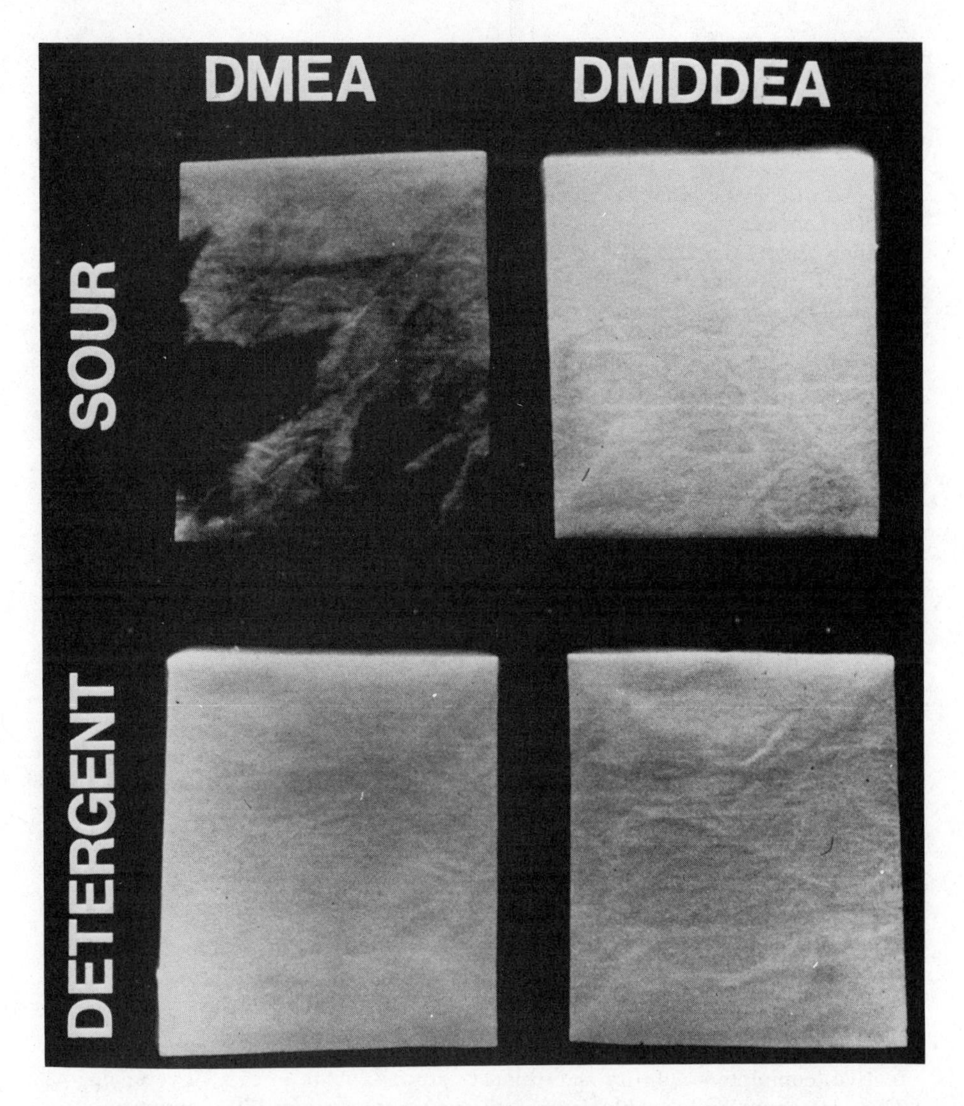

FIG. 18. Effectiveness of DMEA and DMDDEA. From Ref. 62.

distribute itself evenly, will give unlevel effects when applied in an acid sour (see Fig. 18). At low pH this whitener has insufficient solubility to allow a satisfactory fiber/liquor distribution and, therefore, stains the cotton fiber in an unlevel fashion.

Whitener DMDDEA, with four sulfonate and two diethanolamino groups, has sufficient solubility even at low pH to give "level" effects and good buildup. The pH is only one of many factors that govern the distribution of a whitener between the fiber and the wash liquor.

d. Buildup

As with many other technical terms, "buildup" means different things to different people. ASTM* D-12 has adopted the following definition for buildup of an FWA: "The course of change in fluorescence emission intensity or fluorescence shade or both, using specified exhaust procedure: 1) for a specified number of successive applications of FWA, or 2) by varying the FWA concentration in a series of single applications." The maximum buildups obtained by using application methods (1) and (2) above are not necessarily identical. The maximum whiteness reached on repeated usage of a detergent should be as high as possible. On reaching the maximum, it is desirable that an overdose (on continued use) of FWA does not cause fluorescence quenching or fabric discoloration (yellowing). The maximum buildup that can be reached in a one-wash cycle is also of interest. Some consumers want to see a high effect after one application of the product. Allen [55] discussed the reason why FWAs reach a maximum in buildup. The reason is that almost all the available ultraviolet energy in commonly used light sources is absorbed by high concentrations of whiteners. Additional FWA, if applied, will not function because there is no more ultraviolet light (required to excite the whitener) to absorb.

Between 0.3% and 0.4% whitener on the weight of the cotton fabric is the level beyond which no further increase in fluorescence was measured by Allen. The active-ingredient content of the whiteners used, however, is not indicated. Based on Allen's findings, a rough calculation can be made to determine the number of wash cycles and/or maximum concentrations of whitener one can use before the maximum in buildup is reached. For this calculation it can be assumed that the percentages given by Allen are related to full strength products and also that detergent whiteners exhibit the same type of buildup as the textile whiteners used by Allen. Typically, around 3 oz (or approximately 90 g) of detergent is used for each 8 to 9 lb (or approximately 4 kg) of fabrics in a U. S. home laundry cycle. It can be calculated then, assuming that the whitener present in the detergent is completely effective (complete solubility and exhaust onto the fabric in every cycle; no FWA destroyed by hypochlorite bleach; no destruction of FWA during use, etc.), that maximum buildup will be reached after 30 to 40 cycles if the detergent used contains 0.5% full-strength whitener. If less than 100% of the whitener applied ends up on the fabric, if some of the whitener is destroyed by bleach, and if the fabrics are worn and exposed between cycles, it is expected that at least 1% FWA on the weight of the detergent must be applied in 30 to 40 cycles in order to reach the maximum in buildup. Naturally, the maximum level of a whitener that will improve the whiteness of a substrate depends, among other things, on the spectral characteristics of the FWA and, as mentioned, on the amount of ultraviolet radiation available.

*American Society for Testing and Materials Committee

Figure 19 shows the relative spectral radiance curves (against $BaSO_4$) of increasing levels of whitener DMDDEA on cotton. The values indicate next to the curves refer to the visual whiteness rating of the fabrics relative to a whiteness scale (0 to 260). The higher the concentration of whitener on the substrate, the higher the light absorption by the whitener. More and more ultraviolet radiation can be absorbed until, at a certain concentration of whitener, it is almost completely absorbed. At high whitener concentrations, the absorption moves into the violet-blue visible region. At still higher levels, whiteners may form aggregates or microcrystals with new and different optical properties. The quantum yield may also change. On buildup (the level depends on the FWA) either one or a combination of these factors, or the presence of impurities, may result in a yellow appearance of fabrics containing very high levels of whiteners.

e. Concentration

Use concentration affects the selection of a whitener. For example, one whitener may have better properties and, at a low concentration, exhibit a hue superior to another; at a higher whitener level, the same compound may be inferior. It is, for example, feasible to use a higher concentration of bleach-stable FWA with a relatively unstable product at a low total whitener level in a heavy-suty detergent. At high levels, part of the relatively belach-unstable whitener will remain intact on exposure to bleach, and a relatively smaller amount of the bleach-stable FWA is required.

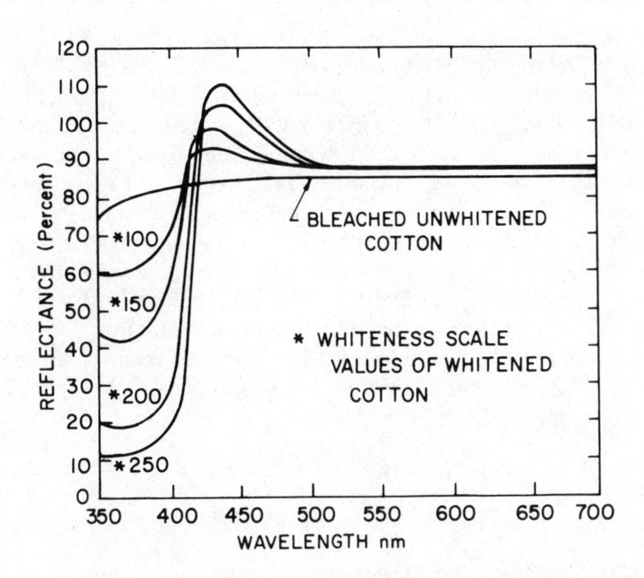

FIG. 19. Relative radiance of increasing levels of whitener DMDDEA on cotton. From Ref. 3.

5. Wash Liquor Conditions

a. Surfactants

Anionic system. Most of the large volume heavy-duty detergents are based on an anionic or a nonionic system. Some detergents are anionic and nonionic mixtures which, in some cases, are "spiced" with a cationic, a zwitterionic, or a semipolar modifying compound. Whitener performance properties, such as affinity for fiber substrates, bleach stability in solution, and also detergent powder whitening, are affected by the surfactant system used [3, 23, 24, 92]. Most of the commonly used stilbene-type cotton whiteners exhibit comparable affinity to cotton in anionic wash liquors, but may differ in effectiveness in nonionic wash liquors.

Nonionic system. Haeusermann and Keller [81] reported on the relative effectiveness of some cotton whiteners of the 4,4'-bis (1,3,5-triazinyl-6-amino)-stilbene-2,2'-disulfonate (CC/DAS) type on cotton and nylon in a nonionic (alkylphenolpolyglycol) based system. The authors conclude that CC/DAS whiteners with relatively good solubility in water exhibit better buildup on cotton in the nonionic wash liquor than do compounds with lower solubility. Stensby's discussion [46] of the effect of surfactants on FWA exhaust is based on Wegmann's ion-pair theory. According to Wegmann's theory, the whitener will be absorbed only in a monomolecular, neutral ion-pair form, and not in an aggregated form. It can be assumed that surfactants affect the equilibrium among dissociated whitener, neutral ion pairs of the whitener, and whitener aggregates.

A decreased affinity for the fiber would result, should nonionic surfactants promote formation of whitener-whitener or whitener-surfactant aggregates or micelles. Aspland [89] indeed indicates that the exhaust of cotton dyes (whiteners) will be reduced by the addition of nonionic ethylene oxide condensates to the solution because the nonionics form micelles capable of "dissolving" the dye (whitener) molecules. Large aggregates between single whitener molecules and the surfactant are probably formed. The more hydrophobic the FWA, the greater the size of the aggregates and, therefore, the lower the degree of whitener exhaust. The suggested whitener-surfactant interaction is shown schematically in Fig. 20. Schick [99] and Becher [100], in discussions of micellar properties of nonionic micelles, conclude that the aggregation numbers of nonionic micelles, derived from most commonly used nonionic surfactants, depend on the length of both the hydrophilic and

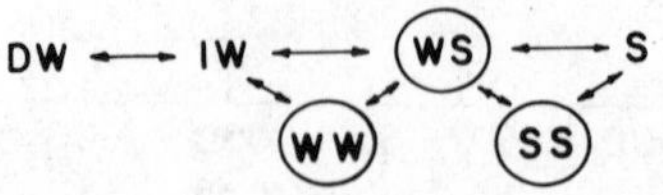

DW ⟷ IW ⟷ (WS) ⟷ S

(WW) (SS)

FIG. 20. Whitener/surfactant interaction: D - dissociated; S - surfactant; W - whitener; I - ion pair; ○ - micelle/aggregate.

hydrophobic groups. At constant size of the hydrophobic group, the aggregation number of nonionic micelles decreases with increasing ethylene oxide chain length. Whitener exhaust is, therefore, likely to increase with increasing length of the hydrophilic group. The reasons why commonly used anionic surfactants form exhaust retarding aggregates with the more hydrophobic anionic whiteners (CC/DAS types, NTS, etc.) to a much lesser extent than many nonionics are probably related to differences in micelle structures of the two surfactant types. An anionic whitener-anionic surfactant aggregate is less likely to be formed than is a whitener-nonionic aggregate because of the charged "cloud" around, and the nonpolar nature of the interior of anionic surfactant micelles. Also, anionics generally form micelles at higher concentration than do nonionics.

Anionic/nonionic/cationic system. The effect of surfactants on four CC/DAS cotton whiteners listed in order of increasing water solubility is discussed by Stensby [46]. Tests were run in unbuilt systems, using LAS-type (anionic) surfactant, oxyethylated-nonylphenol-type (nonionic) surfactant, or dihydrogenated-tallow-dimethyl-ammonium-chloride-type (cationic) surfactant. Findings shown in Table 9 demonstrate the retarding effect of nonionic and cationic surfactants on exhaust of anionic whiteners. The effect is more pronounced on the less soluble compounds. The results confirm findings reported by others. Similar tests were run in anionic- and nonionic-built (1 g/liter STPP) systems. The presence of builder increases whitener exhaust in the nonionic system. The order of effectiveness of the whiteners is the same as in the unbuilt system. The addition of builder, however, affects both the state of the nonionic surfactant and the state of the whitener in the solution. Addition of electrolyte leads to an increased aggregation of the nonionic surfactant, probably because the ethylene oxide chains undergo a salting-out effect [99]. Although higher surfactant aggregation should lower whitener exhaust, the opposite is observed. A suggested possible explanation is that electrolyte addition decreases whitener dissociation and increases the concentration of ion pairs, thereby improving the

TABLE 9

Cotton Fluorescence in Unbuilt and Built Systems

	Unbuilt Systems			Build Systems	
Whitener	Anionic	Nonionic	Cationic	Anionic	Nonionic
TA	150	22	40	150	37
DM	150	60	43	150	123
DMEA	150	72	47	150	137
DDEA	150	85	52	150	141

Source: From Ref. 46.

whitener exhaust. Very high electrolyte concentrations can cause whitener association and decrease the degree of FWA exhaust.

"Igepal" series. Wash tests with "nonionic sensitive" whitener NTS in various "Igepal" CO (polyoxyethylated nonylphenol type) surfactants of the following general structure were run:

$$C_9H_{19} \cdot C_6H_4O \cdot (CH_2CH_2O)_nH$$

The number (n) of ethylene oxide units, and the cloud point of 1% solutions of the Igepal products used are shown in Table 10, together with the cotton fluorescence obtained with NTS at 80°F and 120°F in these surfactants. Fluorescence in "Nacconol" NRSF (an alkylaryl sulfonate) serve as reference points. As can be seen, all Igepal products tested retarded exhaust of the whitener. The degree of influence is probably related either to (1) the surfactant's cloud point, or to (2) its ability to form aggregates or micelles with the whitener in solution. The cloud point of Igepal CO-340 and CO-530 is below the wash temperature and the retarding action on the whitener buildup of these is found to be relatively low. Increase in the length of the ethylene oxide chain results in increased hydration. High solubility lowers aggregation. This is probably why CO-990 (n = 100) shows less retarding action than the other Igepal products. With increasing wash temperatures (from 80°F to 120°F) the degree of hydration of the hydrophilic group of the nonionic surfactant is lowered and, therefore, its solubility is decreased. A decrease of solubility of the surfactants by increasing temperatures should promote formation of aggregates and should lower FWA buildup (lower de-

TABLE 10

Cotton Fluorescence of NTS in Igepal Surfactants

Igepal CO	n	Cloud point (°F)	(F) of NTS 120°F	(F) of NTS 80°F
340	4	<32	131	173
530	6	32	71	59
630	9-10	130	58	46
710	10-11	160	58	50
880	30	>212	67	55
890	40	>212	77	68
970	50	>212	79	68
990	100	>212	120	118
Nacconol NRSF			188	165

Source: From Ref. 46.

gree and/or rate of whitener exhaust). Yet the opposite is observed; whitener exhaust is somewhat better at 120°F than at 80°F in most cases. This is probably because the increased effectiveness of NTS (observed in absence of a nonionic surfactant) with increasing temperatures overshadows the "negative" effect of the surfactant under the chosen conditions. Results in the test with Nacconol indicate this. The results in Table 10 are presented graphically in Fig. 21. The fact that exhaust onto cotton of "nonionic sensitive" whiteners such as NTS decreases with increasing "Igepal" CO-360 concentration is probably due to increases in whitener/surfactant aggregation with surfactant concentration. These results are shown in Fig. 22. Data indicating how the exhaust onto cotton of NTS in a solution containing 0.4% Igepal CO-630 depends on temperature are shown in Table 11. The significant increase in exhaust from 120 to 160°F is expected for this Igepal, since the latter temperature is above the cloud point of the surfactant. As shown in Table 12, pH of the wash liquor containing 0.4% of Igepal CO-630 has no noticeable effect on the exhaust of whitener NTS at 120° F. Results of tests with NTS on nylon run as described earlier for cotton are shown in Table 13 and graphically in Fig. 23. As expected, the overall effectiveness on nylon is lower than on cotton. The influence of the different Igepal types, and the effect of temperature on whitener exhaust are direct parallels to observations on cotton. In other words, the change to nylon substrate does not affect the conclusions as to the whitener-surfactant interaction [46].

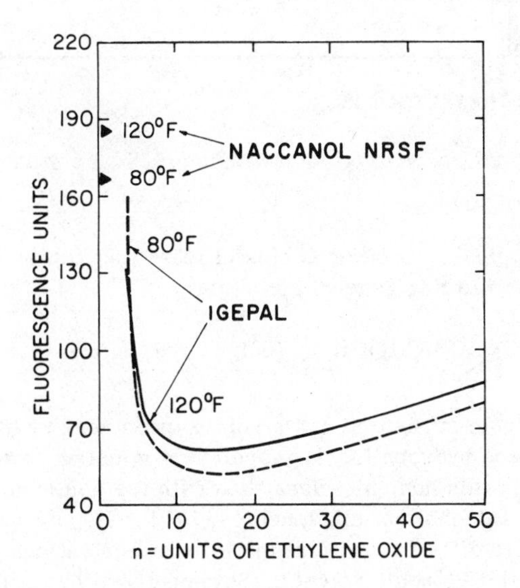

FIG. 21. Igepal surfactants; whitener NTS on cotton. From Ref. 46.

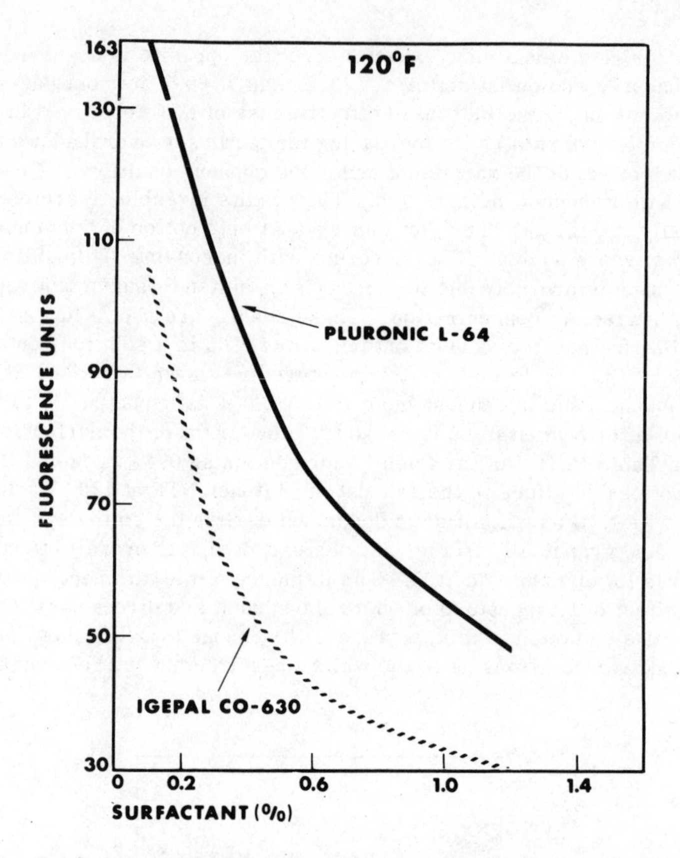

FIG. 22. Surfactant concentration; whitener NTS on cotton. From Ref. 46.

"Pluronic series". Another series of tests was run in "Pluronic" surfactants with the following general structure:

$$HO(CH_2CH_2O)_x[CH(CH_3)CH_2)]_y(CH_2HC_2O)_zH$$

This type of nonionic surfactant has a polyoxypropylene hydrophobe between two polyoxyethylene hydrophils. A more loose micellar structure is probably obtained with this nonionic class than with the commonly encountered oxyethylated fatty alcohols of alkylphenols [101]. Depending on the test method used, quite different critical micelle concentrations have been reported [102] for the Pluronic polyols. Some authors have claimed that certain oxyethylene-oxypropylene block copolymers fail to form micelles, while Cowie and Sirianni [103] reported that the micelles derived from Pluronic products differ and are related to the molecular weight of the

TABLE 11

Effect of Temperature on Cotton Fluorescence of NTS

Surfactant	80°F	120°F	160°F	190°F
Igepal CO-630	46	58	94	98
Pluronic L-64	183	91	95	104

Source: From Ref. 46.

TABLE 12

Effect of pH on Cotton Fluorescence of NTS

	pH			
Surfactant	10	8	6	3
Igepal CO-630	58	57	51	50
Pluronic L-64	91	90	90	87

TABLE 13

Nylon Fluorescence of NTS in Igepal Surfactants

		Cloud point	(F) of NTS	
	n	(°F)	120°F	80°F
Igepal CO				
340	4	<32	76	136
530	6	32	22	20
630	9-10	130	21	18
710	10-11	160	23	19
880	30	>212	28	22
890	40	>212	34	23
970	50	>212	49	30
990	100	>212	85	51
Nacconol NRSF			100	47

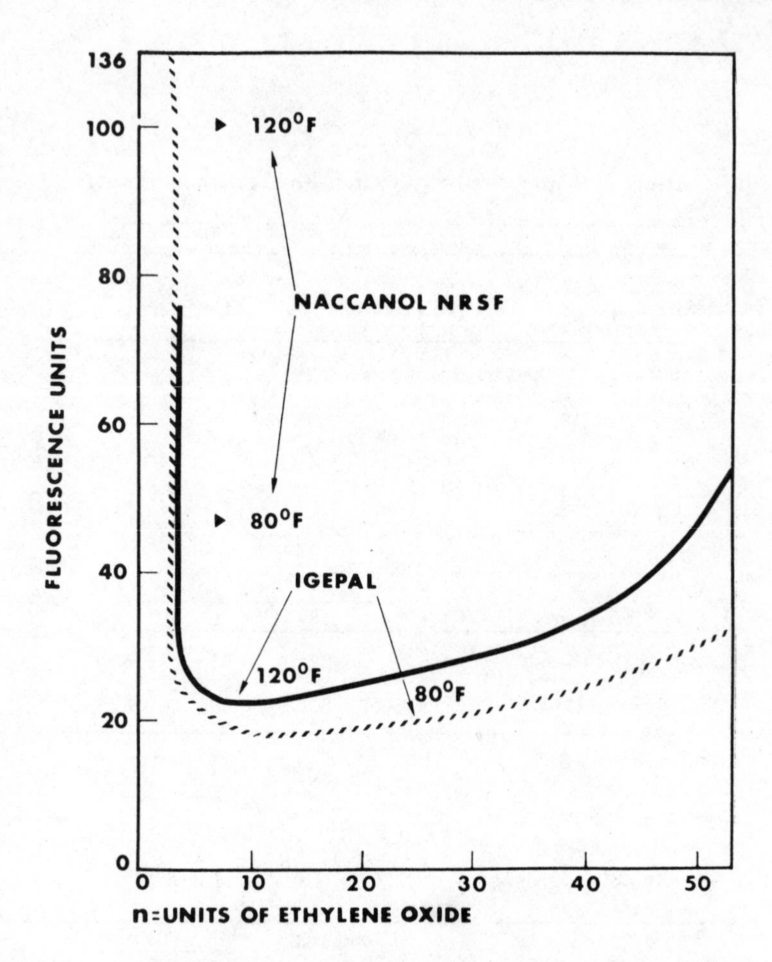

FIG. 23. Igepal surfactants; whitener NTS on nylon. From Ref. 46.

Pluronic products differ and are related to the molecular weight of the hydrophobe. It is believed that products derived from a 950 or 1200 molecular weight hydrophobe (such as F-38 or L-44) do not form aggregates. There appears to be a relationship between the length of the ethylene oxide chains of Pluronic polyols, with the same-molecular-weight hydrophobe, and degree of FWA exhaust. As expected, no direct relationship was observed between whitener exhaust and ethylene oxide chain length of products with different molecular weight hydrophobes. However, a relationship, valid in most cases, has been established between whitener exhaust of low solubility anionic whiteners at 120°F, and the R value is calculated using the following equation:

$$R = \frac{(x+z)10^4}{(y)M_T}$$

where (x + z) is the average number of ethylene oxide units in the molecule, (y) is the average number of propylene oxide units, and M_T is the overall molecular weight. The R values, based on average surfactant composition, and cloud points of 1% aqueous solutions of Pluronic polyols used are shown in Table 14. It should be noted that there is no simple correlation between R values and the CMC values determined by Schmolka and Raymond [102] by the dye-complex method. Results of wash tests run at 80° F and 120°F are shown in Table 15. The degree of exhaust of NTS and TA (both "nonionic sensitive") at 120°F is a function of the R value of the surfactants. A high R value is related to a high degree of whitener exhaust. Surfactants with R values above 5.0 do not significantly retard exhaust of the nonionic sensitive whiteners (low solubility products).

Fabric fluorescence obtained with the above whiteners in solutions of Pluronic polyols with R values above 5.0, therefore, is expected to be approximately the same (close to maximum) under specified test conditions. It should be pointed out in this connection that F-68 is an exceptionally good dispersing agent. Whitener DMEA, a whitener with good solubility, as expected, is much less retarded by the nonionic polyols than are NTS and TA.

A review of results in Table 15 shows that whitener exhaust in the Pluronic polyols is better at 80°F than at 120°F. Exhaust at the lower temperature is more or less independent on the Pluronic type used. The Pluronic polyols apparently form far fewer exhaust-retarding aggregates with the FWA at 80° F than at 120°F. This is probably because the hydrophils become increasingly hydrated with decreasing wash temperature [101]. Whitener exhaust should, therefore, be less affected at low temperatures. Results in Table 15 are shown graphically in Fig. 24 to 26.

TABLE 14

R Value of Pluronic Surfactants

Pluronic	(x+z)	(y)	(M_T)	R	Cloud Pt. (°F)
F-68	150	30	8,350	5.97	>212
L-44	23	21	2,200	4.99	149
F-88	194	39	10,800	4.64	>212
P-66	60	30	4,375	4.52	>212
P-65	40	30	3,500	3.77	180
L-64	26	30	2,900	2.99	136
P-85	53	39	4,600	2.99	185
P-84	44	39	4,200	2.72	165

Source: From Ref. 46.

TABLE 15

Cotton Fluorescence in Pluronic Surfactants

Pluronic	R	(F) of NTS 120°F	(F) of NTS 80°F	(F) of TA 120°F	(F) of TA 80°F	(F) of DMEA 120°F	(F) of DMEA 80°F
F-68	5.97	177	185	148	141	177	173
L-44	4.99	178	186	142	145	178	167
F-88	4.64	140	184	86	134	169	174
P-66	4.52	133	184	75	135	166	165
P-65	3.77	109	184	59	138	152	162
L-64	2.99	91	183	51	143	144	163
P-85	2.99	90	183	48	122	144	174
P-84	2.72	78	180	45	123	142	168
Nacconol NRSF		172	150	152	150	187	178

Source: From Ref. 46.

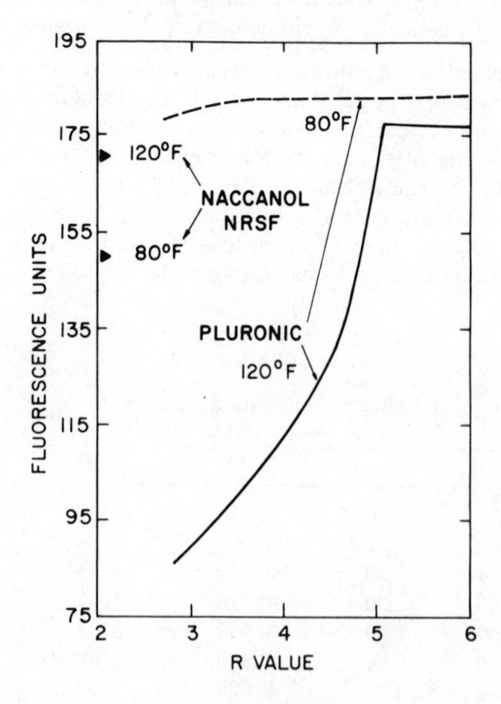

FIG. 24. Pluronic surfactants; whitener NTS on cotton. From Ref. 46.

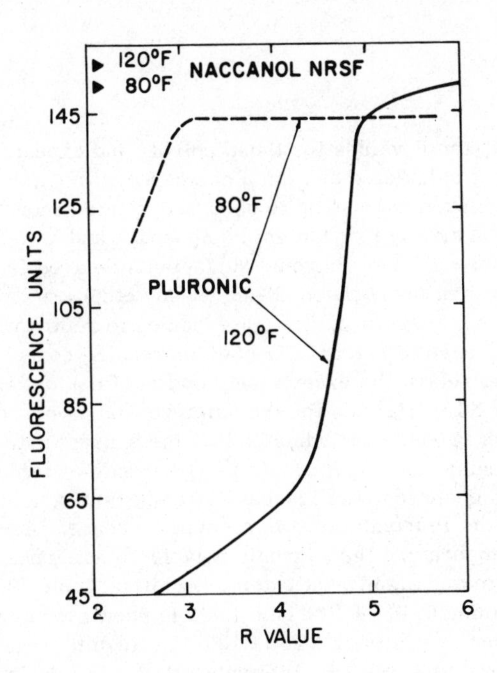

FIG. 25. Pluronic surfactants; whitener TA on cotton. From Ref. 46.

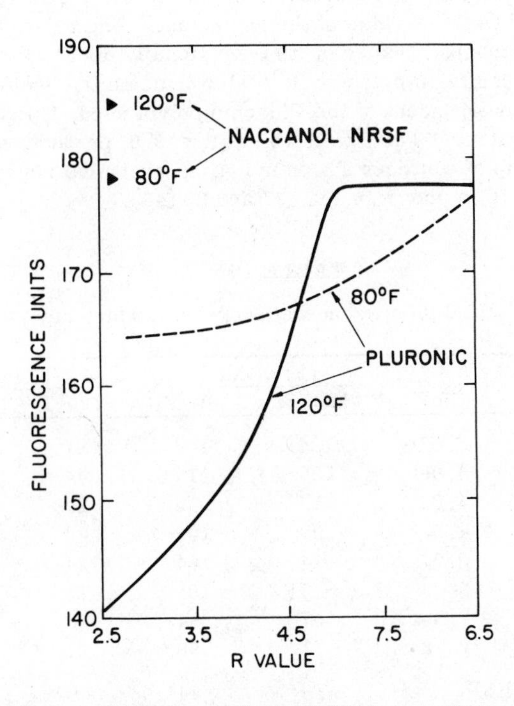

FIG. 26. Pluronic surfactants; whitener DMEA on cotton. From Ref. 46.

In other tests, the R values for these polyols and expected fabric fluorescence were first calculated and the n compared with experimental measurements. The agreement was quite good taking into account that the R values are based on average surfactant composition and also the experimental errors involved. The Pluronic surfactants derived from a 950- or 1200-molecular-weight hydrophobe (F-38, P-46, etc.) are believed not to form aggregates and, therefore, are not expected to retard whitener exhaust. As was the case for Igepal CO-630, increasing concentrations of Pluronic L-64 also retard the exhaust onto cotton of nonionic sensitive whiteners such as NTS (Fig. 22) for the same reason as proposed previously. Data shown in Table 11 indicates how the exhaust onto cotton of NTS in solutions containing 0.4% of Pluronic L-64 depends on temperature. An initial decrease is observed with increasing temperature, and then above the cloud point of the Pluronic there is a small increase. As mentioned earlier, at low temperature the Pluronic polyols do not appear to interact with the whitener to any significant extent. Again in Table 12, pH of the wash liquor containing 0.4% of Pluronic L-64 is shown to have no noticeable effect on the exhaust of whitener NTS at 120°F. Results on nylon at 80°F and 120°F are shown in Table 16. Whitener DMEA, exhibiting no nylon affinity, was excluded from this test. The findings on nylon in the Pluronic polyols confirm the results reported on cotton. Again, effectiveness at 120°F is related to the R value of the surfactant. The relationship does not hold for F-88, perhaps because of its exceptionally high molecular weight (10,800). The results on nylon at 80°F also indicate that whitener exhaust is more or less independent of the Pluronic polyol used, if $R > 3.0$, at the low wash temperature. The low-R-value ($R < 3.0$) products significantly reduce the buildup of whitener TA on nylon. A graphical representation of results in Table 16 is shown in Fig. 27 and 28 [46].

TABLE 16

Nylon Fluorescence in Pluronic Surfactants

Pluronic	R	(F) of NTS 120°F	(F) of NTS 80°F	(F) of TA 120°F	(F) of TA 80°F
F-68	5.97	134	106	67	61
L-44	4.99	120	110	44	53
F-88	4.64	79	98	29	62
P-66	4.52	90	106	29	59
P-65	3.77	74	104	24	54
L-64	2.99	49	103	18	62
P-85	2.99	57	94	18	62
P-84	2.72	51	98	21	37
Nacconol NRSF		88	44	49	38

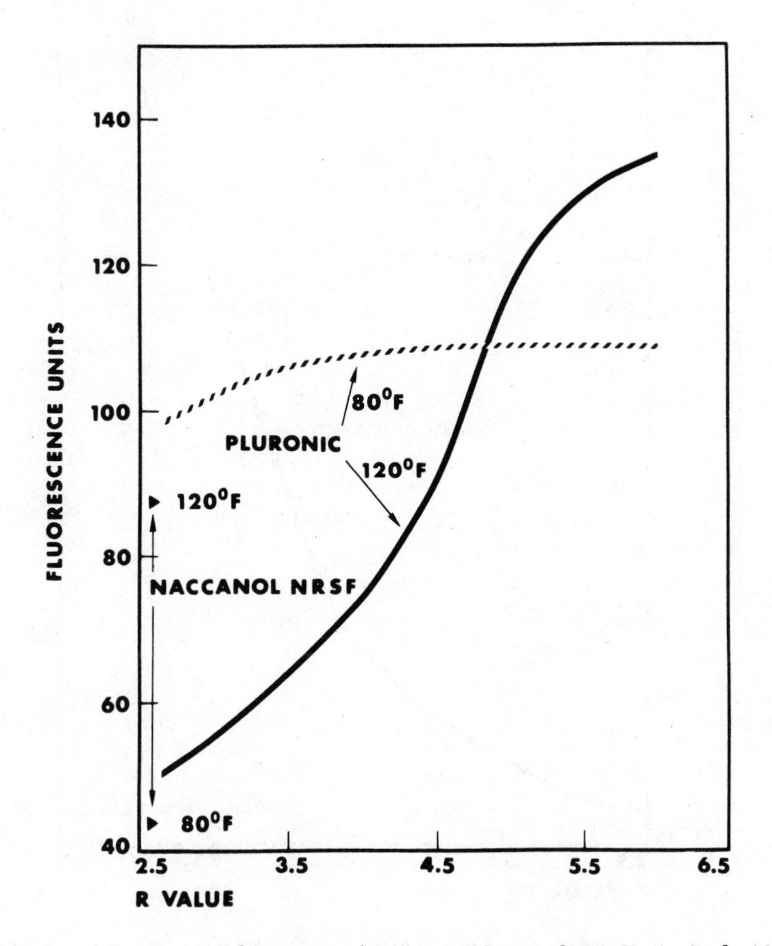

FIG. 27. Pluronic surfactants; whitener NTS on nylon. From Ref. 46.

Cationic surfactants. Whiteners have found a widespread use in cationic fabric softeners, mainly because whitener-free solutions based, e.g., on dihydrogenated tallow dimethyl ammonium chloride, ethoxylated dialkyl dimethyl ammonium sulfate, or imidazolinium derivatives "quench" the effect of FWA on fabrics. Because of the cationic character of these surfactants, only specially selected whiteners may be used successfully in fabric softeners. As indicated by Tables 1 and 2, anionic whiteners with relatively high solubility (such as DDEA and DSBP) are best suited as cotton whiteners for cationic systems. Nonionic whiteners such as BBO and NOS can also be used in softeners for whitening of synthetic fabrics. Only at

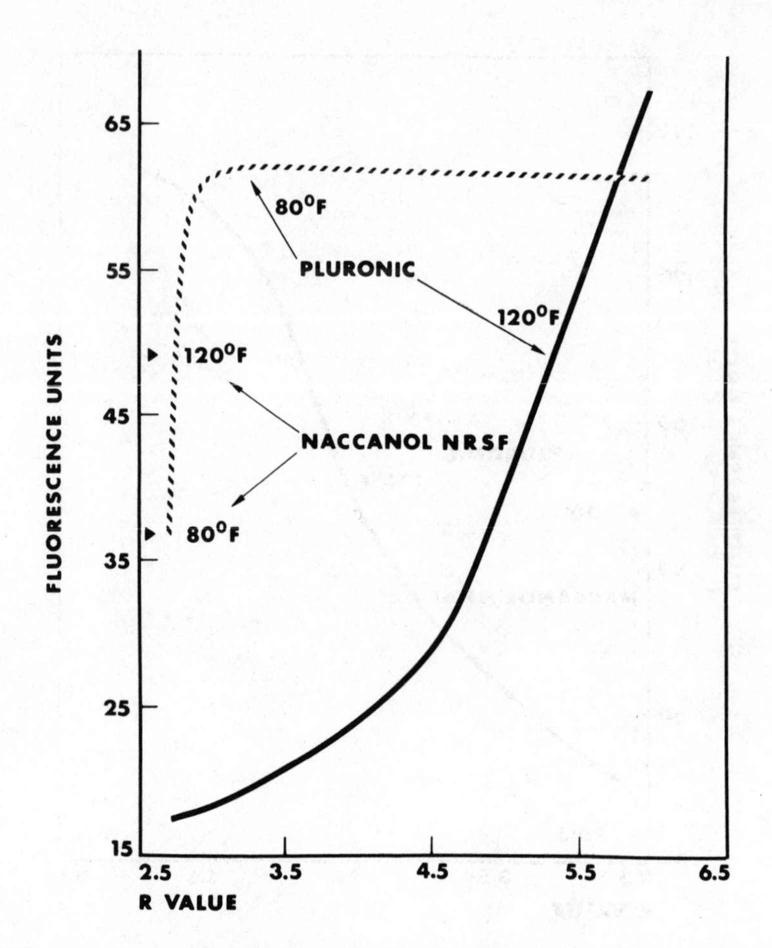

FIG. 28. Pluronic surfactants; whitener TA on nylon. From Ref. 46.

relatively low prewhitening levels on the fabric will the FWA in the softener be effective. This is demonstrated in Fig. 29. Figure 30 shows the effect of an FWA in softener after one, two, and three consecutive rinse cycles.

Zwitterionic and semipolar compounds. It is claimed in U. S. Patent 3, 309, 319 [104] that the effectiveness on nylon of whitener TA in an anionic sodium dodecyl benzene sulfonate-based built detergent, for example, can be improved by replacing part of the ABS with (or by adding to the ABS) a zwitterionic or semipolar detergent compound. Findings are said to be unique for whitener TA. Mentioned as zwitterionic compounds are quaternary ammonium compounds and such compounds as 3-(N, N-dimethyl-N-

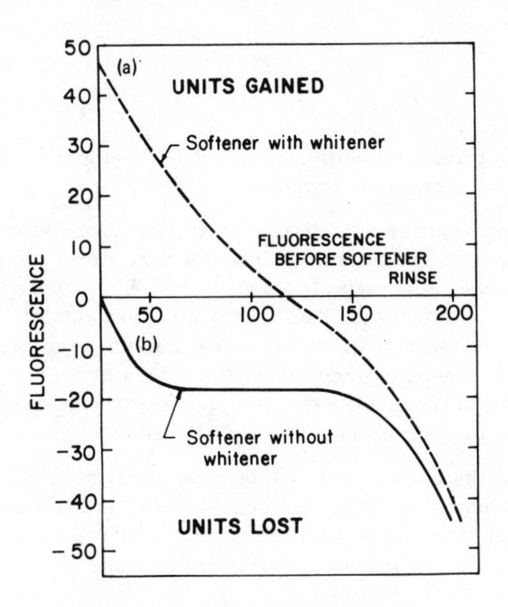

FIG. 29. Effect of softener (a) with and (b) without whitener. From Ref. 48.

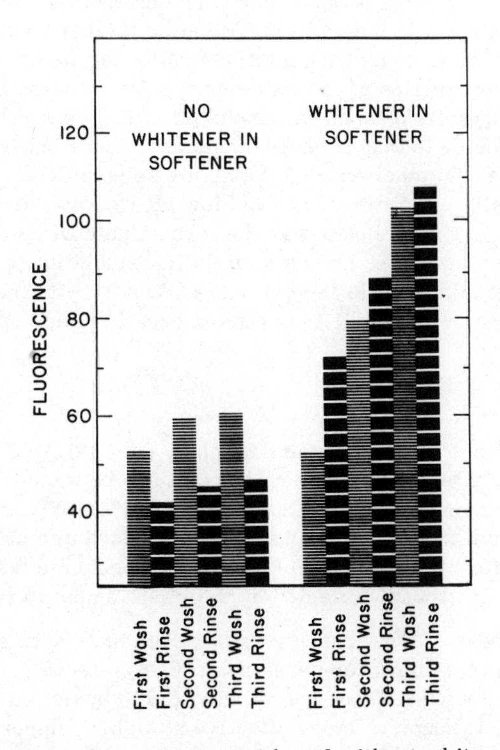

FIG. 30. Effect of softener rinses with and without whitener. From Ref. 48.

hexadecylammonio) propane-1-sulfonate and 3-(N, N-dimethyl-N-dodecyl-ammonio)-2-hydroxy propane-1-sulfonate.

While an anionic-zwitterionic detergent system gives best results, semipolar compounds such as dimethyldodecylamine oxide, dimethyl-dodecylphosphine oxide, and methyl-3-dodecoxy-2-hydroxypropyl sulfoxide are also claimed to be effective. The improved nylon whitening of whitener TA in the presence of these zwitterionic and semipolar compounds might be related to increased ion-pair formation of the FWA or to a higher degree of swelling of the fiber in the presence of the specified compounds. No explanation is offered for this phenonmenon in the patent.

Because one of the main topics of this book is surfactants, the effect of these on the performance of FWAs has been covered in some detail. There are, however, other wash liquor conditions which have equally as important an effect on the behavior of the FWA during application.

b. pH

Changes in pH of the wash liquor can cause changes in the "affinity" of the FWA for the aqueous phase and can also alter its distribution between liquor and fiber. Changes in pH also influence the degree of swelling of the fiber and leveling properties of the whitener. Effectiveness of DMDDEA and DSBP are essentially independent of the pH of the wash bath. The build-up of DMEA is inferior to that of DMDDEA in acid sours but is better in alkaline detergents. In other words, the highly soluble CC/DAS whitener (DMDDEA) is equally effective at high and low pH values, but its affinity for cotton is lower in an alkaline bath than the less soluble DDEA, DMEA, DM, and TA. The latter products, having a fairly low solubility in an alkaline liquor, become partly insoluble in acid sours. Contrary to the low solubility CC/DAS types, whitener NTS is almost equally effective at low and high pH values.

c. Temperature

The cotton affinity of most of the commonly used CC/DAS derivatives is not significantly affected by temperature changes between 70 and 160°F (the range of U. S. home laundering temperatures) if the FWAs are added in predissolved form to the wash liquor. But, if added in a solid form with the detergent powder, the rate of solution (and hypochlorite bleach stability) of some whiteners is strongly affected by the wash temperature.

The effectiveness of slow-rate-of-solution products such as TA (Table 17) increases with increasing temperatures. Fast-rate-of-solution products are only slightly affected (DM) or not affected (NTS) by the wash temperature. Whitener BBO becomes more effective at higher temperatures because polyester fiber is easier to penetrate above its glass transition temperature.

TABLE 17

Effect of Temperature

Whitener (%)	Substrate	Wash temperature		
		80°F	130°F	160°F
		W	W	W
0.4 TA (slow)	C	100.7	114.8	124.7
0.4 DM	C	122.6	126.7	127.7
0.1 NTS	C	107.3	108.7	107.7
No whitener	C	90.3	91.3	94.7
0.1 BBO	P	86.9	91.3	95.7
No whitener	P	—	83.5	—

Source: From Ref. 62.

d. Inorganic electrolyte

The solubility of and, therefore, the distribution between wash liquor and substrates of some whiteners is affected by the concentration of inorganic electrolyte in the liquor. Fluorescent whitening agents that require a certain concentration of electrolyte to exhaust are not suited for electrolyte-free systems. Yet high concentrations of electrolyte may force certain whiteners out of solution, making the product an ineffective precipitate.

e. Agitation

The amount of mechanical work applied during washing will increase the rate of solution of an FWA and also influence its distribution between wash liquor and cloth. High agitation is desirable, of course.

f. Whitener/detergent/fabric/liquor ratio

Variations in ratio will affect the performance of whiteners. For example, at a constant detergent/fabric ratio, a decrease in fabric/liquor ratio will affect the rate of solution. If the CC/DAS whiteners are already dissolved in the liquor, changes in the fabric/liquor ratios commonly used in the laundry will not affect performance significantly. Whitener DSBP will on the other hand, because of its extremely high solubility, exhibit a higher affinity at a 1:10 than at a 1:30 fabric to liquor ratio.

g. Wash cycle duration

Both step 1 and step 3 (Fig. 14) are relatively slow processes, the rates of which depend on the whiteners used. In other words, the products might perform equally well in wash cycles of 20-min duration, yet in a shorter cycle one whitener could be more inferior to the others.

h. Hypochlorite stability

Active chlorine products affect the stability of some FWAs. The extent of attack on the so-called bleach-unstable whiteners depends on their chemical structure, their rates of solution, the amount of bleach used, the timing and order of addition of all ingredients of the wash, and temperature, etc.

The longer the whitener remains in the wash liquor in the presence of active chlorine, the more whitener will be destroyed. It is preferable, therefore, to delay the hypochlorite addition until most of the whitener is adsorbed on the substrate. Protected by the substrate, most bleach-unstable FWAs are not significantly affected by hypochlorite. As previously mentioned, differences in stability to hypochlorite and dichloroisocyanurate bleaches exist even among the CC/DAS derivatives. The bleach stability of these compounds is directly or indirectly related to the amine which is used for the reaction with the second reactive chlorine on the triazine ring [62] during the manufacture of the whitener.

Whitener NTS, which has no secondary amine groups, is quite bleach stable. Whiteners like BBO, NOS, and BBI, which are practically water insoluble, also exhibit good bleach stability. An oxidation of these whiteners by NaOCl is more difficult because the whiteners are not dissolved in the aqueous phase and are thus not as susceptible to attack by the bleach. In the case of DSBP not containing any nitrogen group, the type of amine group is not a factor in regard to bleach stability.

Two chemically equivalent whiteners may react differently to the same bleach, depending on the time required for the whitener to go into solution, the time spent in the aqueous phase and the time it takes for the whitener to exhaust onto the fiber.

Differences in bleach stability of whiteners and the effect of exposure time are shown in Table 18. Because the bleach also can improve the whiteness of the substrate, W (whiteness) values (see Sec. IV) reflect improvements in whiteness caused by both bleach and the FWA. The F values (see Sec. IV) are better approximations for the amount of intact whitener exhausted onto the fabric. The whiter the substrate, of course, the more efficiently the FWA emits radiation.

The reactivity of hypochlorite increases with increasing temperatures. Rate of solution (step 1) and the rate of steps 2 and 3 are also temperature dependent. For the fast-rate-of-solution whitener DM, the bleach stability decreases with increasing temperatures (Table 19). It is interesting to note, however, that the slow-rate-of-solution type TA whitener is more effective at 130 and 160°F than at 80°F, and also that the slow TA type exhibits better whiteness than DM at the higher temperatures. Normally DM is regarded as being more bleach stable than is TA.

TABLE 18

Bleach Stability

Whitener (%)	Substrate	Exposure time: No bleach		20 sec		3 min	
		W	F	W	F	W	F
0.4 TA (fast)	C	124.9	403	104.0	174	99.3	98
0.4 DM	C	124.9	407	113.8	265	101.1	126
0.1 NTS	C	107.2	299	111.5	284	109.1	270
No whitener	C	91.5	20	93.2	—	93.4	—
0.1 BBO	P	91.3	122	90.8	119	91.1	113
No whitener	P	83.5	10	—	—	85.0	—

Source: From Ref. 62.

TABLE 19

Effect of Temperature on Bleach Stability
(3 min of Exposure)

Whitener (%)	Wash temperature: 80°F		130°F		160°F	
	W	F	W	F	W	F
0.4 TA (slow)	100.1	99	105.8	147	104.1	139
0.4 TA (fast)	—	—	99.3	98	—	—
0.4 DM	119.6	242	101.1	126	93.8	37
No whitener	93.1	20	93.1	—	94.3	—

Source: From Ref. 62.

At the higher temperature in the case of slow-rate-of-solution TA, the hypochlorite bleach reacts with other components in the wash liquor and becomes deactivated before all the TA (slow) has gone into solution. During the last minutes of the wash cycle, the whitener deposits on the fabric and finally exhibits a higher fluorescence than do the fast-rate-of-solution products.

As mentioned previously, the bleach stability of a whitener is to some extent affected by the type of surfactant used. Whiteness values obtained in an anionic- and a nonionic-built detergent with frequently used detergent whiteners, are shown in Table 20. Whitener DM is equally effective in both systems without bleach so that the better whiteness in the nonionic detergent

TABLE 20

Effect of Surfactant on Bleach Stability
(3 min of Exposure)

Brightener (%)	Substrate	Anionic	Nonionic
		W	W
0.4 TA (fast)	C	99.3	100.6
0.4 DM	C	101.1	104.2
0.1 NTS	C	109.1	103.6
No whitener	C	93.4	93.4
0.1 BBO	P	91.1	88.5
No whitener	P	85.0	85.0

Source: From Ref. 62.

is believed to be related to the protective action of the latter surfactant. The protective action of the nonionic may be related to the way this surfactant forms a micelle or aggregate with the whitener of a type different from that obtained in an anionic system. Another possible explanation for the apparently better bleach stability of the FWA is an interaction between the oxidant and the nonionic surfactant resulting in a less active bleach. A comparison of effectiveness of the bleach stable whiteners NTS and BBO, respectively, in the anionic and nonionic systems in the absence of bleach, indicates that the lower effect observed with bleach in the nonionic system is related to inferior exhaust properties of the whiteners in the latter system, and not to reduced bleach stability.

E. Whitener on Fabric

1. Whiteness

The ultimate objective of employing FWAs in detergents is to improve the whiteness of the textiles being washed. In addition to factors already mentioned, the selection of a specific FWA is, of course, mainly dictated by its contribution to improving the whiteness of the substrate. This contribution is a function of the whiteness of the substrate to be treated. The "base" whiteness varies among textiles, not only among fiber types, but also as a function of pretreatment, by the degree and type of prewhitening, by the degree of yellowing and soiling, and by the type of FWA previously applied to the substrate.

Most textile garments reach the consumer with high initial whiteness. As mentioned earlier, most polyester fibers, for example, are mass whitened during their manufacture. Cotton, polyester/cotton blends, and nylon are whitened in the textile mills during finishing. Lowering of the initial whiteness is caused mainly by soil buildup and loss of whitener effectiveness on exposure to light during wear of the textiles or during drying. Light-, wash-, and bleach-fastness of mass-whitened and textile mill-whitened synthetic fibers are good. The CC/DAS textile mill cotton whiteners are not as fast to light. Table 21 shows light fastness of typical mass and textile mill FWAs on different fibers and also shows the graying/yellowing tendency of white garments on washing and wearing as a function of fiber substrates.

Whiteners used on cotton are those least stable to light. Fortunately, cotton is the fiber most economically and easily whitened in the home laundry. In general, light fastness of good quality textile mill whiteners on nylon is better than for those on cotton.

2. Fastness

The light fastness of polyester whiteners is excellent. The fabric fluorescence of polyester in a 65:35 polyester/cotton fabric, for example, was reduced only by an amount equivalent to a 15% decrease in FWA concentration on exposure to 20 average days of sunlight [47]. Consequently, the mass whitener in polyester will, to a large extent, remain intact during the life of a garment. The goal of using the fluorescent whitener in a detergent must be to improve the whiteness of as many as possible of different bases of white. Not one, but a combination of FWAs (e.g., one bleach-unstable cotton FWA, one bleach-stable cotton and nylon FWA, and one product for fine fabrics) may be used. The combination of FWA with multifunctional properties should be so selected that it contributes to optimum whiteness of all fabrics laundered. The lower the base whiteness, the more significant is the contribution of the FWA in the detergent.

TABLE 21

Whiteness of Washload

Substrate	Typical light fastness[a]	Ease of laundry whitening	Graying/yellowing tendency
Cotton	2	Easy/low cost	Low
Nylon	3-4	↓	↓
Cotton/polyester	4-5		
Polyester	6	Difficult/high cost	High

[a]Wool blue scale rating.
Source: From Ref. 3.

The effectiveness of laundry whiteners decreases with increasing prewhitening; addition of some detergent whiteners can, at high levels, even result in an unfavorable change in the hue of whiteness. Interference between hue contribution ofthe detergent whitener and the textile-mill whitener can also occur. The limited fastness of cotton whiteners will create a greater reliance on detergent whiteners for this substrate. No unfavorable hue change, therefore, is expected due to the buildup of the cotton FWAs. Consequently, when testing and selecting a detergent whitener, cotton can be regarded as an unwhitened fabric.

Whiteners used to obtain initial whiteness of nylon and especially of polyester fabrics, on the other hand, exhibit good to excellent fastness. For correct testing and selection of a nylon and polyester whitener, therefore, the detergent FWA should be evaluated on fiber or textile-mill prewhitened fabrics (containing a typical product) as well as on unwhitened fabrics.

Since the detergent FWA does not build up to any significant degree on mass-whitened polyester fibers, no interference with the hue of the mass whitener is to be expected.

IV. EVALUATION OF WHITENESS

Just as individuals differ in their opinion of what smells or tastes good, they also differ in which kind of white fabric is the most preferred or which white fabric is the better white. We are dealing here with a psychological as well as a physiological phenomenon. Physiological differences, too, exist among individuals. It is well known that the yellow or brownish pigmentation of the human eye increases with age [105]. In the blue region of the spectrum, therefore, the response of individuals of different age groups will differ. In addition, the yellow-blue sensitivity is a function of retinal position and luminance level [106].

Whiteness can be evaluated by visual comparison of (1) one white substrate with another or (2) by comparison of a white substrate with an imaginary standard "stored" in the memory system of the observer. The opinion of a large number of observers is required to arrive at a valid average observer response in regard to the preferred white.

Comparisons of visual and instrumental findings have enabled investigators to correlate observer preference with measured whiteness. There are many pitfalls, however, in correlating visual and instrumental whiteness results. These have been discussed in Refs. 107-149.

The available instrumental evaluation techniques more or less accurately reflect visual results. Some of the important factors affecting both visual and instrumental whiteness evaluation will be discussed next.

A. Physical Factors

As mentioned above, an individual's preference for the whiteness of one fabric over another is related to a number of factors, illustrated in Fig. 31 and discussed below in detail.

1. Light Source

The light energy reaching the eye of an observer reflected from an unwhitened substrate is a function of both the substrate and the relative spectral energy distribution of the light source in the visible region. The "white" observed when magnesium oxide (with close to 100% reflectance between 400 and 700 nm) is illuminated, for example, is the color of the light source itself. If bleached, whitened cotton fabric is substituted for MgO, the observed color under the same light source will appear yellower because the cotton fabric absorbs light in the blue region. On the other hand, the observed color of the near-white fabric will change if the spectral energy distribution of the light source is altered. An unwhitened cotton fabric may appear to have quite different hues when observed under different light sources. Therefore, it is important, when evaluating unwhitened substrates, to have a standard light source with constant relative energy distribution in the visible region.

In evaluating whitened (containing an FWA) fabrics, it is important to have a light source with a known and constant spectral energy distribution, not only in the visible but in the ultraviolet region as well. At high levels,

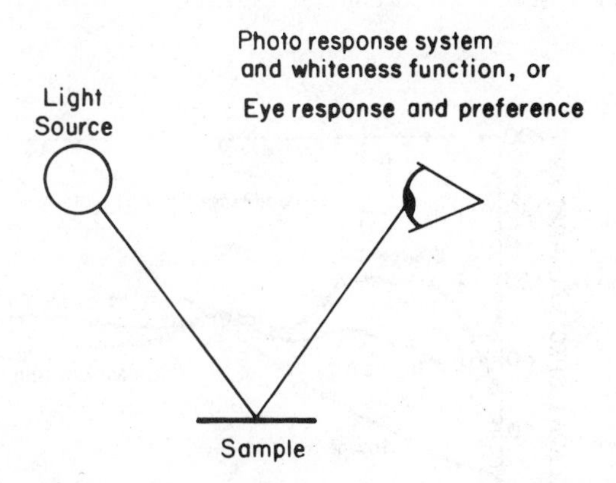

FIG. 31. Whiteness evaluation. Factors: 1. light source; 2. geometry of illumination and observation; 3. substrate (light absorption, surface properties, gloss, transparency, and moisture content of fabric); 4. photo response of observer or of instrumental recording system); 5. preference (visually by observer, instrumentally by whiteness function).

the effectiveness (yield) of the whitener depends on the amount of energy available in the near-ultraviolet region. Therefore, the hue of white as seen by the observer, is a function of the quality and quantity of the light emitted by the light source in both the visible and the ultraviolet region. A light source with a standardized, constant light emission between 300 and 750 nm, therefore, is required to obtain reproducible findings when comparing FWA-treated white substrates.

Standard light sources for color evaluation were recommended in 1931 by C.I.E. (Commission International de l'Eclairage). Figure 32 shows the relative spectral irradiance of illuminants A, B, and C. Source A is typical of the gas-filled incandescent lamp; source B is typical of noon sunlight; source C is typical average daylight [105]. Source D_{65} with spectral radiance between 300 and 750 nm, a region important to whitener evaluation, was recently defined [150-152]. The relative energy distribution for D_{65} is shown in Fig. 33. The best artificial daylight source for whitener evaluation has been discussed by several authors [129, 131, 133, 153, 154]. The energy distribution of average daylight in the ultraviolet region can be approximated by using a tungsten lamp (3100°K), if the energy distribution of the lamp in the visible part of the spectrum is corrected with filters. For another approach, a combination of two tungsten lamps can be employed. The visible light of one of the lamps can be excluded by filters and the ultraviolet portion added to the light emitted by the other. This procedure allows the use of artificial daylight with varying but measured amounts of ultraviolet energy. However, the tungsten lamp tends to "darken" (The tungsten filament sublimes and condenses on the glass) so that the results obtained are not reproducible. The tungsten lamp, therefore, is not suitable as a standard light source for whitener evaluation.

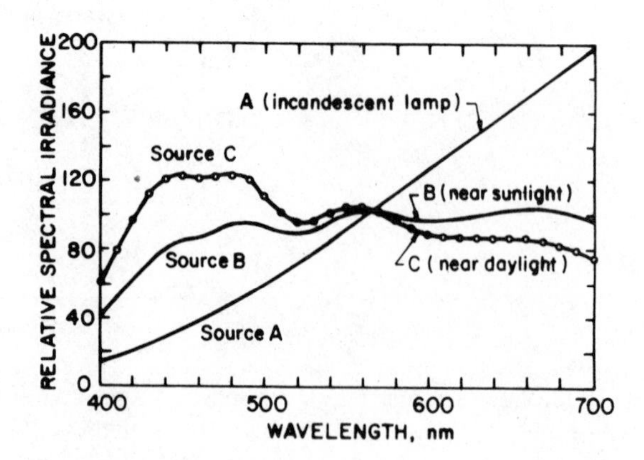

FIG. 32. C.I.E. light sources A, B, and C. From Ref. 3.

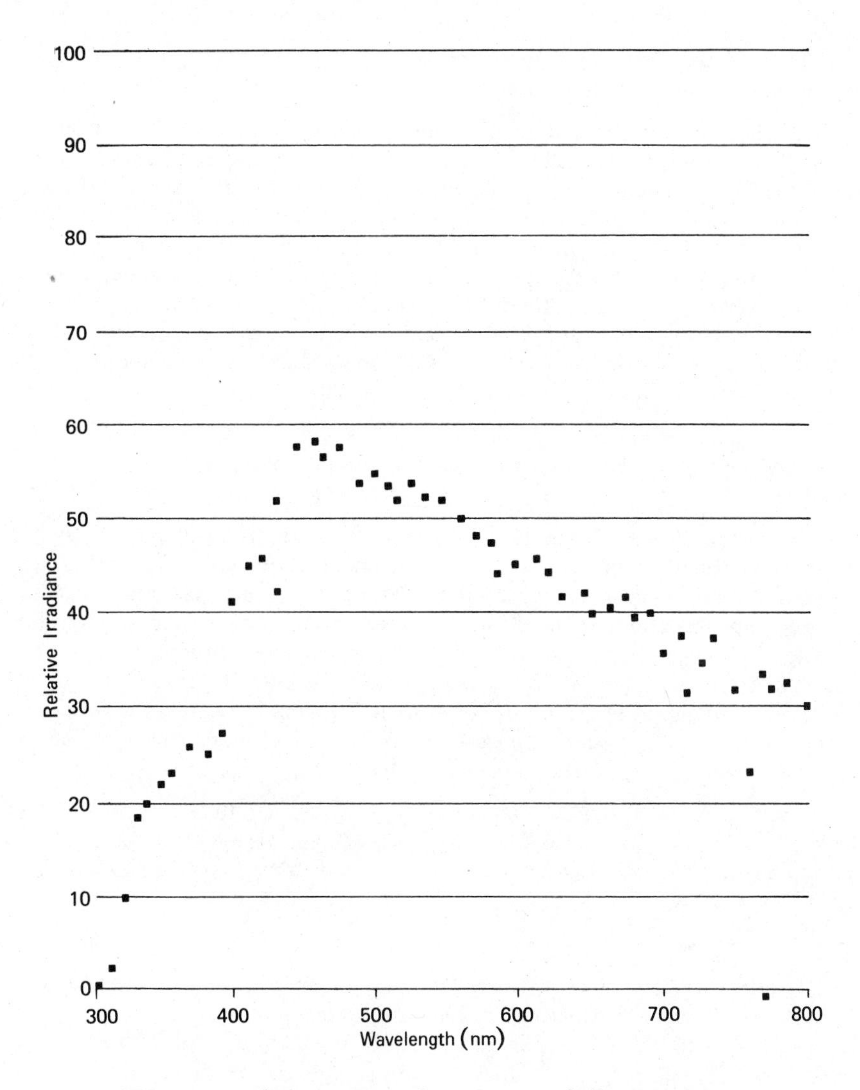

FIG. 33. Relative energy distribution of illuminant D_{65}.

The quartz-iodine lamp, relatively richer in ultraviolet radiation than is the tungsten, appeared at first to be well suited for whitener evaluation. Recent studies, however, indicate that the spectral irradiance of the currently available quartz-iodine lamp does not remain as constant during use as was initially believed. Further work must be done to establish the best conditions for proper operation of the quartz-iodine lamp before this light source can be put into general use.

Fluorescent light tubes are being used as D_{65} simulators but are hardly a good choice for viewing whitened fabrics.

Artificial "daylight" can also be approximated by the use of a xenon lamp/filter combination. The spectral properties of different xenon lamps vary somewhat in quality. A slight shift in the position of the arc will cause erroneous results if the xenon lamp is used as a light source in front of a narrow slit during instrumental measurements. In addition, intensity variations of improved xenon lamps are still a problem when measuring a fluorescent fabric against a nonfluorescent standard [133]. The xenon lamp/filter combination although probably the best simulator, is not the ideal light source either. In other words, a perfectly stable, reproducible light source is not yet available for whitener evaluation.

Instruments used for colorimetric FWA evaluation measurements are most often equipped with either a tungsten lamp (Hunter, Color-Eye, Gardener, etc.) or a xenon lamp (e.g., Elrepho).

In the United States whitened substrates are usually compared by eye under North daylight through a window; or under artificial daylight obtained by mixing light of incandescent lamps, fluorescent lamps, and black lights; or under special lamp combinations supplied by companies such as MacBeth Corp., Newburgh, N.Y. Sometimes, only incandescent light is used. The combination of whitener can be overemphasized if black light is included in an artificial daylight source. If FWAs with different excitation spectra are compared under such light, the whitener with maximum absorption close to the peak emission of the lamp will be favored.

One should keep in mind that although daylight, whether natural or artificial, is preferred for FWA evaluation, this is not necessarily the light quality most frequently available when the housemaker inspects the laundry.

2. Geometry

An ideal white surface reflects diffusely in all directions. Goniophotometric curves show that cotton broadcloth, for example, is an almost perfect white diffusor [131]. Therefore, the angle of viewing is not too critical in the case of such material. Broadcloth samples can be inspected or measured under direct or diffuse illumination or under a combination of direct and diffused light without any significant effect on findings. However, instruments with different geometries can give different results when fabrics with some directional reflectance are tested. A standardized geometry should be prescribed in such cases. The effect of illumination and viewing geometry on the whiteness evaluation of whitened fabrics should not be overlooked.

Illuminating and viewing angles most commonly used in the United States are (1) direct 45° illumination and 0° viewing such as in Hunter Color and Color Difference Meter, Model D25; and (2) diffuse illumination and 0° or near 0° viewing angle as in the Beckman DK-2 spectroreflectometer.

3. Substrates

Ultraviolet light is required to excite the fluorescence of FWAs. Soil and fibers (naturally yellowed or pigmented during manufacture) that absorb light in the near ultraviolet range will compete with a whitener that is absorbing light energy in the same wavelength region; this, of course, diminishes the effectiveness of the whitener. Ultraviolet and visible reflectance curves (relative to $BaSO_4$) of some fibers appear in Fig. 34. Bleached cotton exhibits a reflectance of more than 90% between 700 and 500 nm; the reflectance decreases somewhat as the wavelength drops between 550 and 300 nm. Polyamide, polyester, and acrylic textile fibers are pigmented with TiO_2 and will absorb strongly below 400 nm.

Depending on the type, soil can also absorb both ultraviolet and visible light. Figure 35 shows reflectance curves for both soiled and unsoiled cotton. Heavy metal stains, dyes bleeding from colored fabrics, oxidized fatty oils, etc., are typical light-absorbing soils. If fiber and soil absorb strongly in the visible region as well as in the ultraviolet, the total reflection of light will be low and the addition of emitted light contributed by the whitener will not be enough to produce a satisfactory white. Because of this, an FWA cannot cover up dirt to any significant extent. An FWA can produce a white of good quality only if the detergent in the wash liquor, perhaps with the assistance of bleach, has done a good job of removing the soil from the fabric. Whiteners with different excitation and emission properties can be affected differently by the specific spectral characteristics of the soil present, a point which must be considered in whitener evaluation.

For both visual and instrumental comparison, the moisture content of swatches should be kept constant, since whitening power increases with the

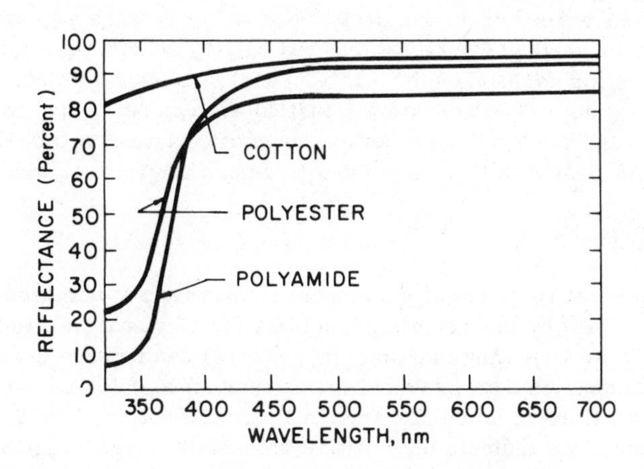

FIG. 34. Reflectance; unwhitened fibers.

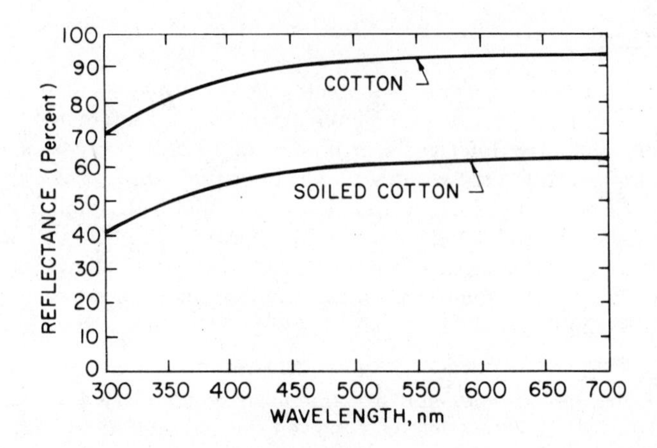

FIG. 35. Reflectance of soiled and unsoiled cotton. From Ref. 3.

moisture content of the substrate. The effectiveness of most FWAs increases with rising moisture content; however, the reverse is true of some FWAs. Heating of fabrics, therefore, should be avoided during evaluation.

The effect of surface properties, gloss, and transparency on observed whiteness are additional factors to consider when fabrics of different fibers or fabric construction are compared.

B. Physiological/Psychological Factors

While physical factors involved in whitener evaluation can be defined and conditions agreed upon, the preference of the average housemaker for one white over another can be established only by submitting a large number of whites for comparison by a large panel of observers. Physiological and psychological differences among individuals can be studied in this manner and an "average" observer response (if existent) established. The following will deal with factors related to human response to whiteness.

1. Photo Response

The spectral response of the standard observer recommended by C. I. E. in 1931 is defined by the weighting functions for 2° visual field shown in Fig. 36. These weighting functions ($\bar{x}_\lambda$, $\bar{y}_\lambda$, $\bar{z}_\lambda$) can be used to reduce spectrophotometrics data (physical measurement) to colorimetric terms (psychophysical expression) like tristimulus values X, Y, and Z. The weighting functions indicate the positive amounts of three imaginary primaries, X, Y, and Z, required to match unit amount of monochromatic

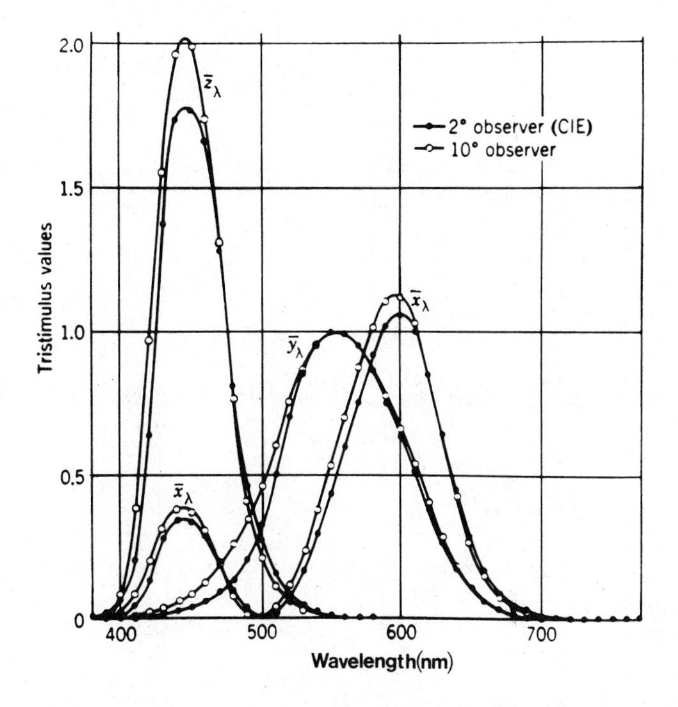

FIG. 36. Color-matching functions. From Ref. 3.

light. The weighting functions are related to the spectral sensitivity of the human eye. The spectral radiance of whitened fabrics illuminated by a defined light source can be analyzed by a tristimulus colorimeter. This instrument has a red, green, and blue photo-response system which more or less closely (depending on how well duplication of the C.I.E. color matching function has been achieved) measures the X, Y, and Z tristimulus values of the light radiated by the sample. These values can be converted into one of the four chromaticity systems listed in Table 22.

The x, y coordinates of samples can be plotted in the C.I.E. chromaticity diagram (Fig. 37) to obtain information on the dominant wavelength, percentage purity, and lightness (Y axis perpendicular to x, y plant). The location of illuminants A, B, and C are indicated in the figure. The diagram shows that incandescent light is, of course, yellower than sunlight and that sunlight is yellower than daylight. Coordinates for two whitened cotton swatches (1 and 2) are also plotted. Sample 2 contains a much higher whitener level than sample 1; addition of whitener generally results in smaller x and y values. The relative hue of a white sample can be obtained easily from the chromaticity diagram. But the diagram gives no information on whiteness because the relative importance of luminosity and chromaticity

TABLE 22

Chromaticity Systems

1. 1931 C.I.E. Chromaticity Coordinates
$x = X/X + Y + Z)$
$y = Y/(X + Y + Z)$
Y = luminosity
Nonuniform chromaticity system
2. 1960 C.I.E.-U.C.S. Chromaticity Coordinates
$u = 4x/(-2x + 12y + 3)$, $v = 6y/(-2x + 12y + 3)$ — Uniformity Chromaticity Scale
Y = luminosity — Nonuniform
3. 1963 Wyszecki Chromaticity Indexes
$W = 25Y^{1/3} - 17$ — Lightness index
$U = 13W(u - u_0)$, $V = 13W(v - v_0)$ — Chromaticness index
u_0, v_0 = coordinates of achromatic color
Tridimensional uniform system
4. Hunter Uniform Color System
$L = 10Y^{1/2}$
$a = 17.5\ (1.02\ X - Y)/Y^{1/2}$
$b = 7.0\ (Y - 0.847Z)/Y^{1/2}$

Source: From Ref. 3.

is not indicated. The C.I.E. chromaticity diagram is illustrated in color in Ref. 155.

2. Preference

Whiteners are added to overcome a deficiency in blue reflectance of a fabric. Ideally, the light emitted by the whitener should, when added to the spectral radiance curve of the substrate, result in a hue of white close to that preferred by the observer. The degree and spectral character of the

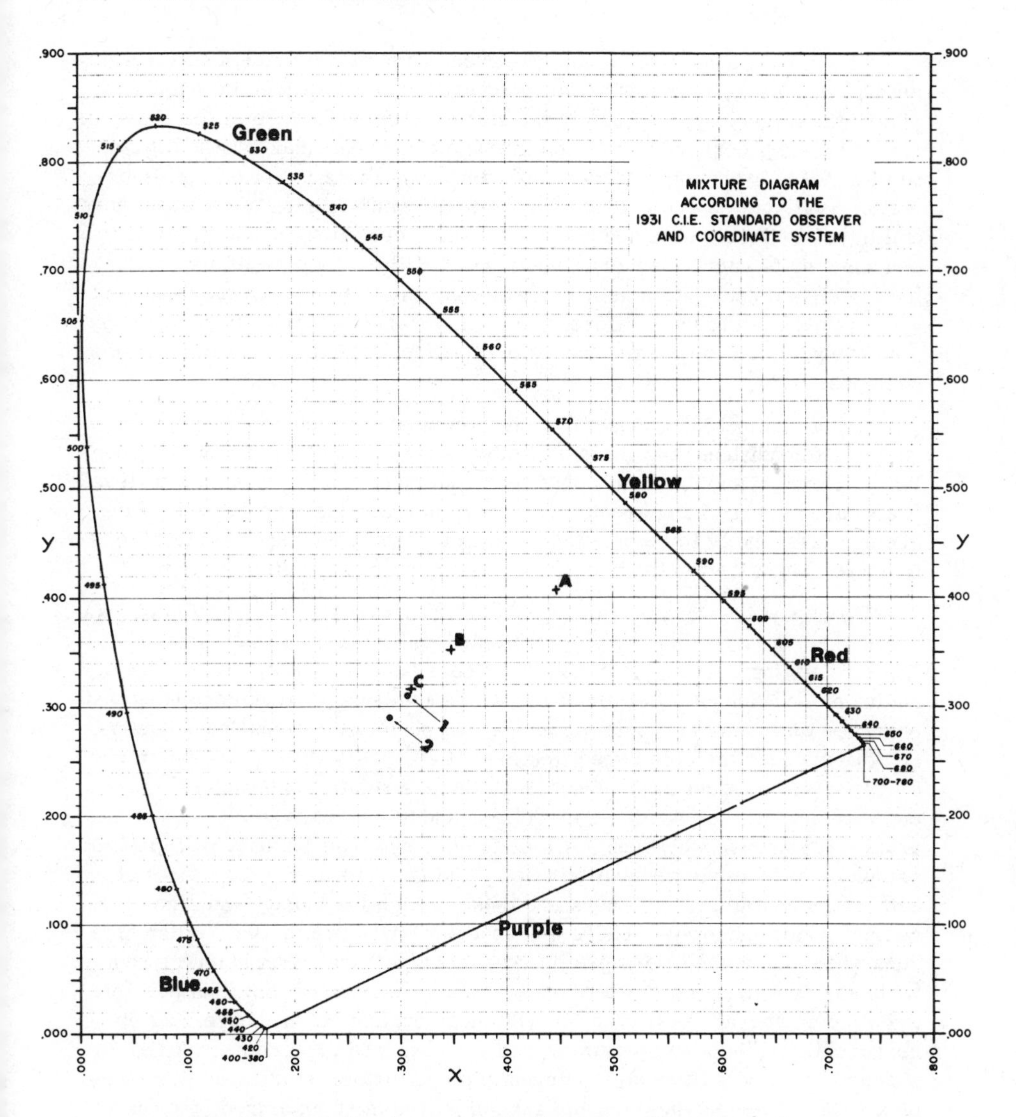

FIG. 37. C.I.E. chromaticity diagram. From Ref. 3.

deficiency in blue reflectance will vary mainly as a function of light source, geometry, surface and light absorbance of substrate. The same factors determine, to a certain extent, the quantitative contribution of the whitener. Consequently, only in some cases will the whitener be able to contribute, in combination with substrate, the type of emission required to approach the color preferred by the observer.

The eye responses to human observers with no eye defects (color blindness), with the exception of differences caused by variations in macular pigmentation, are approximately the same. The preference of the observer (we are mainly concerned with the average U.S. housemaker) when it comes to whiteness, however, differs. Unfortunately, there is not enough information available on preferences for whitened white fabrics of different hues and lightness levels obtained under controlled conditions. The average U.S. housemaker is said to prefer a bluish white with a slight violet cast [133], but little published data exist to substantiate this. However, the "preferred" white is believed to be a different color in different geographical locations; also, people tend to change, from time to time, their preference regarding whiteness [120].

Goldwasser [135] conducted a study comparing preferences (under a variety of conditions) for laundered fabrics treated with "reddish," "greenish," "bluish," and "neutral" brighteners at equal energy of emission levels. It appeared that red and green were either most or least preferred. Blue seemed least likely to be on either extreme of the preference scale. Neutral was ranked better than, equal to, or inferior to the other hues.

Vaeck and co-workers discussed [111-113, 144-146] correlation of visual and instrumental evaluation of nonwhitened and whitened white substrates. After reviewing results published by authors such as Selling and Friele, Sugiyama and Fukuda, Berger, Coppock, and Jeanne, Vaeck concludes that observed whiteness ". . . is a complex but perfectly rational phenomenon, and that besides the color specifications of the near-white surfaces, many other factors have an influence." Such factors relate to personal preference, mainly in the green/red direction; and an increase in blueness appears always to increase whiteness. Other factors are transparency and gloss of samples, and sample separation when viewing. The preferred whites of substrates containing whiteners are those of high luminosity and high blue content (limit unknown). There are observers who prefer the reddish blue hues while other observers prefer greenish blue hues. Based on his own findings, as well as on those of others, Vaeck concludes that it is possible to define an average observer for whiteness evaluation, although real observers differ, especially with regard to green/red preference. A C.I.E. subcommittee on whiteness is currently doing further studies on this complex area. The conclusion of the subcommittee is summarized by Ganz [156].

C. Visual Evaluation

The importance of correct choice of light source, illumination, and viewing geometry has already been discussed. For meaningful results, the relative positions of fabric samples must be exchanged during evaluation [157]. The distance between the samples does not seem too important [113], provided that the background color does not introduce an effect. Usually, laundered white fabrics are compared for whiteness under artificial daylight in a gray room (neutral gray, Munsell value 7.5 to 8.5) or under natural North daylight inside a window. Surrounding colors can affect what the observer sees. For example, a certain yellow may look greener against a red background; a green background makes the same yellow appear redder. Similarly, a green lawn or a red brick wall could affect the white seen, for example, by a housemaker.

Many laboratories use specialists to evaluate whiteness. Indications, however, are that specialists are influenced by their experience and will rate whites differently than will untrained observers. The U.S. laundry industry must please the U.S. housemaker; randomly selected female observers appear to be the best judges [24]. Female observers also have other advantages. They are rarely color blind and have better self-correlation [109] than men (individual self-correlation coefficients for women ranged from 0.65-0.92 and for men they were in the 0.24-0.72 range). In all types of panel testing, of course, the greater the number of observers, the higher the reliability of the test.

The relative effectiveness of whiteners or of whitener systems can be determined by panel testing. Fabrics washed under standard conditions, dried, and reconditioned, can be evaluated under controlled humidity conditions in paired comparison, by ranking from most white to least white, or by any proven statistically based method.

Several companies determine the whiteness of laundered swatches by means of a whiteness scale. Whiteness scales (in intervals of 10 units between, for example, 0 and 260) consist of a series of whitened fabrics. The relative visual increase in whiteness between each step in the series is the same throughout the scale. The whitener concentration, on weight of the fabric, increases roughly exponentially with increasing whiteness. Samples, preferably on the same substrate as the fabrics of the scale, are compared (by specialists) with the scale under standard lighting conditions. The sample is assigned the whiteness number of the comparable fabric on the scale. The weakness of this method is related to the specialist's preference for a certain hue, as noted above, which may not necessarily be the housemaker's preference.

D. Instrumental Evaluation

1. Instruments

Three general types of instrument are used to evaluate whitened fabrics: fluorometers, tristimulus colorimeters, and spectroreflectometers. Accuracy and reliability generally increase with cost of the instrument.

Fluorometers (such as those supplied by Jarrell-Ash, Engineering Equipment, Photovolt, and Turner, etc.) measure only gross fluorescence values. These do not give any indication of spectral radiance and, strictly speaking, can be used only to compare the effectiveness of whiteners having the same excitation and emission characteristics and even those only within a concentration range in which the hue remains constant. Fluorescence values (logarithmic functions of whitener concentration) can serve as a first approximation and are quite useful when buildup, light fastness, and hypochlorite stability, etc., are being studied. Such readings are also of value in determining whether differences in the appearance of whitened fabrics are related to differences in buildup of the whitener on the fabric or to differences in hue.

A better approximation of the relative whiteness exhibited by whiteners or whitener systems can be obtained with a high quality tristimulus colorimeter (such as those supplied by Hunter, Gardner, Instrument Development Laboratories, and Carl Zeiss, etc.). However, tristimulus colorimeters are not particularly well suited to recording small differences in emission spectra of whiteners, differences recorded by the observer, because of the large half-widths of the tristimulus filters. An additional weakness of tristimulus colorimeters is that the currently used $\bar{z}$ function (2° field) does not put enough weight on the region close to 400 nm, the region in which many whiteners absorb and emit light. Use of a more violet tristimulus $\bar{z}$ function has been suggested [131].

The $\bar{z}$ function for the 10° field is about 4 nm to the violet side. Hunter [131, 158] found that the latter $\bar{z}$ function gave good correlation of instrumental with visual ratings when a series of cotton fabrics with increasing concentrations of whitener was evaluated. No data are available, however, on the best blue function for comparing fabrics treated with whiteners of different hue. At high levels, such whiteners have different spectral absorption and emission in the 400-nm region.

Spectral radiance curves for unwhitened (A) and whitened cotton (B) are superimposed on the $\bar{x}_\lambda$, $\bar{y}_\lambda$, $\bar{z}_\lambda$ weighting function curves in Fig. 38. The addition of whitener to unwhitened fabric will cause an increase mainly in the Z value, to some extent in the X value (secondary maximum), and very little, if any, in the Y value.

Instrument Development Laboratories' Color Eye Model LS (Fig. 39), an abridged spectrophotometer with integrating sphere, permits measurements of tristimulus values and reflection values at 16 different wavelengths, using filters distributed evenly in the visible region. Some of the filters in the long-wavelength region can be exchanged for additional filters in the short-wavelength region. Additional readings then can be made below 550 nm in the region of main interest. Color Eye LS, used as an abridged spectrophotometer, permits a more detailed measurement of spectral

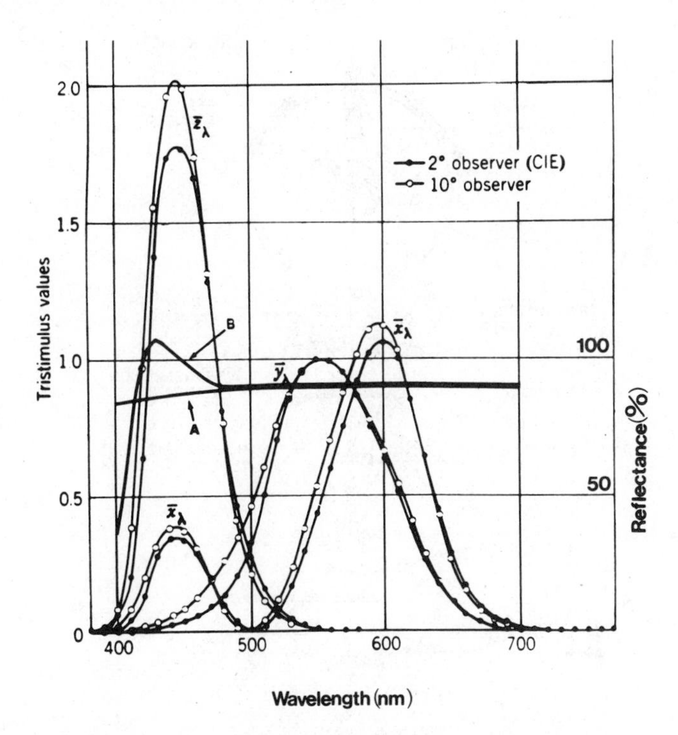

FIG. 38. Tristimulus whiteness measurements. From Ref. 3.

radiance than does a regular tristimulus colorimeter. Newer, more modern instruments are also available.

Only a continuous-recording instrument such as the Beckman DK-2A reflectometer with reversed optics (Fig. 39), however, can furnish information on the total spectral radiance of whitened fabrics.

2. Whiteness Functions

Many whiteness functions for transferring measurements into meaningful whiteness values have been suggested and are used. Some are valid only for a single plane in color space, however, or in the limited area for which they were developed. Experience has shown that most whiteness functions developed for unwhitened substrates fail when applied to FWA-whitened samples. Other functions have failed when fluorescent whiteners of different hues are involved. Several of the most recently suggested and frequently used whiteness functions are listed in Table 23; references to authors of other whiteness functions appear in the footnote. A more updated review on and evaluation of proposed whiteness functions has been written

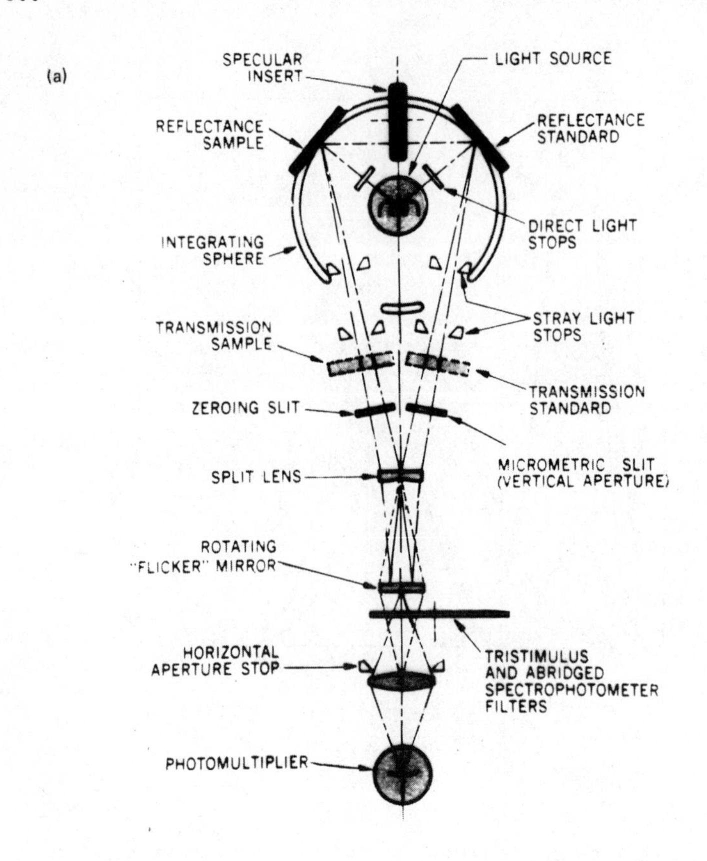

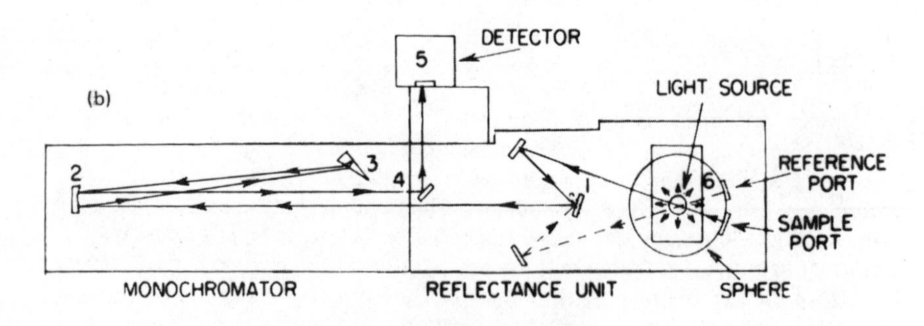

FIG. 39. (a) Color-Eye LS and (b) Beckman DK-2A spectroreflectometer.

TABLE 23

Whiteness Functions

Functions	Remarks	Refs.
1. $W = \sum_{400}^{700} P(\lambda)A(\lambda)\Delta\lambda + K$	$P(\lambda)$ = relative radiance measured at wavelength λ $A(\lambda)$ = spectrum weighting function at wavelength λ K = constant	Grum [140]
2. W = Y + kE	E = chromaticity value from graph k = factor weighting luminosity against chromaticity	Vaeck [111, 113, 146]
3. W = 2B - A	The relation between X, Y, Z and A, G, B depends on	Stephansen
4. W = G/3 + (B-A)	the instrument used. In one case, for example:	Berger [119]
5. W - 4B - 3G	X = 0.782A + 0.198B; Y = G; Z = 1.181B	Taube
6. W = L - 3b		Hunter [159]
7. W = L + 3a - 3b		Stensby [3]

Source: Fron Ref. 3.

by Ganz [156]. Although the C.I.E. subcommittee on whiteness concluded that there are no universal whiteness functions applicable in all industries, Ganz [156] suggested one which, depending on selection of values of constants, can be adapted to fit the requirements.

Grum and co-workers [140, 150, 151] proposed a very interesting approach to a whiteness value based on measurements. He selected various tints of near-white samples, all except four of which contained fluorescent whiteners. The samples were ranked visually under a mixture of cool white, fluorescent, and daylight illumination. The spectral radiance of each sample was also measured on a Beckman DK-2 spectroreflectometer with reversed optics. The relative radiance at each 10 nm between 400 and 580 nm, and each 20 nm between 580 and 700 nm, was multiplied by a weighting factor.

The sum of the products was calculated and a constant (in this case negative) was added to bring the rating of MgO to a whiteness value of 100. The weighting function (Fig. 40) selected for transformation of radiance data into observed whiteness was one which gave the most favorable correlation between calculated values and visual whiteness. This function exhibits some deficiencies when whitened fabrics of different hue are compared or when bluing dyes are involved. Other improved weighting functions are being considered.

Vaeck and co-workers [111-113, 146] found that a near-perfect correlation with average visual whiteness (North daylight) can be obtained by a linear combination of the luminosity Y and a value E derived from the chromaticity of the sample:

$$W = Y + kE$$

In the formula, k, a function of transparency and gloss of the sample, is a factor weighing luminosity against chromaticity. The factor will depend on the conditions under which the whiteness is evaluated. In Vaeck's experiments, k is equal to 1 for normal, nonglossy fabrics, and papers. E is taken from a graph (Fig. 41) representing the "equiwhiteness" contours (at constant luminosity = Y) in a chromaticity diagram (u, v coordinates). Vaeck's measurements were made with an Elrepho instrument equipped with a xenon lamp. Using this method, whiteness of a fabric can be determined by measuring X, Y, Z, by calculating and plotting u, v values, by determining the chromaticity value E, and by following with a summation of luminosity (Y) and chromaticity value.

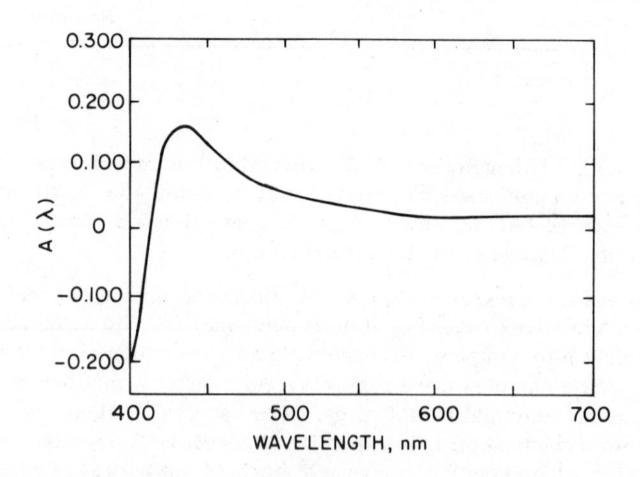

FIG. 40. Spectrum-weighting function. From Ref. 3.

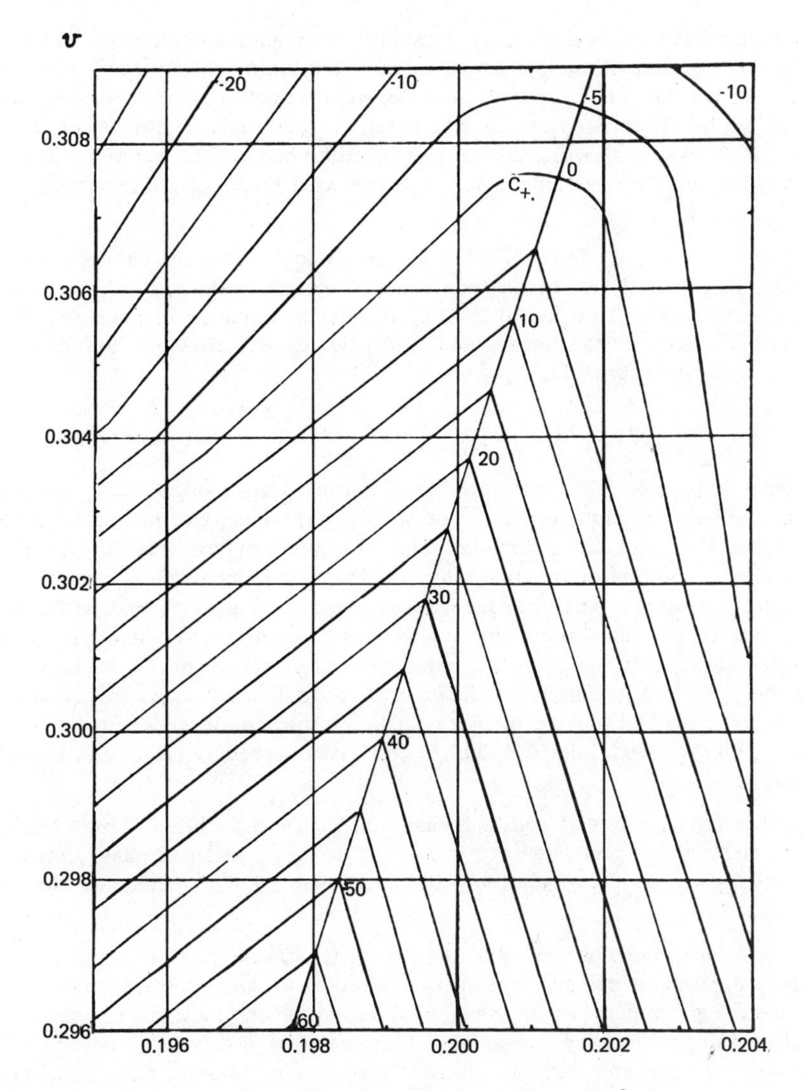

FIG. 41. Equiwhiteness diagram. From Ref. 3.

One must keep in mind that Grum's weighting function, as well as Vaeck's work, were developed on the basis of visual evaluation by a limited number of people and only under one type of light. The validity of the methods, therefore, may be limited.

Other functions frequently referred to in the European literature are those suggested by Stephansen and Berger. Whiteness values calculated from these functions are based primarily on measurements made on the Elrepho instrument equipped with a xenon arc lamp. Kling [134] and Vaeck and Van Lierde [144] reported good correlation between whiteness calculated by the Stephansen and Berger functions and visual whiteness in the case of fabrics with whiteners of very similar hues. The functions are unsatisfactory, however, if fabrics whitened with FWAs of different hues are compared.

In the U. S., whitened fabrics frequently are measured on a Hunterlab Model D40 (two-filter) instrument and whiteness is calculated by the function developed by Taube. Seve [121] and Zimmerman [122] point out the similarity between the Berger and Taube functions. Both can be covered by the following formula:

$$W = G - n(G - B)$$

When $n = 3$, a function very similar to that of Berger is obtained; when $n = 4$, Taube's function results. The $W = L - 3b$ equation (measured using a Hunterlab model D25 instrument, for example) approximates the Taube equation. The Hunter function indicates that a unit increase in blueness (negative b values) will improve whiteness three time more than will a unit increase in L value (luminosity of a fabric). Neither of the above functions differentiates between greenish blue and reddish blue hues, since only blueness and not greenish red differences are taken into account. Vaeck's work and results reported by others indicate that small hue differences among fabrics containing high brightener levels strongly affect visual preferences.

The function introduced by Stensby indicates that whites of high luminosity and reddish blue hues are preferred [3, 160]. In most cases, Stensby's equation gives good correlation with visual evaluation of laundered whitened fabrics.

All the above whiteness functions have limitations, but will furnish a good indication of relative whiteness in defined areas. Because they are time saving (compared with visual evaluation), such functions are satisfactory for preliminary screening. Final results, however, should be confirmed by visual comparison. It must also be kept in mind when using any of these whiteness functions that at a certain, as yet undetermined, distance from the achromatic point in the chromaticity diagram—at least in the green, yellow, and red direction—the functions lose their validity. The white converts into a color at this point. The same is true for the vertical luminosity (Y) axis going from white toward black. At one value of Y (around 70) the sample no longer appears white but gray. One must keep in mind that one whiteness equation which may be suited for internal application in a textile or paper mill, and which may correlate well with what the investigator feels is a preferred white, might easily fail if compared with the preference of the average U. S. homemaker.

Calculation of whiteness from the relative spectral radiance curve (obtained under standard conditions) using a weighting function may well be the best procedure. Weighting functions might differ with surroundings and conditions, but so will the correlation (to visual ranking) coefficient for any whiteness function. Whiteness should not be expressed by only one number, however, because it depends on both luminosity and chromaticity of the substrate.

Some investigators rate whitener performance after the first application as the most important; others judge the whitener on its effectiveness after repeated washings. Because whiteners frequently are used in combinations, the contribution of the "package" is often evaluated rather than that of the single components. Also, the overall effect on a mixed, soiled, prewhitened washload may be considered a more realistic measure than the effect of a single whitener on a clean, unwhitened, one-fiber substrate.

V. PRODUCT SAFETY

Another property which must be considered prior to adopting an FWA is its toxicological status. Before releasing whiteners for sale, reputable suppliers will examine them carefully for consumer and environmental safety. Laundry aid manufacturers, in turn, will examine the whiteners in their own formulations to ascertain that no cross-sensitization is caused by other ingredients in their products.

The human and environmental safety of the more important FWAs currently used by the major U. S. detergent producers have been investigated extensively [161-166]. In their paper on the current status of the human and environmental safety aspects of FWAs used in U. S. detergents, Burg et al. [161] concluded that the concentrations of FWAs which are observed in the environment in the United States appear to be quite small when compared to toxic quantities.

Considerable information is available on the potential mammalian toxicity of FWAs [161, 162, 167-171]. Tests of oral toxicity (including mutagenicity, teratogenicity, and carcinogenicity) which have been completed using high daily doses do not indicate any hazard. Studies of effects resulting from the topical application of FWAs, including phototoxicity, carcinogenicity, and photocarcinogenicity experiments, have been negative. Considering the low concentrations of the FWAs with which humanity comes into contact, no current human health hazard from these materials can be projected.

Most of the different FWAs currently used in U. S. detergents produced by the major detergent manufacturers can be identified in sewage. Only trace quantities have been found in drinking waters. The concentration in various natural waters and in fish from polluted waters is usually about 0.000001%. The compounds all have low fish toxicity.

The FWAs currently used in detergents in the United States do not exhibit biomagnification through either food chains or bioaccumulation. Hence, greater amounts of FWAs are not appearing as a body burden.

Most FWAs demonstrate a low biodegradability. The limited breakdown of FWAs that has been demonstrated in laboratory experiments lessens concern as to the toxicity of a family of metabolites. The whiteners, however, are to a large extent removed from sewage in treatment plants as the FWAs attach themselves to the sludge.

When assessing the overall human safety aspects, it is important to relate more specifically exposed maximum daily oral concentrations to which humans would reasonably be exposed to the toxicity of the compounds. A millionfold safety factor has been established between the lowest concentrations at which any toxic manifestations have been observed in animals and human exposure. The maximum daily human exposure of 40 $\mu g\ kg^{-1}\ day^{-1}$ percutaneous and 0.3 $\mu g\ kg^{-1}\ day^{-1}$ oral has been calculated. Consequently, the current status of human safety is no cause for concern. Likewise, the environmental aspects of the current use of FWAs in detergents demonstrate no problem areas.

REFERENCES

1. B. Wendt (to I. G. Farben AG). German patent 731558, 1940.
2. P. Krais, Melliand Textilber., 10, 468 (1929).
3. P. S. Stensby, Soap Chem. Spec., 43, pp. 41 (Apr.), 84 (May), 80 (July), 94 (Aug.), and 96 (Sept.) (1967).
4. M. Scalera and D. R. Eberhart (to American Cyanamid). U.S. patent 2,563,795, 1948.
5. O. Troesken (to Casella Farbwerke). German patent 850,008, 1950.
6. H. Gold and S. Petersen (to Farbenfabriken Bayer AG). U.S. patent 2,668,777, 1948.
7. E. R. Keller, R. Zweidler, and H. Haeusermann (to J. R. Geigy AG). U.S. patent 2,784,183, 1951.
8. K. Weber, P. Liechti, H. R. Meyer, and A. E. Siegrist, (to Ciba-Geigy). U.S. patent 3,984,399, 1967.
9. K. Weber (to Ciba-Geigy). German patent 2,201,857, 1972.

10. W. I. Lyness, R. T. Amel and G. E. Booth. U.S. patent 3,711,474, Proctor & Gamble, 1969.

11. N. N. Crounse (to Sterling Drug Co.). U.S. patent 3,642,642, 1967.

12. L. R. Hamilton (to Procter & Gamble Co.). U.S. patent 3,646,015, 1969.

13. H. Bloching and G. Walther (to Henkel and Cie). German patent 2,557,783, 1975.

14. E. Ohtaki, R. Mita, K. Ohsawa, and I. Ohkubo (to Mitsui Toatsu Chemicals, Inc.). Japanese patent 7681826, 1975.

15. A. Boeck, D. Jung, and W. Wuest (to Henkel and Cie). German patent 2,354,096, 1973.

16. Y. Yamakawa and M. Suda (to Sumitomo Chemical Co., Ltd.). Japanese patent 73 03889, 1969.

17. C. Luthi (to Ciba-Geigy Corp.). U.S. patent 4,105,399, 1978.

18. H. Schlaepfer (to Ciba-Geigy Corp.). U.S. patent 4,009,994, 1977.

19. R. B. Barbee and E. C. Taylor (to Eastman Kodak Co.). U.S. patent 3,830,804, 1974.

20. A. Wagner, C. W. Schelhammer, and S. Petersen, Angew. Chem. (Int. Ed.) 5, 699 (1966).

21. R. Zweidler and H. Haeusermann, Ency. Chem. Tech., 3, 737 (1964).

22. H. W. Zussman, Ency. Chem. Tech., 2, 606 (1965).

23. F. G. Villaume, J. Am. Oil Chem. Soc., 35, 558 (1958).

24. H. W. Zussman, J. Am. Oil Chem. Soc., 40, 695 (1963).

25. D. A. W. Adams, J. Soc. Dyers Col., 75, 22 (1959).

26. J. H. P. Tyman, J. Soc. Dyers Col., 81, 102 (1965).

27. E. Uehlein, Optische Aufheller, Moser Verlag, Garmisch-Partenkirchen, W. Germany, 1957.

28. A. E. Siegrist, Soap Chem. Spec., 31, 44 (Nov., 1955).

29. A. E. Siegrist, Soap Chem. Spec., 31, 58 (Dec., 1955).

30. H. W. Zussman, W. Lennon and W. Tobin, Soap Chem. Spec., 32, 35, (Aug., 1956).

31. R. Zweidler, Specialties, 1, 14, (Feb., 1965).

32. A. E. Siegrist, J. Am. Oil Chem. Soc., 55, 114 (1978).

33. J. Dostmann, Dtsch. Textiltech, 7, 575 (1957).

34. A. Schlachter, Fette-Seife Anstrichm., 56, 9 (1954).

35. Y. Sayato, J. Food Hyg. Soc. Jpn., 2, 56 (1961).

36. T. H. Morton, J. Soc. Dyers Col., 79, 238 (1963).

37. K. Schoenol, Seifen Oele Fette Wachse, 91, 25 (1965).

38. H. Bloching, in Waschmittelchemie (Henkel & Cie GmbH, eds.), Hüthig Verlag, Heidelberg, 1976, p. 137.

39. Y. Sakai, H. Kato, Sen'i Kako, 25(2), 88 (1973).

40. H. K. Banerjee and K. G. Pai, Colourage, 22(7), 25 (1975).

41. D. S. Rao, Colourage, 21(25), 23 (1974).

42. R. Von Ruette, Seifen Oele Fette Wachse, 100(3), 55 (1974).

43. J. Lin, D. E. Rivett, and J. F. K. Wilshire, Aust. J. Chem., 30(3), 629 (1977).

44. I. H. Leaver, Aust. J. Chem., 30(1), 87 (1977).

45. M. Tanaka and S. Sakuma, Nippon Kagaku Kaishi, 5, 885 (1975).

46. P. S. Stensby, Deterg. Age, 5, 24 and 26 (Jan. and Feb., 1968).

47. P. S. Stensby, Deterg. Age, 3(9), 20 (Feb., 1967).

48. P. S. Stensby, Soap Chem. Spec., 41, 85 (May, 1965).

49. P. S. Stensby, W. R. Findley, and C. W. Liebert, Deterg. Spec., 6(5), 29, (May, 1969).

50. P. S. Stensby and W. R. Findley, Soap Cosmet. Chem. Spec., 48, 52 (Oct., 1972).

51. R. Anliker, H. Hefti, and H. Kasperl, J. Am. Oil Chem. Soc., 46, 75 (1969).

52. A. E. Siegrist, Papier, 8(7/8), 109 (1954).

53. A. Berger, Papier, 13(1/2), 5 (1959).

54. F. S. Mousalli and C. L. Browne, Text. Chem. Color., 3, 202 (1971).

55. E. Allen, Am. Dyest. Rep., 48(14), 27 (1959).

56. D. H. Powers, Am. Dyest. Rep., 55, 532 (1966).

57. R. Williamson, Int. Dyer, 157(8), 359, (9), 408 (1977).

58. N. A. Evans, D. E. Rivett, and P. J. Waters, Text. Res. J., 46(3), 214 (1976).

59. H. L. Needles, Text. Res. J., 46(1), 39 (1976).

60. S. D. Mehendale and V. E. Inamdar, Text. Dyer Print., 2(1), 153 (1970).

61. I. Soljacic and K. Weber, Textilveredlung, 9(5), 220 (1974).

62. P. S. Stensby, J. Am. Oil Chem. Soc., 45, 497 (1968).

63. A. Wagner, Fachorg. Textilveredl., 19, 466 (1964).

64. J. Lanter, J. Soc. Dyers Col., 82, 125 (1966).

65. H. Theidel, Melliand Textilber., 39, 61 (1958).

66. M. Pestemer, A. Berger, and A. Wagner, Fachorg. Textilveredl., 19, 420 (1964).

67. J. Kurz, Fette-Seifen Anstrichm., 67, 792 (1965).

68. J. Kurz, Fette-Seifen Anstrichm., 68, 42 (1966).

69. Anonymous, Ciba Rev., 3, 10 (1966).

70. Anonymous, Ciba Rev., 3, 24 (1966).

71. J. Lanter, Fachorg. Textilveredl., 19, 469 (1964).

72. N. Seiler, G. Werner, and M. Wiechmann, Naturwissenschaften, 50, 643 (1963).

73. J. Schulze, T. Polcaro, and P. S. Stensby, Soap Cosmet. Chem. Spec., 50(11), 46 (1974).

74. D. Kirkpatrick, J. Chromatogr., 139(1), 168 (1977).

75. R. Van Ruette, Manuf. Chem. Aerosol News, 46(1), 36 (1975).

76. O. Koch and K. Bunge, Chem. Ing. Tech., 32, 812 (1960).

77. V. G. Bochard, A. Z. Karimova, and V. V. Makarov, Zh. Prikl. Spektrosk., 15(4), 636 (1971).

78. J. Eisenbrand and A. Klauck, Dtsch. Lebensm. Rundsch., 61, 370 (1965).

79. H. Joeder, Melliand Textilber., 40, 1190 (1959).

80. L. E. Weeks, J. C. Harris, and J. T. Lewis, Soap Chem. Spec., 35, 66 (May, 1959).

81. H. Haeusermann and R. Keller, Text. Rundsch., 16, 176 (1961).

82. J. Wegmann, Melliand Textilber, 48, 59 and 183 (1967).

83. W. Postman, Dyestuffs, 42(5-8) (1958) (Reprint).

84. H. Zollinger, Am. Dyest. Rep., 49, 142 (1960).

85. H. Bach, E. Pfeil, W. Philippar, and M. Reich, Angew. Chem., 75, 407 (1963).

86. F. L. Tewksbury, Dyestuffs, 45(2), 72 (1965).

87. C. H. Giles, Chem. Ind., pp. 92 and 137 (1966).

88. E. Merian, Text. Res. J., 36, 612 (1966).

89. J. R. Aspland, Specialties, 2, 3 (Nov., 1966).

90. R. McGregor, J. Soc. Dyers Col., 82, 450 (1966); 83, 52 (1967).

91. P. B. Weisz and H. Zollinger, Melliand Textilber., 48, 70 (1967).

92. P. S. Stensby, Paper presented at Am. Oil Chem. Soc. Short Course, Pocono Manor, Pa., June, 1967.

93. F. G. Villaume, J. Am. Oil Chem. Soc., 37, 188 (1960).

94. F. Fortress, Text. Chem. Col., 3, 103 (1971).

95. Textile Organon Monthly Reviews, Textile Economics Bureau, Inc., New York (June, 1970).

96. Man-Made Fiber Fact Book Annual Reviews, Man-Made Fiber Producers Assoc., New York (1967-68).

97. Anonymous, Mod. Text., 51, 31 (March, 1970).

98. Current Industrial Reports—Woven Fabrics: Production, Inventories and Unfilled Orders (Series M22A, Monthly Reports), USDC/Bureau of Census.

99. M. J. Schick, J. Am. Oil Chem. Soc., 40, 680 (1963).

100. P. Becher, in Nonionic Surfactants (M. J. Schick, ed.), Dekker, New York, 1967, pp. 478-513.

101. I. R. Schmolka and L. R. Bacon, J. Am. Oil Chem. Soc., 44, 559 (1967).

102. I. R. Schmolka and A. J. Raymond, J. Am. Oil Chem. Soc., 42, 1088 (1965).

103. J. M. G. Cowie and A. F. Sirianni, J. Am. Oil Chem. Soc., 43, 572 (1966).

104. T. L. Coward and N. R. Smith (to Procter & Gamble Co.). U.S. patent 3,309,319, 1967.

105. D. B. Judd and G. Wyszecki, Color in Business, Science and Industry, Wiley, New York, 1965.

106. S. Weissman and J. A. S. Kinney, J. Opt. Soc. Am., 55, 74 (1965).

107. E. Allen, Am. Dyest. Rep., 46, 425 (1957).

108. Washington Section of AATCC, Am. Dyest. Rep., 50, 812 (1961).

109. W. A. Coppock, Am. Dyest. Rep., 54, 343 (1965).

110. Rhode Island Section of AATCC, Am. Dyest. Rep., 55, 997 (1966).

111. S. V. Vaeck and F. Van Lierde, Ann. Sci. Text. Belg., 7(1) (Mar., 1964).

112. S. V. Vaeck and F. Van Lierde, Ann. Sci. Text. Belg., 95(1) (Mar., 1966).

113. S. V. Vaeck, Ann. Sci. Text. Belg., 95(1) (Mar., 1966).

114. ASTM, Proposed Method of Test for Indexes of Whiteness and Yellowness of Near-White, Opaque Materials (ASTM 59S). Preprint of Report, 12, 1965.

115. A. Berger, Bayer Farben Rev., Spec. Ed. no. 7E, 48 (1965).

116. F. Mueller, Bayer Farben Rev., Spec. Ed. no. 1/3, 2 (1966).

117. A. Fargues and E. Bonte, Bull. Inst. Text. Fr., 18(111), 249 (1964).

118. L. R. C. Friele, Farbe, 8(4/6), 171 (1959).

119. A. Berger, Farbe, 8(4/6), 187 (1959).

120. A. Berger and O. Koch, Farbe, 9(4), 259 (1960).

121. R. Seve, Farbe, 12(1/6), 187 (1963).

122. W. Zimmerman, Fette-Seifen Anstrichm., 67, 946 (1965).

123. H. Hemmendinger and J. M. Lambert, J. Am. Oil Chem. Soc., 30, 163 (1953).

124. E. Allen, J. Opt. Soc. Am., 47, 933 (1957).

125. E. Allen, J. Opt. Soc. Am., 53, 1107 (1963).

126. M. L. Jeanne, J. Opt. Soc. Am., 54, 1046 (1964).

127. R. Seve, A propos de la Blancheur et de sa Conception, Kodak-Pathe, Vincennes (Seine), France (1963).

128. R. Seve, Les Problemes de Blancheure dans l'Industrie Papetiere, Kodak-Pathe, Vincennes (Seine), France (1963).

129. R. Seve, Essais pour Determiner la Blancheur de Papiers Contenants des Produits Fluorescents, Kodak-Pathe, Vincennes (Seine), France (1963).

130. F. O. Sundstrom, Mod. Packag., p. 168 (Sept., 1957).

131. R. S. Hunter, Paper presented to Annual Convention of AATCC, 1966.

132. A. Berger and H. Unterbirker, Papier, 13(1/2), 5 (1959).

133. W. A. Coppock, Pulp Pap. Mag. Can., 67, 499 (1966).

134. A. Kling and J. Kurz, Seifen Oele Fette Wachse, 89, 427 (1963).

135. S. Goldwasser, Soap Chem. Spec., 38, 47 (Aug.), 62 (Sept.) (1962).

136. F. Grum and T. Wightman, J. Tech. Assoc. Pulp Pap. Ind., 43, 400 (1960).

137. J. M. Patek and F. Grum, J. Tech. Assoc. Pulp Pap. Ind., 45, 74 (1962).

138. R. S. Hunter, J. Tech. Assoc. Pulp Pap. Ind., 45, 203A (1962).

139. T. W. Lashoff and J. M. Patek, J. Tech. Assoc. Pupl Pap. Ind., 45, 566 (1962).

140. F. Grum and J. M. Patek, J. Tech. Assoc. Pulp Pap. Ind., 48, 357 (1965).

141. O. Oldenroth, Tenside, 3(1), 9 (1966).

142. K. J. Nieuwenhuis, Text. Rundsch., 13, p. 82 (Feb., 1958).

143. A. Kling and J. Kurz, Text. Rundsch., 15, 297 (1960).

144. S. V. Vaeck and F. Van Lierde, Waescherei-tech. Chem., 17, 26 (1964).

145. S. V. Vaeck and G. Brouwers, Waescherei-tech. Chem., 17, 898 (1964).

146. S. V. Vaeck and G. Brouwers, Waescherei-tech. Chem., 18, 155 (1965).

147. K. J. Nieuwenhuis, Was-ind., 11(1), 4 (Aug., 1961).

148. J. Kurz and W. Lebensaft, Z. Ges. Textilind., 67, 555, 659, 718, 803, and 975 (1965).

149. J. Kurz and W. Lebensaft, Z. Ges. Textilind., 68, 29, 300, and 679 (1966).

150. F. Grum, R. F. Witzel, and P. S. Stensby, J. Opt. Soc. Am., 64(2), 210 (1974).

151. F. Grum, S. Saunders, and T. Wightman, J. Tech. Assoc. Pulp Pap. Ind., 53, 1264 (1970).

152. G. Wyszecki, Farbe, 19, 43 (1970).

153. A. S. Stenius, Letter to Secretariat of ISO/TC6/SC2/WG1, Cellulosaindustriens Centrallaboratorium, Stockholm, July 1966.

154. J. M. Adams, J. Soc. Dyers Col., 77, 670 (1961).

155. H. Bloching, in Waschmittelchemie (Henkel & Cie GmbH, eds.), Hüthig Verlag, Heidelberg, 1976, p. 140.

156. E. Ganz, J. Col. Appear., 1(5), 33 (1972).

157. J. Lanter, Ciba Rev., 12(140), 26 (1960).

158. R. Hunter and W. Schramm, J. Opt. Soc. Am., 59, 881 (1969).

159. R. Hunter, J. Opt. Soc. Am., 50, 44 (1960).

160. P. Stensby, J. Col. Appear., 2(1), 39 (1973).

161. A. W. Burg, M. W. Rohovsky, and C. J. Kensler, Crit. Rev. Environ. Control, 7, 91 (1977).

162. R. Anliker and G. Müller, in Environmental Quality and Safety, supp. to vol. IV (F. Coulston and F. Korte, eds.), Thieme, Stuttgart, 1975.

163. C. Eckhardt and R. Von Rütte, in Environmental Quality and Safety, supp. to vol. IV (F. Coulston and F. Korte, eds.), Thieme, Stuttgart, 1975, p. 59.

164. C. R. Ganz, C. Liebert, J. Schulze, and P. S. Stensby, J. Water Pollut. Control Fed., 47(12), 2834 (1975).

165. C. R. Ganz, J. Schulze, P. S. Stensby, F. L. Lyman, and K. Macek, Environ. Sci. Technol., 9(8), 738 (1975).

166. R. N. Sturm, K. E. Williams, and K. Macek, Water Res., 9(2), 211 (1975).

167. C. Gloxhuber, H. Bloching, and W. Kaestner, Environ. Qual. Saf. (Suppl.), 4, 202 (1975).

168. C. Gloxhuber and H. Bloching, Clin. Toxicol., 13(2), 171 (1978).

169. J. McPhilip, Miljoevardscentrum Reports, 2, 123 (1973).

170. P. E. Osmundsen, Miljoevardscentrum Reports, 2, 69 (1973).

171. J. F. Griffith, Arch. Dermatol., 107(5), 728 (1973).

Chapter 21

DISHWASHING

William G. Mizuno

Economics Laboratory, Inc.
St. Paul, Minnesota

I. INTRODUCTION AND HISTORY 816

A. Eating Utensils 816
B. Dishwasher Development 817
C. Detergent Development 823

II. THEORY OF DISHWASHING 825

A. Kinetic Energy 827
B. Thermal Energy 828
C. Hydration and Solvation 830
D. Surface and Interfacial Tension 830
E. Surfactants 831
F. Hypochlorite 832
G. Filming and Spotting 833

III. DETERGENT EVALUATION 838

A. Wash Action 839
B. Selection of Test Pieces 840
C. Soil and Soiling Methods 840
D. Soil Residue Measurements 841
E. Film and Spot Measurements 842
F. Corrosion 844
G. Finalizing on Test Methods for Evaluation 845

IV. FUNCTIONS AND PROPERTIES OF VARIOUS DETERGENT COMPONENTS 848

A. Polyphosphates 848
B. Chelates other than Polyphosphates 861
C. Silicates 863
D. Alkaline Builders 867

E. Active-Chlorine Compounds 872
F. Neutral Salts and Urea 878
G. Corrosion Inhibitors 879
H. Low-Foaming Surfactants and Defoamers 881
I. Enzymes 883

V. RINSING AND RINSE ADDITIVES 885

A. Polyphosphates 885
B. Surfactants 885
C. Acids 887
D. Iron Staining 887

VI. SUMMARY 888

REFERENCES 890

I. INTRODUCTION AND HISTORY

Dishwashing became a necessity with the advent of reusable eating utensils. However, mechanical dishwashing such as we are concerned with in this chapter is a more recent development, and follows the industrial revolution and the development of the electric motor as a source of mechanical power.

A. Eating Utensils

Vases and pottery were known to exist at least 500 to 1000 years B. C. and perhaps as long ago as 5000 years B. C. Anthropologists have used the refinement of pottery as a measure of the advancement of culture. The forerunners of the present-day chinaware, undoubtedly were introduced into Europe from China in about the 13th century by Marco Polo: hence the derivation of the generic name "china" for this porcelainware. By about the 18th century the manufacture of porcelain began to take hold in Europe [1]. The china differed from the conventional earthenware then prevalent in Europe in being light colored and translucent and was accepted at once.

Pottery is defined generally as ware made from clay and fired to render it stable to the action of water. It must be coated with low-melting glass to make the surface nonporous and readily cleanable. Porcelain, on the other hand, is generally considered to be intermediate between pottery and glass, for it may be considered either as pottery rendered translucent by the addition of clay to glass and fired, or it may be considered as glass rendered translucent by the addition of clay. As will be discussed later, the surface and condition of a dish or eating utensil is an important factor in dishwashing.

There are various types of glaze as far as chemical composition and properties and physical properties are concerned. Some are more resistant to shock, wear, and chemical action (from foods and detergent) than others. Generally, the glaze compositions that melt at higher temperature thus require a higher firing temperature and are more resistant to physical and chemical damage in use.

With regard to patterns, there are underglaze and overglaze patterns. In the underglaze pattern, the decorative colors (various metal salts or oxides) are applied to the dish or "bisque" before the glaze is put on, then glaze is put on and the dish is fired to fuse the glaze. Since the pattern is now covered by a glass or glaze, the pattern is well protected from chemical and physical wear. In the overglaze pattern, the glaze is put on first, fired to fuse the glass, cooled, and then the pattern is put on over the glaze and refired at a lower temperature to cure the pattern but not remelt the glaze. The overglaze process produces more delicate colors and shadings and is the process used for producing so-called "fine china." Because the pattern is not protected by a glaze, the overglaze pattern is susceptible to damage by chemical (food or detergent) and mechanical action. The wide acceptance of fine china took place about the same time that the home dishwashing machines came into wide use.

Institutions and restaurants use underglaze patterns almost exclusively, but overglaze patterns are commonly found in home use.

Plastic dishware, often called molded-ware, is a relatively new innovation and has assumed an important position among dishware. The main advantages of plasticware are lightness and freedom from breakage in handling and/or dropping. For use in serving meals on aircraft, light weight is important; and in home use, where the dishes are handled by young children, plasticware has gained wide acceptance. Most of the plastic dishware is made of melamine resin. The earlier melamineware had a high percentage of alpha cellulose which made it susceptible to staining by wicking action when the surface became worn, but the filler-free or relatively filler-free melamine of current vintage has a much greater stain and chip resistance.

B. Dishwasher Development

Mechanical dishwashing is believed to have been documented first in 1865 with the issuance of U. S. patent 7, 365 to J. Houghton of Ogden, New York. His device was a crude wooden affair but was destined to be the forerunner of the modern dishwasher.

Several people and companies were later involved in the manufacture of early dishwashing machines. It is reported [2] that in 1880 a Mrs. Josephine Garis Cochrane built a dishwashing machine as a labor-saving device for her own use. The interest among restaurateurs was very strong, and by the time of the Columbian Exposition in Chicago in 1893 a total of about nine machines were used in the concessions at this World's Fair. These machines were all sold immediately at the conclusion of the Fair by Crescent Washing Machine Company, now a part of Hobart Manufacturing Company.

Some 44 years and more than 30 patents later, Walker Brothers Company of Syracuse, New York, in 1909 sold a dishwashing machine in the form of a free-standing model with a cast-iron tub on legs [3]. The hydraulic action came from hand-cranked paddles which splashed water on the dishes. The next major improvement came in 1913 when an electric motor replaced the hand crank. Further improvements were made with the introduction of a porcelain tub in 1918 and a cabinet-type dishwasher in 1928. Both commercial and home dishwashers were in the making during this period. The two most active developers, the Crescent Washing Machine Company and the Walker Brothers Company, soon merged with Hobart Manufacturing and General Electric Company, respectively.

In principle, the current home and commercial dishwashing machines are alike in many respects. However, in mechanical details, operation, and dishwashing capacity they differ widely. Also there are considerable differences among the makes and models.

The top-line home machines may have six or more different cycles designed to give wide flexibility for different needs. These variations are in the number of washes and rinses, wash temperature, etc. Typically, these include rinse-and-hold, short wash, long wash, 145° or 150°F sanitize wash, pot and pan, and plate warm cycles.

In a typical "built-in", "front-load" machine, the detergent dispenser is located on the door. It consists of an "open" cup for the first wash and another cavity with a cover for the second cup. The cover of the second detergent cup is held in place during the first wash, and then it is automatically released by the timer at the beginning of the second wash to provide a detergent charge for the second wash.

"Portables" differ from built-in machines in that they are not permanently plumbed-in and are usually set on wheels. Portables may be front loading or "top loading." The top-load machine, as the name suggests, has a swing-up door on top of the machine and the dishes are loaded from the top. The detergent dispenser is located on the front wall and performs the same function as in the front-load but designed somewhat differently. Most portables are of the front-load type and many of these are known as "convertibles," i.e., they may be converted to built-in dishwashers by removing the wheels and top and by making a permanent installation. Convertibles are by necessity front loading.

The detergent charge is volume measured and determined by the cavity volume. The water volume is regulated by a time-fill by means of a solenoid valve controlled by a timer. The detergent concentration thus becomes a function of the bulk density and water of hydration present in the detergent for a given machine. The detergent concentration may vary from 0.30 to 0.35% concentration for most machines.

The wash action is obtained by spraying the detergent solution over the dishes which are racked to obtain most efficient utilization of the space available. The solution is recirculated by means of a centrifugal pump which distributes the detergent solution via a manifold to a rotating two- or four-arm sprayer similar in principle to a rotating-arm garden sprinkler. The wash arms are rotated in order to get uniform and complete coverage and to minimize "wash patterns." Most machines have double wash arms, one for the lower rack and one for the upper rack. The duration of the wash may vary from about 3 to 10 min. After the wash, the detergent solution in the reservoir is discharged into the drain. Where a second wash is employed, the same sequence is repeated. To minimize food redeposition, the sump of some machines has a fine mesh strainer which prevents particles of food from being recirculated. Any food particles trapped on the strainer are flushed down the drain after each cycle.

The rinse, which may number from two to four times depending on the cycle chosen, is carried out in a similar manner to the wash except that no detergent is added. The duration of the rinse is usually of the order of 2 to 3 min. The top-line model provides injection of rinse aid to the final rinse to minimize filming and spotting. A cutaway diagram of a typical home machine is shown in Fig. 1. A complete wash, rinse, and hot-air dry cycle may require about 1 hr.

The acceptance of dishwashers in the home is indicated by the sales figures. It is reported that in 1964 sales reached over 1,000,000 units. In January 1965 11.8% of homes had dishwashers, and by January 1971 one out of every four homes had a dishwasher. In 1977 40% of U.S. homes had dishwashers.

A breakdown of manufacturers' shipment of home machines from 1958 to 1978 is shown in Table 1. This table shows that manufacturers' shipment roughly doubled each 5 years from 1958 to 1968 and the total shipment for 1978 is expected to top 3,500,000 units. The table also indicates that built-in machines are currently outselling the portables by more than four to one.

In contrast to the 1977 figure for the dishwasher market of about 40% saturation for the United States, it is reported that West Germany had reached about 14% saturation by 1977. Most homes in West Germany do not have central hot-water systems and those that do usually have small hot-water storage capacity. In European dishwashing machines the water

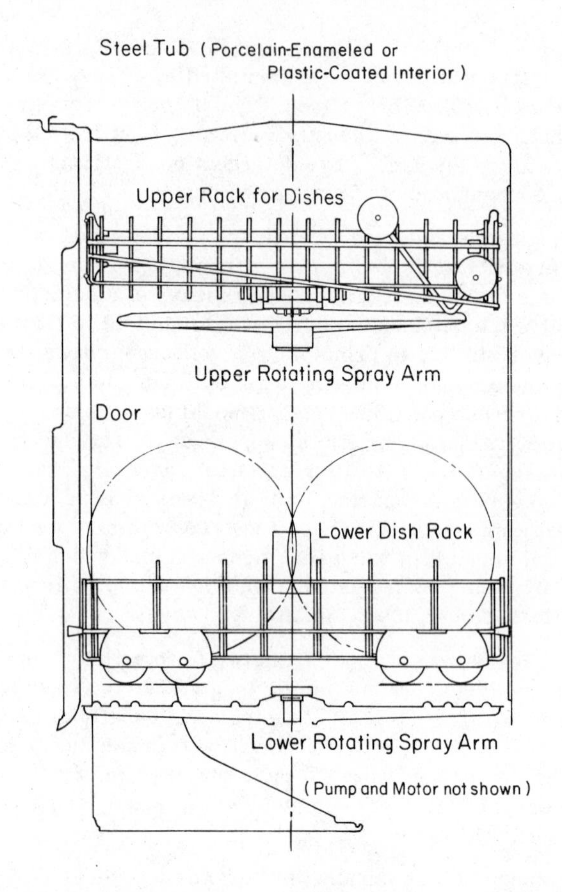

FIG. 1. Cutaway diagram of a typical home dishwashing machine.

is heated directly in the machine, similar to our 145 or 150°F sanitized cycle but often starting with cold water.

The small commercial-type machine is similar to a home machine except that the complete wash and rinse cycle is reduced to about 5 min. A higher wash and rinse temperature is used and the dishes are air dried outside the machine.

In the intermediate-sized, single-tank commercial machine, a rack of dishes is washed for 40 to 50 s; this is followed immediately by a 12-s freshwater spray rinse at line pressure. The wash and rinse is completed in about 1 min. In order to make up for the shorter wash and rinse and to provide sanitizing action, the wash temperature is 160°F and the rinse temperature is 180°F. The wash pump is much larger than a home machine, providing about 175 gal/min volume.

TABLE 1

Manufacturers' Shipment of Home Dishwashers

Year	Portable	Built-in	Total units
1958	—	—	424,700
1959	—	—	547,000
1960	232,000	323,000	550,000
1961	240,000	380,000	620,000
1962	260,000	460,000	720,000
1963	305,000	575,000	880,000
1964	340,000	710,000	1,050,000
1965	445,000	817,000	1,262,000
1966	545,000	983,000	1,528,000
1967	599,000	991,000	1,590,000
1968	677,500	1,283,000	1,960,500
1969	685,900	1,432,200	2,079,100
1970	679,900	1,435,700	2,115,600
1971	711,900	1,764,800	2,476,700
1972	842,000	2,357,000	3,199,000
1974	860,000	2,460,000	3,320,000
1976	803,000	2,337,000	3,140,000
1977	651,000	2,705,000	3,356,000
1978[a]	643,000	2,900,000	3,543,000

[a]Estimate
Source: From Ref. 159.

The detergent solution is reused many times before it is drained. The detergent tank has a capacity of 15 gal or more and each rinse may dilute by adding about 1 to 2 gal of water. This dilution serves to freshen the wash solution. The wash tank is provided with a food strainer as well as an overflow standpipe to skim off floating solids and to keep the wash solution as free of solid food particles as possible.

Because of this progressive dilution each time the rinse is turned on, the commercial machines are usually equipped with an electronic detergent concentration controller. The controller consists of a conductivity cell which is located in the detergent tank and monitors the solution conductivity continuously; this cell is connected to the bridge network of the controller so that, as the detergent concentration drops, the control opens a solenoid water valve. The water flows into a hopper reservoir containing the solid detergent powder. As the flow continues, the saturated detergent solution overflows into a standpipe. This overflow is directed into the detergent tank and raises the detergent concentration (conductivity). When the preset concentration is achieved, the control which receives the signal from the conductivity cell shuts off the solenoid water valve, thus cutting off the detergent feed into the wash tank. The detergent concentration may be controlled within narrow limits. The concentration of the detergent is usually set in the range of 0.15 to 0.25% depending on soil load, water hardness, and other factors.

The loaded dish racks are manually pushed through the machine (double door) and are supported by two "L"-shaped rails inside the machine.

The larger commercial machines consist of a number of sections joined together and are often engineered to the kitchen, work load, space available, and the desired end result. A large machine may have a prescrap (prewash) section, one or two wash sections, and one recirculated (power) rinse section and a final freshwater rinse. The temperature of each tank is maintained by a thermostat and the heat is provided by a steam coil or electrical heater. Some smaller machines have direct gas-fired heaters. Because the machine is made up of a number of sections joined together to form a tunnel, it is sometimes referred to as a tunnel washer. The rack of dishes is moved through this tunnel by means of a conveyor.

Water replenishment takes place as follows. The spent freshwater rinse is collected and diverted to the power rinse. In some machines a part of this rinse may be diverted directly to the prescrap section. The overflow from the power rinse in turn is diverted to the wash (or series of washes) and into the prescrap section. Thus maximum utilization of the rinse water is achieved. The hot-water consumption is essentially that of the freshwater rinse requirement except for the initial filling of the machine and is usually of the order of 5 or 6 gal/min during its actual operation.

The dishes are usually sorted, "scrapped" to remove gross food, and sometimes given a quick preflush to remove additional soil. They are then racked and mechanically conveyed through the machine. Some machines are equipped to provide a back-and-forth motion while the rack is in the wash section to minimize wash pattern or blind spots. The wash arm may be slotted rows of stationary pipes or rotating wash arms. The conveyor moves the rack through the tunnel at a speed of 1 to 2 in./s.

After the final freshwater rinse, the conveyor moves the racks out of the machine and onto a drain table where the racks may be lifted off the table and stacked onto a dolly if they consist of cups or glasses. Dishes are usually allowed to dry briefly, they are then stacked on carts, and the empty rack returned to the soiled dish area. Silverware is best washed in special racks which holds it vertically.

A variation of this tunnel machine is an arrangement where the machine and the tables are joined together to form a large oval racetrack. The machine is located at the straight part of the track and the remainder of the curved table forms an oval. In such an arrangement the racking is done on the table at one end (soiled end) and the clean dishes are removed at the other end. The racks need not be lifted off the table unless dishes are to be stored in them. The empty rack is pushed back to the soiled dish area to be reracked with soiled dishes. This arrangement reduces rack handling and the need to lift the racks.

There are also rackless machines known variously as "flight type," "rackless conveyor," or "auto-racking conveyor." These machines have endless belts having numerous rows of pegs. The pegs will directly support cups and most dishes and trays and will do so in such a position as to obtain the best cleaning. Glasses, silverware, and other small pieces still must be put in racks, placed on the belt, and sent through the machine. This arrangement does away with the need to rack most dishes and thus saves labor. In such machines the peg-belt usually extends 5 to 10 ft beyond the final rinse to allow enough drying time so that most dishes are essentially dry when removed at the clean end of the machine. The rinse aid greatly assists in shortening the drying time.

The commercial machines vary in size from about 2 ft × 2 ft to as large as 2-1/2 ft × 60 ft for flight-type machines and in washing capacity from 300 to 15,000 dishes per hour.

A schematic diagram of a typical conveyor-type machine is shown in Fig. 2.

C. Detergent Development

The dishwashing machine came into general use both in the home and in institutions such as restaurants, hospitals, etc., starting rather rapidly in the early 1940s. This rapidly expanding market for the machines led to a corresponding expanded market for detergents.

From the previous brief description of the various dishwashing machines, it may be apparent that soap or foaming detergents cannot be used because of the intense mechanical action which would create intolerable amounts of foam. Moreover, these surface active agents are more difficult

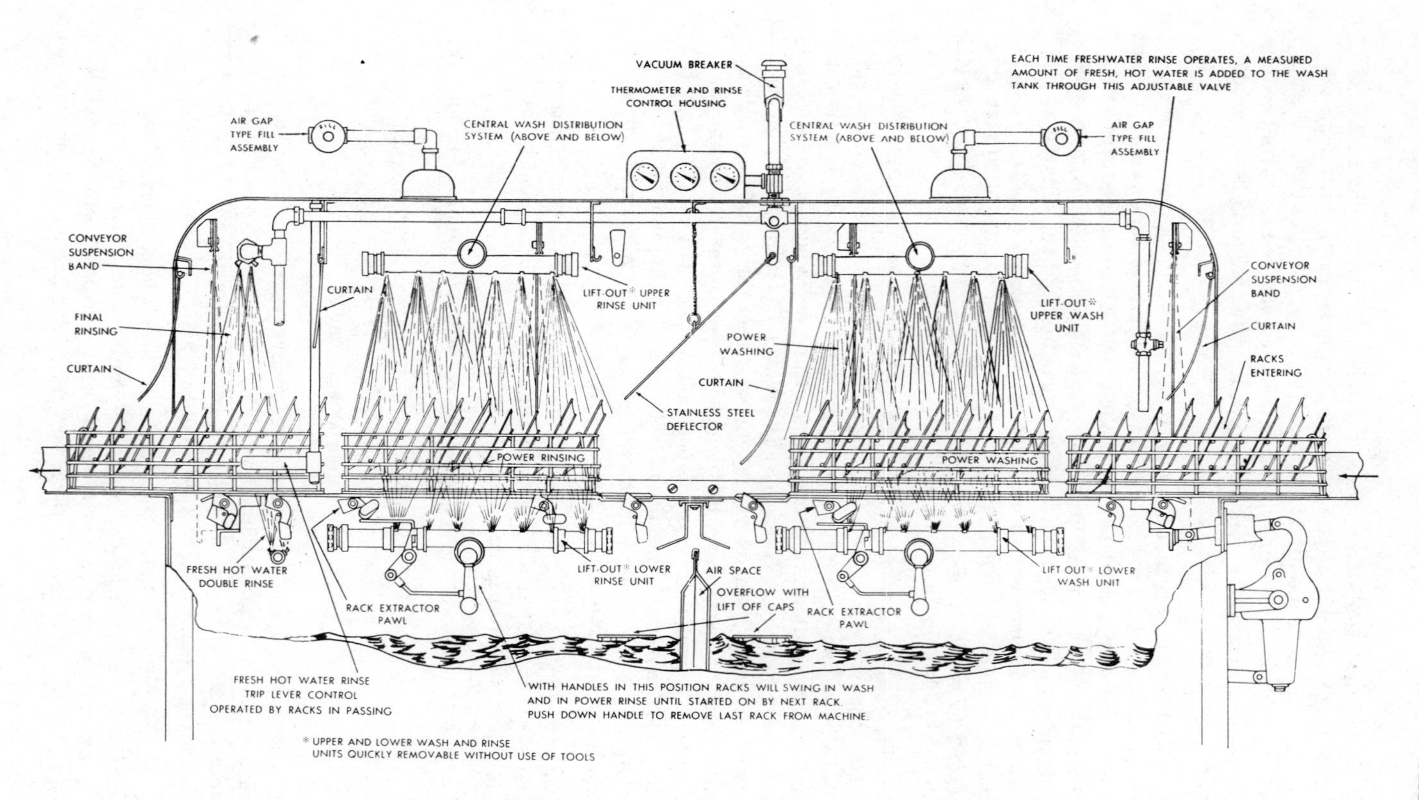

FIG. 2. Schematic diagram of a conveyor-type commercial dishwashing machine. Courtesy of Hobart Manufacturing Co., Troy, Ohio.

to rinse than are conventional dishwashing detergents. This is a critical factor in commercial operations because of the cost of hot water and the heating equipment.

Early detergents were soda ash and trisodium phosphate purchased locally. Though trisodium phosphate represented a real improvement over soda ash alone, these products softened water by precipitation and left a good deal to be desired, especially when used in harder waters. These products left a semipermanent film on glasses and dishes which would eventually build up to an unacceptable heavy film.

Substantial improvement in detergents came about in 1934 when Calgon-type polyphosphate was introduced. This was a glassy phosphate having an Na_2O/P_2O_5 ratio of 1.1:1.0. Subsequently, a glassy phosphate with an Na_2O/P_2O_5 ratio of 1.5:1.0 and known as sodium tetraphosphate was introduced. Still later a glassy polyphosphate with a ratio of 1.33:1.00 and known as sodium hexaphosphate, sodium pyrophosphate and sodium tripolyphosphate were introduced. The sodium tripolyphosphate eventually became the most widely accepted sequestering phosphate for detergent use. These phosphates all form soluble complexes with calcium and magnesium and greatly reduce the film formation tendency of the detergent due to water hardness and alkalinity. In addition, they have other desirable properties which contribute toward overall performance of the detergent.

The next major event was the use of chlorinated trisodium phosphate [4,5]. Subsequently, other chlorine donors were introduced such as the alkali metal dichloroisocyanurates [6-8] which resulted in a product with an improved shelf life.

The introduction of nonionic surfactants and defoamers took place during the period 1950-1960. It was recognized that foaming due to food soils such as egg (protein) resulted in a loss of wash pressure; thus a search for defoamers compatible with other detergent ingredients was sought for and found [9,10]. The search for better defoamers still continues. The various problems, methods for laboratory evaluation, field evaluations, stain removal, rinse additives, development of dishwashing equipment standards, and recent developments in detergent formulations are discussed by Podas and Crecelius [10].

II. THEORY OF DISHWASHING

Much has been written on the theoretical aspects of laundering. Primary attention has been directed on surface and interfacial tension, critical micelle concentration, redeposition or prevention thereof, and water conditioning, as well as other factors, such as effect of various builders. In view of the fact that laundering products are heavily dependent on organic surfactants and involve a large specific surface (fiber-soil interface), the great attention paid to surface and interfacial tension is not surprising. On the other hand, machine dishwashing products by necessity contain little or no surfactants because foaming cannot be tolerated; consequently, they are essentially "builders" with or without a minor proportion of low-foaming organic surfactant. Although the builder route of formulating dishwashing products was forced by necessity and/or design of the equipment, it also follows that the specific surface in dishwashing is very much smaller than in laundering. Thus the lesser attention given to surface and interfacial tension is understandable.

There is less written on the theory of dishwashing than on laundering, apparently because of lack of "handle" on which to grasp. In laundering studies, surface and interfacial tensions and the critical micelle concentrations are rather convenient handles which can be used to advantage because some degree of correlation may be found between theory and results. Theories developed for laundering may also hold for dishwashing but the relative priorities are not of the same order.

Warewashing, as we prefer to call it, is the removal of the food soil from a ware surface. Such removal requires the overcoming of the adhesive forces between the soil/substrate interface and the cohesive forces which hold the soil together. These are largely Van der Waals' forces and depend not only on the nature of the substrate surface (glass, metals, plastic, etc.) but on the condition of the surface, such as scratches, erosion or corrosion, and crater formation; these are sometimes found on worn ceramic glaze due to erosion of the surface and exposure of gas bubbles in the glaze formed during firing and subsequent cooling.

The condition of the soil affects adhesion as well as cohesion. A soil which is dehydrated is generally more difficult to remove than is a hydrated soil because the net Van der Waals' forces are greater due to close packing. These attractive forces decrease as the square of the distance so that close packing will greatly increase the difficulty with which the soil may be removed and/or dispersed. Moreover, with many soils such as proteinaceous types, drying may cause denaturation and loss of solubility or at least increase the difficulty of rehydration. Burnt cooking oils are examples in which heating may cause two- or three-dimensional cross-linking of unsaturated or polyunsaturated fats and are relatively independent of dehydration/rehydration mechanism for their removal. The resulting varnish-like coating or, if carried to the extreme, a carbonaceous material may be nondispersible by ordinary means.

In general, the adhesional forces between substrate and common food soil decrease if the substrate is hydrophobic and conversely increase if the substrate is relatively hydrophilic. Thus, in general, a relatively hydrophilic surface such as glass will hold soils more tenaciously than will a relatively hydrophobic surface such as Teflon. This is due to lack of polar sites on the Teflon surface and, consequently, to reduction of the Van der Waals' forces to attract and hold the soil. Relatively nonpolar soils such as tristearin or triolein (triglycerides) adhere lightly to glass surface, especially at temperatures above their melting point and are relatively easily removed. One reason for this is probably a combination of lack of polar groups for mutual attraction between the soil and substrate; another reason for this is the tendency for the electrolyte detergent to displace oil preferentially [11, 12] by a combination of mechanical (spray) action and the physical-chemical phenomenon of adsorption of the electrolyte onto the surfaces.

The various factors involved in soil removal will be considered next.

A. Kinetic Energy

So far as gross soil removal is concerned, kinetic energy is one of the most important factors in dishwashing. (Redeposition, filming, etc., may be important factors in overall performance.) The applied kinetic energy may be described by the well-known equation:

$$\text{K. E.} = \frac{\text{mass} \times \text{velocity}^2}{2}$$

Everything else being equal, one might expect soil removal to increase directly proportional to mass (volume of wash water) and as a square of the velocity (pressure). Therefore, a high-pressure, low-volume wash spray ought to be more effective than a high-volume, low-pressure spray. However, in practice this is complicated by a number of factors such as the angle of impingement, the water film or layer which must be moved before the applied K. E. acts on the soil, the interaction between the upper and lower wash arm, or the cascading water which interacts with the lower wash spray, the spray pattern, etc. A certain minimum volume of water is necessary usually to eliminate the spray pattern. Moreover, there is a practical upper limit to the wash pressure governed by the tendency of light plasticware to fly off the racks if the wash pressure is increased beyond a certain point. Kinetic energy acts directly by erosion (and/or "peeling") to remove the soil as well as indirectly by providing new surfaces to further hydration and chemical action. Kinetic energy may also be instrumental in the "roll-back" process of oily-soil [13, 14] removal and interface-activated mechanism involving the air-solution-solid interfaces. Any soil-removal mechanism involving preferential wetting or emulsification would be greatly assisted by kinetic energy. Schwartz [15] discusses recent advances in detergency theory and suggests breaking down the complex phenomenon of detergency into simpler, basic model systems for study.

Although mechanical action is known to be the key factor in soil removal, the relationship between this mechanical energy and chemical additive we call detergent is still not clearly agreed upon. As pointed out by Loeb and Shuck [16], Bourne and Jennings proposed the definition: "A detergent is any substance that, either alone or in a mixture, reduces the work requirement of a cleaning process." Loeb and Shuck also discussed a short communication by Viswanadham and Rao who commented on this definition of detergent. Loeb and Shuck studied soil removal in relation to total energy input using a Terg-O-Tometer. By providing a thermally isolated system containing a known volume of wash solution and known weight of soiled fabric, and then delivering a given amount of mechanical energy, they measured the heat change dependent only on the mechanical energy delivered. At constant detergent concentration, soil removal increased with increase in energy input as measured calorimetrically (energy increased by increasing stroke rate). In another study where detergent concentration was the prime variable and where the work input (stroke rate) was maintained constant, the soil removal increased with increasing detergent concentration up to about 0.15% concentration, and then the rate of change decreased. The studies were carried out for several fabric-load sizes. In order to relate soil removal and work input to detergent concentration, they calculated the efficiency ratio defined as the quotient of soil removal/work input. They then plotted these values against detergent concentration and obtained a family of curves in which a correlation was obtained between efficiency ratio and detergent concentration. From these studies Loeb and Shuck proposed a new definition for detergent as follows: "A detergent is any substance that, either alone or in a mixture, increases the efficiency with which mechanical energy is utilized in a cleaning process." Loeb and Shuck [17] have recently reported a follow-up study of this calorimetry method using a full-sized clothes washer.

While this work was done on laundry studies, it would appear that in principle these findings and this definition would apply to dishwashing.

B. Thermal Energy

The temperature of the wash has some major effects on soil removal as well as providing a sanitary treatment. Fatty soils and lipstick are much more efficiently removed at temperatures above their melting point because their viscosity changes dramatically at the melting point. The wash temperature should be above the melting point of the fatty soil being removed in order to take advantage of the phase change at their melting point. Below their melting point, fat removal is poor because cohesional forces are strong and interfere with the "roll-back" or displacement effect and emulsification.

The rate of rehydration and solvation may be increased by increasing the washing temperature. Where a chemical reaction is involved in soil removal, the general rule of $Q_{\Delta t} = 10°C = K_2/K_1 = 2$ (where Q is the temperature coefficient, K_1 is the reaction rate at temperature T_1, and K_2 is the reaction rate at $T_1 + 10°C$) may be expected to hold. If a soil can be related to the solubility behavior of a salt, temperature will increase the solubility of the salt (soil) according to the theorem of LeChatelier-van'tHoff [18]:

$$BA + \text{heat} \leftrightarrows B^{+} + A^{-}$$

While generally an increase in temperature is favorable for soil removal, there are some economic and practical limits. Proteinaceous soils such as egg are denatured at above 160°F, and excessively high temperature is found to be detrimental to such soil removal because denaturation renders the soil much less soluble (and irreversibly insoluble) and causes a phase change.

Home machines are operated usually at a lower wash temperature than commercial machines. For safety reasons the home water supply is usually set between 140 and 150°F but may be even lower than this in some cases. The actual wash temperature in a home machine is 5 to 10°F below the supply temperature because of heat loss in the pipe and heat loss in the machine itself. Many newer home machines have the 145 or 150°F sanitize cycle which assures adequate heat treatment for sanitation purposes.

The wash and rinse temperatures for the commercial machines must comply with the local health ordinance. Usually the local ordinance conforms to the recommendations of the U. S. Public Health Service, and according to the Food Service Sanitation Manual [19] the wash water must be at least 140° F and single-tank conveyor machines must be at least 160°F. The final rinse-water temperature must be at a temperature of at least 180° F at the entrance of the manifold. This temperature requirement is set up as a part of the heat treatment for sanitization of the dishes but also aids materially in getting better soil removal results.

Mallman et al. [20] reported on heat penetration studies from front to back of chinaware using a thermocouple to measure the rate of penetration at several spray temperatures. More recently, Scalzo et al. [21] reported findings on dish temperature in relation to wash temperature and rinse temperature using a single-tank conveyor-type commercial machine and comparing thermocouple readings to those obtained with "paper thermometers," i. e., paper strips that change from white to black at specific temperatures.

C. Hydration and Solvation

Solvation effect actually cannot be separated from erosion or kinetic energy, since the two forces work together in many situations. Solvation is one of the more important actions of the detergent solution. Proteinaceous soils such as casein and albumin may form a soluble sodium salt in alkaline solution, probably in part due to mass-action effect. A sequestering agent such as sodium tripolyphosphate is known also to dissolve or disperse milk solids. This may be due to calcium-sodium exchange (displacement), since calcium caseinate is known to be less soluble than sodium caseinate and since the polyphosphate has a greater affinity for calcium than for the sodium ion. The shift in pH away from its isoelectric point (pH 4.6) for casein is also a factor in solvation of casein by alkaline salts.

In theory, neutral salts could form complex ions, as is well known from the effect of salts on the solubility of a precipitate in quantitative analysis [22]. Thus, in such gravimetric analysis excess salt is avoided to prevent complex ion formation leading to the solvation of the precipitate. Whether this plays an important role in detergency is not really known, since in detergency studies one is concerned with the overall result and such factors as these may be well overshadowed by other factors. Neutral salts such as sodium chloride or sodium sulfate are used sometimes in detergent formulation but, for the most part, are considered as inactive fillers or processing aids in agglomerated detergents.

D. Surface and Interfacial Tension

Surface and interfacial studies are complicated by the fact that solid-substrate, solid-soil, liquid-soil, liquid-detergent solution, and air interfaces are simultaneously involved in combinations in spray-type washing. Most of the studies relating interfacial tension to dishwashing detergency have been made on fatty soil such as tryglycerides and mineral oil. Anderson et al. [11] studied the effect of fatty soil removal from glass by electrolyte detergent builders and concluded that the mechanism of the radio-tagged tristearin removal from a glass substrate by sodium tripolyphosphate is primarily one of preferential displacement. That is to say, the sodium tripolyphosphate being more polar than tristearin displaces the tristearin from the glass surface. This mechanism could possibly be substantiated by using radio-tagged sodium tripolyphosphate and untagged tristearin. Van Wazer and Tuvell [23] state that the negative zeta potential of oily droplets as well as inorganic particles is increased by the adsorbed chain phosphates, thus favoring desorption of soil. It is a little more difficult to assign adsorption configuration of a molecule such as chain phosphate than for a typical surfactant which has a hydrophobic and hydrophilic end. Moreover, the Helmholtz double-layer effect would tend further to complicate matters so far as relating charge and structure to explaining the phenomenon involved. However, it is clear that while sodium tripolyphosphate is not thought of as a surface active agent in the usual sense of lowering the surface tension of a solution, these observations suggest that sodium tripolyphosphate is definitely involved in the interfacial phenomenon and is surface active, though probably not to the same extent or in the same sense as the conventional long-chain heteropolar organic surfactants. The phosphates are surface active in the sense that evidence suggests that the phosphates change the electrokinetic potential of oil and inorganic particle and that there is evidence for the adsorption of phosphates on the glass surface [11,23]. Any adsorption which tends to make both the substrate and the soil assume like charge will result in desorption of the soil as adhesional forces are reduced.

E. Surfactants

Surfactants used in mechanical dishwashing formulation are of either the low-foaming or defoaming type. Description, function, properties, and uses of low-foaming nonionic surfactants of the polyalkylene-oxide-block copolymers are given in detail in Surfactant Series, Volume I, Chapter 10.* Because of the limitation on foaming, the surfactants usually used have a surface tension in the range of 40 to 45 dynes/cm at 0.1% concentration in water. This is relatively high surface tension compared with a value of about 26 to 30 dynes/cm for the surfactants used in laundry application, but is considerably lower than surface tension for pure water which is about 72 to 73 dynes/cm at 25°C. Although surfactants generally are said to be beneficial, the emphasis currently is more on defoaming than on other properties normally associated with surfactants.

Food soils, particularly eggs and other proteinaceous soils, cause foaming with loss in wash pressure as measured by a Pitot tube placed in the path of a spray nozzle, or as determined by a manometer connected to the pump manifold, or by means of counting the revolutions per minute of the wash arm. Loss in pressure could come about by several mechanisms. Foam may, for example, cause a loss in volume of wash water since most commercial machines are kept at constant volume by an overflow standpipe. Loss in solution volume may cause air intake in the pump manifold in some instances. Even without this direct air intake, air dispersed in the water will cause a loss in kinetic energy due to loss in mass per given volume of solution and will cause the loss in application of energy on impact of the wash stream due to the cushioning (elastic) effect of the foam.

---

*Martin J. Schick (Ed.) Nonionic Surfactants in Surfactant Science Series, Volume I, Dekker, New York, 1966.

Some defoamers are relatively nonspecific and may defoam a number of different kinds of foam, while others are relatively specific and defoam only certain foams. Silicones, for example, are widely used for defoaming detergents, defoaming in food processing, paint, etc. Defoamers for dishwashing detergents usually have a much narrower spectrum but have greater water solubility and wetting properties. There are several theories on the mechanism of defoaming and they may be complementary. Foam stabilizers are said to act in some instances by entering the bubble film, increasing the film viscosity (film strength), and slowing down the drainage rate; Ross and Young [24] found that some foam inhibitors acted in reverse manner by promoting the liquid drainage, thus destabilizing the foam. In other studies, Ross and Bramfitt [25] found that above their critical micelle concentration, foam inhibitors increased the specific conductance of the colloidal electrolyte solutions which they interpret as a result of reduced micellization caused by the inhibitors. Ross and Haak [26] obtained further information that the addition of a foam inhibitor accelerated the attainment of surface tension equilibrium, thus hastening the foam breakdown. Fineman [27] postulated the theory that foam inhibitors are themselves surface active agents that are absorbed into foam films and tend to displace the foaming agents but, being poor stabilizers, cause defoaming. The specificity of some defoaming agents would suggest some sort of foam stabilizer-foam inhibitor interaction.

F. Hypochlorite

The use of active chlorine came into general use beginning in about 1950. Two of the most common sources of active chlorine today are chlorinated trisodium phosphate and alkali metal dichloroisocyanurate. Chlorine has a two-fold purpose (1) to destain food soils such as tea and coffee stains which appear to be especially troublesome in hard water and in water containing iron, and (2) to minimize spotting and filming which is particularly noticeable on glass tumblers. The destaining mechanism of the hypochlorite is probably similar to those involved in destaining fabric. The marked tendency toward minimizing spotting is a less-understood mechanism.

Several important observations to be considered are: that at the normal pH range of the use solution of the detergent (pH 10 to 13), over 99% of the chlorine exists in the OCl^- form [28]; the active chlorine compound must have a relatively large hydrolysis constant in order to be effective; proteinaceous materials and other nitrogen-containing compounds, and especially the albuminoid proteins, form N-chloro compounds with very small hydrolysis constant [29]. This may explain why, in dishwashing solutions containing proteins, the OCl^- becomes rapidly nonavailable. Schmolka et al. [30] studied fabric-bleaching properties of a number of different active chlorine compounds. When they are compared with the hydrolysis constant of various N-chloro compounds as listed by Kirk and Othmer [29], there appears to be a very distinct correlation between the bleaching property and the hydrolysis constant.

One of the most susceptible linkages in a protein is the disulfide linkage. It is believed by Briggs [31] that the key in the reaction between protein and OCl^- might be the disruption of the protein via the disulfide linkages into more soluble products.

Tracer studies [32] have indicated that the addition of 50 ppm hypochlorite in detergent solution greatly retarded both the fat and casein deposition on glass surface compared to the same detergent without the hypochlorite. The tagged soil was introduced into the wash solution. These data are shown in Fig. 3.

Baker [33] has studied the interaction between albumin and hypochlorite; the rate of protein degradation was followed by determining the amount of intact protein remaining after varying time intervals; the intact protein was determined as that precipitable by phosphotungstic acid. His data, summarized in Fig. 4, indicate that there is a good correlation between the rate of disappearance of the precipitable protein and the loss in available chlorine, and that the reaction is rapid, being substantially complete in a few minutes. A similar study was made [32] in the loss of available chlorine in several proprietary detergent formulations when dried milk solids were added to 0.2% concentration of the detergent at 130°F. These data are shown in Fig. 5 and are similar to the data shown in Fig. 4. These data are in agreement with field observations, namely that the available chlorine found in a wash solution is a fraction of the total added. It indicates that the chlorine reacts readily with the protein soil; but despite the small amount of available chlorine demonstrable at any time, it does have a marked effect on dishwashing results. Thus, it may have greater influence on redeposition than on soil removal itself since it is expected that chlorine will react more readily with soil in the wash solution than with soil on the dishes. It is also possible that the chlorine may affect the "monolayer" soil, i.e., the last trace of soil on the dish or glass surface.

Tannins react with protein to produce marked changes in the leather-tanning process. Tannin-protein combinations are believed to be involved in tea staining. It might be speculated that hypochlorite would attack the protein part of the complex; this might be a possible mechanism in destaining.

G. Filming and Spotting

Filming and spotting may be due to soil residue or deposition and/or redeposition of insoluble matter. Frequently, it appears to be most strongly associated with solids in the final rinse. Waters may contain

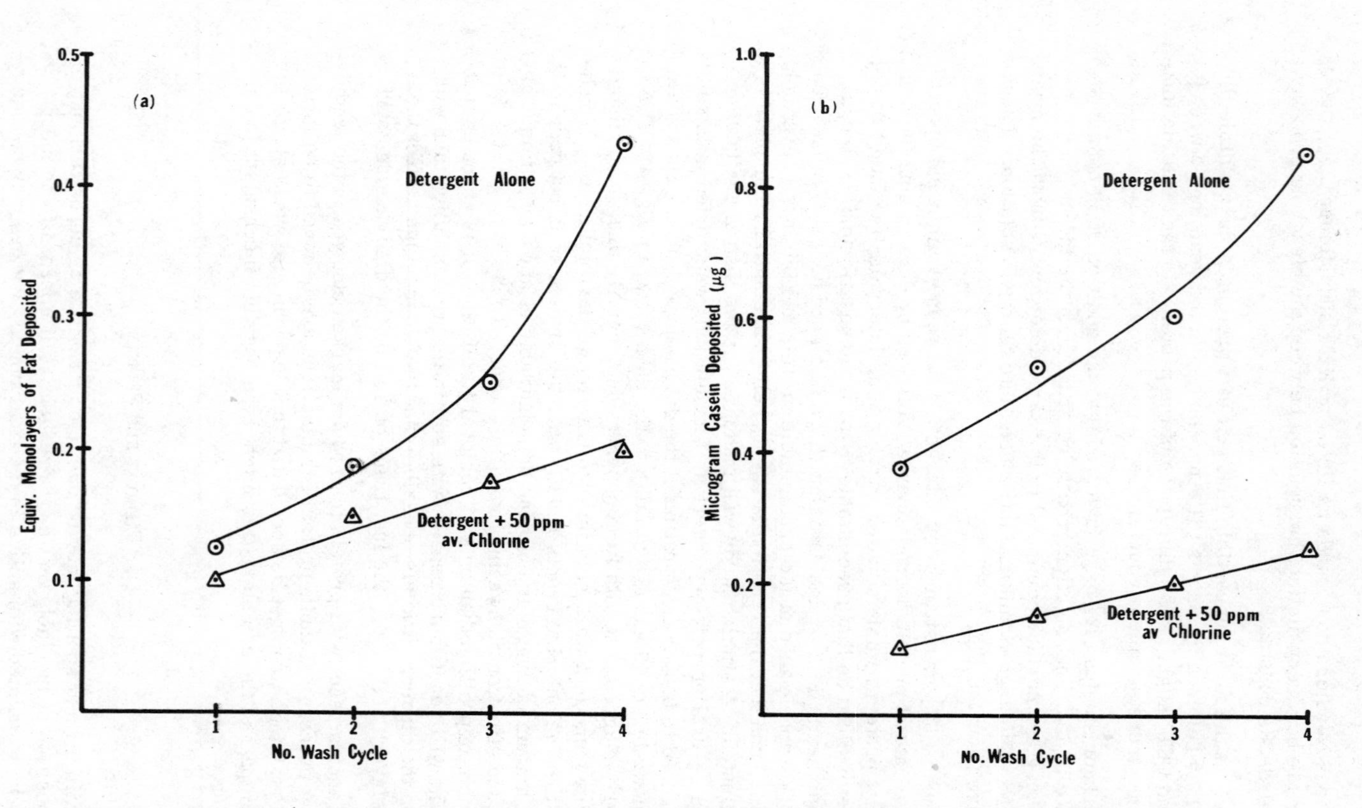

FIG. 3. Effect of available chlorine in detergent on minimizing (a) fat and (b) protein deposition on glass surface. Soil tagged with radioactive tracer. From Ref. 32.

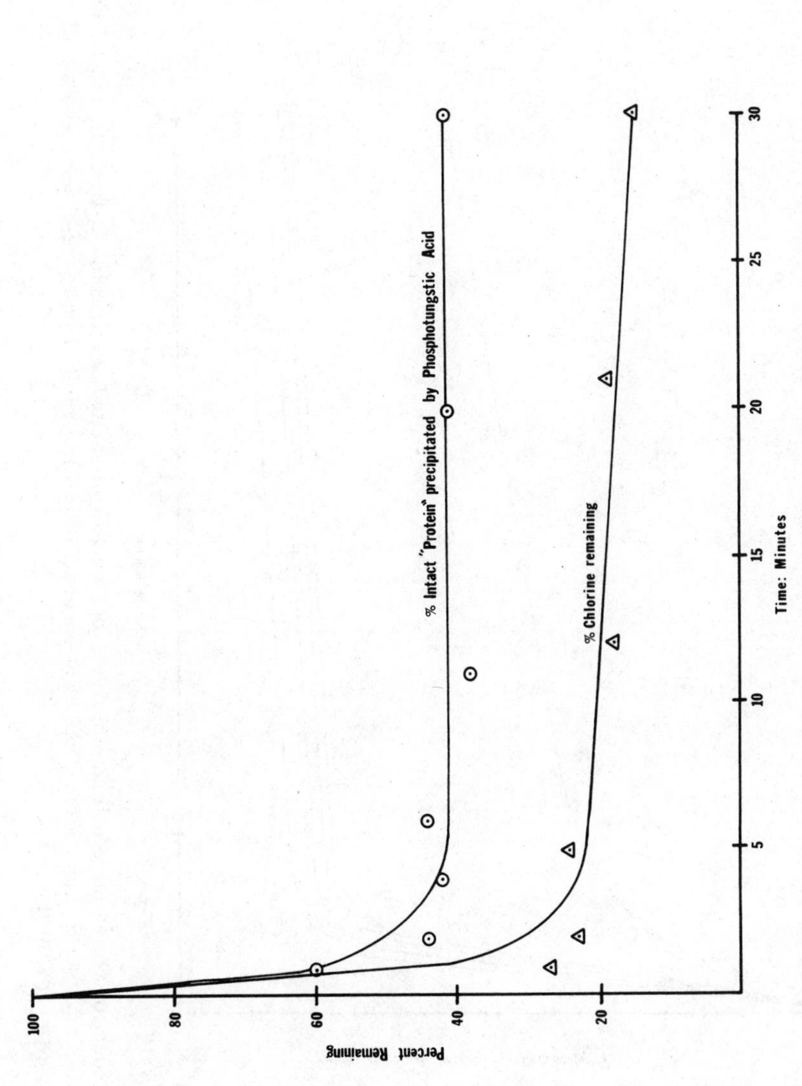

FIG. 4. Studies on effect of hypochlorite/albumin interactions: effect on percent intact protein precipitable by phosphotungstic acid after varying reaction time (⊙% albumin; △% hypochlorite). From Ref. 33.

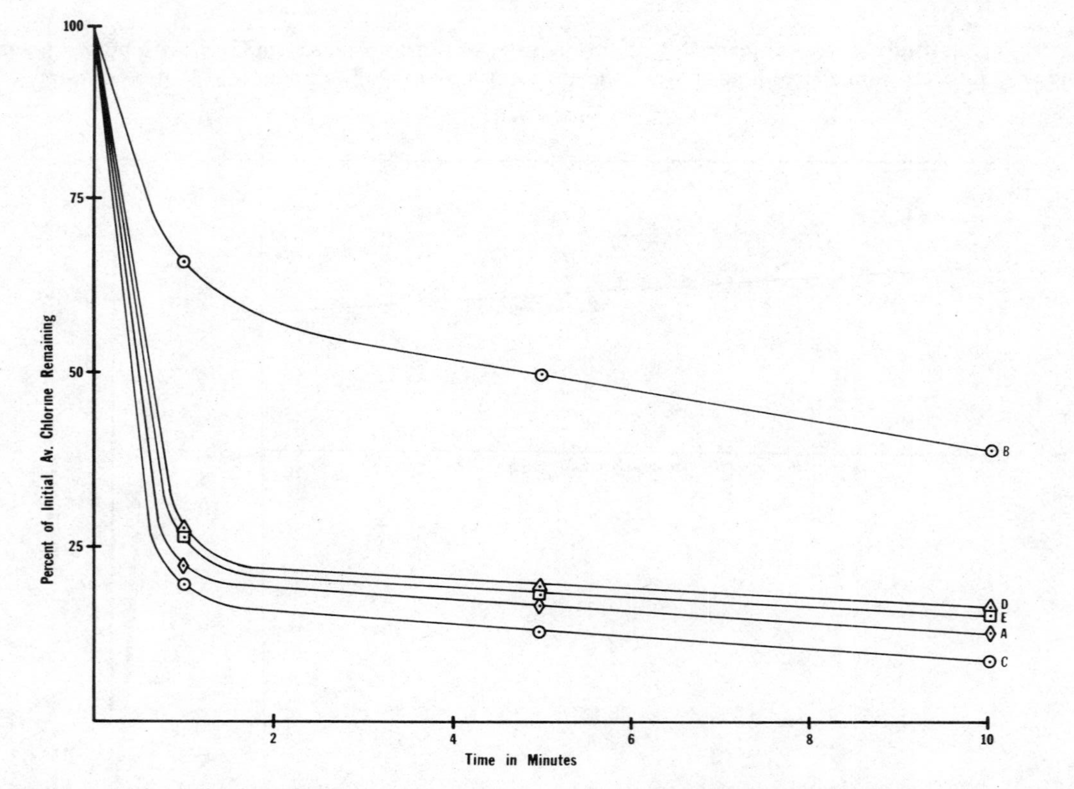

FIG. 5. Reduction of available chlorine in five proprietary chlorinated detergents, A, B, C, D, and E, when 0.12% dried skim-milk solids are added. Detergent concentration—0.2%; initial chlorine- 20-35 ppm av. cl.; temperature—130° F. From Ref. 32.

from less than 100 ppm to over 1000 ppm solids. Thus, any water which is left on a surface will eventually evaporate and deposit its solid. In general, water containing less than 100 ppm hardness as $CaCO_3$ does not give visible filming or spotting problems. Waters containing over 300 to 400 ppm hardness as $CaCO_3$ may be expected to give problems. Experience indicates that the use of a conventional salt regenerated zeolite or resin softener may help in situations where the hardness is high. In waters where the total solids may be 500 to 700 ppm and higher, the residue which remains when the water is evaporated may cause spotting and streaking even though the water may be softened. Under laboratory conditions [34], a distilled water or deionized water appears to correct completely this water problem.

As indicated previously, the filming and spotting is not completely dependent on the hardness of the rinse water alone; proteinaceous and fatty soil are also involved. Tracer studies [32], as well as the general observation that the addition of chlorine to the detergent will minimize filming, do indicate that the rinse water is not the sole factor in filming under practical conditions.

Detergent components may affect the film formation. Studies by Wilson and Mendenhall [35], Madden [36], and independent studies at Midwest Research Institute [37] indicate that in hard waters the film buildup on glasses follows the general pattern from the heaviest to the lightest film: sodium carbonate, sodium metasilicate, trisodium phosphate, chlorinated trisodium phosphate, and sodium tripolyphosphate. The last three components gave very light films. The film formation in the studies by Wilson and Mendenhall and the Midwest Research Institute were measured with a photometer. In the Madden studies, the amount of film on glasses was determined gravimetrically by weighing glasses before and after washing. The chemical composition of the film was determined separately by making up the detergent or component in hard water, collecting the precipitate formed and by analysis of this precipitate by the "wet-methods." It is assumed that the film was formed on glasses by the deposition of the precipitate from the detergent solutions. Of significance here is that in the Madden studies, the films (precipitates) due to the sodium carbonate and sodium metasilicate were both identified as calcium carbonate. The film (precipitate) caused by the sodium tripolyphosphate was identified as primarily calcium tripolyphosphate; and films (precipitates) caused by the combination of trisodium phosphate and tripolyphosphate were identified as calcium phosphate and calcium tripolyphosphate; these films (precipitates) occur when insufficient sodium tripolyphosphate is present to sequester completely the hardness in water.

In addition to the above hard-water films, Madden describes a permanent film or "soft-water etching." This etched appearance occurs on glasses washed in softened water over a number of cycles and with formulations containing low-alkali silicates and high-sodium tripolyphosphate. The film is reported to be practically identical to the infrared spectra of hydrated silica or silica gel. He reports that the tendency of formulations to produce this permanent filming can be minimized by using the more alkaline silicates and sodium carbonate along with minimum sodium tripolyphosphate. This permanent film cannot be removed by ordinary acids and typically starts out as an iridescent film which is later transformed into an etched-appearing film.

Vance [38] studied detergent-hard water interaction by measuring the turbidity of a solution in which the $CaCl_2$ was varied from 12 to 0 mM and $Na_5P_3O_{10}$ was varied from 0 to 4.8 mM. Maximum turbidity was found at a Ca/P_3O_{10} ratio of 5:2, suggesting that the precipitate possesses a formula of $Ca_5(P_3O_{10})_2$. He further noted that substantial precipitate was formed on the $P_3O_{10}^{-5}$-deficient side of this stoichiometric point but, on the $P_3O_{10}^{-5}$-rich side, the precipitate redissolved with a slight excess of $P_3O_{10}^{-5}$.

To relate this to film formation on a glass surface, Vance repeatedly dipped glass slides under controlled conditions into solutions of a detergent and gravimetrically determined the film buildup. Systematic variations of both detergent concentration and water hardness allowed the construction of a response surface upon which contour lines were drawn connecting points of equal weight gain (film buildup). Instead of a family of rectilinear hyperbolic curves expected if no specific interaction occurred, Vance found two areas of distortion to the curve, one indicating more deposit than anticipated and the other less deposit than anticipated, paralleling the turbidity study described previously. This study showed that film formation is minimized when the Ca/P_3O_{10} ratio is 1:1 and maximized when the ratio is 5:2. These ratios correspond to the 1:1 chelate and the uncharged precipitate, $Ca_5(P_3O_{10})_2$.

Vance further discusses the implication of these ratios on the precipitation of $CaSiO_3$ in detergent formulations containing tripolyphosphate and silicate and feels that detergent interactions are a major cause of glassware filming.

III. DETERGENT EVALUATION

The purpose of laboratory detergent evaluation is for product optimization and for projection of what results might be anticipated under actual use. Because use conditions may differ from one extreme to another in regard to various factors, the investigator has the choice of selecting the test conditions to encompass "95% of the universe" or of attempting to set up a condition to simulate the conditions dictated by a given market for which the product is intended. A frequent pitfall is to rely too heavily on a fixed set of conditions and to optimize around this set of conditions which may or may not be truly representative of the true parameters.

Evaluation may be subdivided into soils and soiling methods, wash methods, soil residue measurements, and filming and spotting measurements.

A. Wash Action

Selecting the type of wash action is the first step in designing an evaluation method. For various reasons investigators have selected immersion wash [39,40] with revolving holders for glass slides, adaptation of Terg-O-Tometer [41], specially designed spray machines [42] and the use of modified or unmodified machines that are in wide use [20,36,43-46]. The latter approach is the one being followed most widely today, not only because it does not require engineering development but also because an ideal laboratory designed machine is not indicative of the realistic condition that exists under actual use condition. The investigator ought to have some insight into how representative a particular machine is to the "universe." Some of the variables that need to be considered in selecting a machine are the wash pressure, carryover of the wash into the rinse, susceptibility to foam out, food strainer as it may affect redeposition of food particles, the number of washes and rinses, and their durations, etc.

For certain evaluations such as evaluation for detergent "cup caking" in multiple-wash, home-type machines, the design which gives the most problems may be chosen on the assumption that, if the detergent does not cake in this machine, it will function as well or better in other machine types.

In commercial-type machines, the selection of a machine is more limited. It would be impractical to select a large "flight-type" conveyor machine for laboratory evaluation because of the large amount of hot water, steam, and soiled dishes needed to run the evaluation. Usually, a single-tank, manually operated machine or a simple double tank machine is chosen, and operating conditions are varied to simulate various field conditions such as the effect of temperature, types and amount of soil load, water hardness, wash and rinse duration, etc. The recognition and evaluation of spray-wash pattern is very important as indicated by various investigators [43,46]. In fact, specially engineered laboratory machines were designed by some investigators to eliminate or minimize such patterns. But the overall washing efficiency of a system is influenced by the design of the machine as well as by other factors. Thus there is currently a tendency to evaluate with several widely used machines rather than to overdesign a laboratory machine which would not be truly representative of any in use; this would be true for home as well as commercial machines.

B. Selection of Test Pieces

The selection of test pieces may vary with the object of evaluation. Usually, new pretreated test pieces are used, particularly in the case of glass tumblers because the surfaces are more reproducible and the results less likely to be due to variations in surface conditions. Scratches adversely affect photometer readings. These glasses are prewashed or handwashed in a uniform manner previous to use as a test glass. Plain china is usually preferred to patterned china because, whether it be visual or photometric evaluation, it lends itself to more uniform reading. Various types of dinnerware, tumblers, etc., should be included, and test pieces as well as "dummy" pieces should be included to simulate normal home or commercial operation in regard to loading. Flatware (i.e., knives, forks, and spoons) should also be included. Glasses and flatware give the most problems in normal use, so that any evaluation should weigh these factors rather heavily. The condition of ceramicware surface may often be determined qualitatively by writing on it with a lead pencil. Surface in good condition will not mark, whereas surface roughened with use and wear will mark. It will be noted usually that, after some use, dinner plates will mark on the eating surface but not on the back; this is probably wear due to abrasion with knife, stacking, etc. Plasticware is in common use and should also be considered. Gold- and platinum-rimmed dishes or tumblers normally should be washed by hand but special evaluation should be made to assess the effect of detergent solution on them.

When setting up a detergent evaluation, normal loading of the machine should be followed as recommended by the dishwashing-machine manufacturer. However, overloading and improper loading so as to block off spray action to some areas of the machine is a common problem in the home.

C. Soil and Soiling Methods

Soil and soiling methods are important as performance is a direct measure of the difficulty with which the soil is removed; if the soil is too easily removed, all products tested would be given a high rating and, conversely, if the soil is too difficult to remove, all products would be rated poor. Thus soil should be selected according to the objective of the test and in relation to the normal soil encountered in a dishwashing operation. For optimization purposes, a soil should be selected so that a gradation in removal would be found between the poorest and best detergent but with incomplete removal with the best detergent.

A number of different soils have been tried by various investigators. These range from tagged "pure" triglycerides [11, 12, 32], tagged proteinaceous soils [12, 32], tagged bacteria [47], and starchy soils such as homogenized oatmeal [48] to complex blends of peanut butter, lard, dried milk solids, oleomargarine, mineral oil, and egg [39,40,43,44,49]. A modification of the Hucker soil is suggested by Leenerts et al. [41] and classified as protein-carbohydrate and greasy soil. The degree of cure must be closely controlled in order to get reproducible soil removal. Many soils are cured from room temperature to 150°C [41] and from a few minutes to overnight [20], depending on the soil and degree of tenacity desired. It is advisable regularly to run a standard reference detergent for control to check on the uniformity of soil.

The method of soil deposition varies from placing soil in the wash tank [32,42,46] for deposition studies, to dip application [41], to spraying on soil with a special spray applicator [43], to manual application by rolling with a rubber roller [39]. Each has its merits and shortcomings. Spray methods are applicable to flat glass or ceramic plates or to dinner plates. The dipping method is applicable to small test pieces while the mechanical leveling device (paint-leveling doctor-blade or rubber roller) is suited only for very flat test pieces such as plate glass and for relatively homogeneous soils. The spray methods, using a special spray head, offers a fairly wide degree of flexibility, although it may require homogenization of the soil in a food blender prior to spraying.

Some laboratories soil glasses and dishes with milk, fruit juices (especially orange or grapefruit) that deposit particular residue, gravy, fat, egg, etc., and allow them to air dry to simulate a typical mealtime situation. This is an excellent method but somewhat time consuming for routine use and difficult to standardize.

D. Soil Residue Measurements

The soil remaining on test pieces either as residual or due to deposition and/or redeposition, may be rated in various ways. Visual inspection and feel of grittiness has been used by a consumer testing organization [50]. Visual graded standards have been used, ranging from completely clean to no removal, with numerical values from 1 to 33 using oatmeal, India ink soil [10]. The difficulty here is maintaining such a standard and setting up interlaboratory uniformity. An optical scanning device was described by Mallman et al. [20]. In this device the dinner plate is rotated on a turntable while a reflectometer continuously scans the surface in a spiral path from the center to the periphery. The reflectance reading is averaged to get a rating. Tracer methods have been used by Armbruster and Ridenour [47], Harris and Satanek [12], and Benton et al. [42]. Tracer methods give a high degree of sensitivity but the limitation is on disposal of the radioactive wastes if a large number of tests are to be carried out. Another limitation is that, due to waste problems, investigators [42] have resorted to miniaturized wash-machines; this not only poses a mechanical design problem, but also limits the type of surface that can be washed to small glass slides. Light-transmittance measurements have been used by Mann and Ruchhoff [39] and others. For such measurements, glass slides of high optical quality must be used. The use of polished glass surfaces (plate glass) is believed by some investigators to affect soil binding due to alteration of the charge on the glass surface during the polishing operation [12]. Generally, in such transmittance measurement the soil removal is calculated from Lambert's and Beer's Law:

$$R = \frac{\log Iw - \log Is}{\log Ic - \log Is} \times 100,$$

where R = percentage soil removed,
Iw = transmitted light through washed slides,
Is = transmitted light through soiled slides, and
Ic = transmitted light through clean slides

Photographic record or radio autographs have been used by some investigators as a means of making a permanent record of performance tests.

Qualitative identification of soil is sometimes important, particularly if various soils are used simultaneously or if redeposition is being studied. Such identification is also important when soils are to be identified on dishes or glasses that are brought in from field studies or operations. Starchy foods such as mashed potato granules, oatmeal, rice, etc., may be identified readily by flooding the dish with Lugol's iodine solution. Milk soils contain riboflavin which fluoresces under ultraviolet light. Some soils absorb fluorescent dyes and the dish may be flooded with dilute fluorescent-dye solution containing a small amount of wetting agent, rinsed and drained, and examined under ultraviolet light [51]. Protein residues may be identified by spraying with a ninhydrin solution which reacts with protein to form a lavender-to-purple-colored complex. Armbruster and Ridenour [52] have suggested a technique of sprinkling talc-safranin dye mixture to reveal traces of soil on dinnerware.

There are a number of other techniques possible by removal of the material from the dishes and using one of many spot test techniques of Feigl [53].

E. Film and Spot Measurements

In general, the situation is similar to the measurement of soil residue except that film is due to deposition and/or redeposition; also, the investigator is usually interested in measuring low-optical-density range, and some distinction must be made between film and spots. For film deposition studies one must start with clean, transparent, plain-surface glass using the initial reading as the clean value or setting the initial value as optical density of zero (100% transmission). One of the early photometers described by Wilson and Mendenhall [35] is a true measure of light transmission although light scattering was undoubtedly involved as the glass slides were placed at an angle in reference to the light plane between the source and the phototube. A refinement of the photometer is described by Kimmel et al. [46] to measure films of much lower intensity. Due to the design of the optical system, the photometer measures Tyndall effect; thus the final readings cannot be related to either optical density or light transmission, but in terms of meter reading. This system also requires the use of a plate-glass test piece. This photometer has been used by several laboratories for a number of years with some degree of success. Another method is visual rating of glass tumblers in a light box again using the Tyndall effect and rating according to the degree of filming such as in the CSMA Test Method [44] or by comparing against a series of graded standard tumblers from clean to heavily filmed glasses rated arbitrarily from 1 to 10. Pryor [54] has applied a statistical method for rating tumblers by a ranking method which may be used to establish reliability of the technician evaluating filming and spotting. The problem with all visual grading methods is the variability between operators and between laboratories.

In 1968, a photometer was developed by E. D. Berglund (Economics Laboratory, Inc) in which the film on glass tumblers may be measured using the Tyndall effect. Basically, this is a light-tight box in which the tumbler is thrust through a circular felt-lined slit with the lip edge down. The upper two-thirds of the box is blacked out and isolated optically from the lower third which contains the light source. The only light reaching the upper compartment of the box enters by conduction of light through the tumbler wall itself. The light is then scattered by any film on the glass and the scattered light is picked up by six photoconductive cadmium sulfide cells (peak spectral response at 5500 Å, stable, low memory) located about midway up the glass and around the periphery in the upper part of the box and at right angles to the wall of the tumbler. The current flow through the cadium sulfide cell is balanced by an opposing current in a bridge system and the potentiometer reading at the null point is used as a tumbler reading.

No methods are without drawbacks but each successive generation of photometers is usually an improvement over the previous one.

Radioactive tracer methods have been used to record water spots on X-ray film [42] by wrapping a film around a tumbler, exposing the film, and developing in a conventional manner. However, this method is time consuming and cumbersome and suited only for studying the development of spots from one cycle to another rather than for the routine measurement of spots. Such studies have shown that spots do indeed continue to be present from one cycle to another rather than being washed off between each cycle; and as cycling continues, spots increase in numbers and intensity from one cycle to another.

On occasion it may be desirable to identify qualitatively films or spots. Use may be made of the already-mentioned spot tests for identifying food residues. In addition, it is often desirable to determine whether an ordinary film or an "etch" (permanent softened water film) exists. Usually deposits may be removed readily by a light scraping action of a pin or spatula. Etch will not be removed in this fashion. Hard water deposits may be detected by applying a drop of 5 N HCl and blotting the drop after a minute or so of contact. If the film is due to hardness, a clear area will result due to removal of the deposit. If a successive application of concentrated sulfuric acid and rinsing does not remove the deposit, it is usually an indication of an etch. This can be confirmed by treating a small area with 10 to 20% hydrofluoric acid, which will remove silica film. Silica film can only be removed with the HF treatment, whereas most other materials may be removed with HCl or charred with concentrated sulfuric acid if an organic film or residue is present.

F. Corrosion

Corrosiveness of detergent solutions may be determined in several different ways. China corrosion may be measured by the CSMA test method [44] wherein a cut or broken piece of overglaze pattern is exposed to various concentrations of detergent in a hot-water bath for a specified period of time and then examined for pattern fading. This is a relatively quick and simple method. However, it is not possible to relate this data directly in terms of the number of cycles a pattern may be washed under normal conditions before perceptible fading occurs. Cycling of dishes under normal wash conditions is not practical because of the enormous amount of time required. In several instances, a formulation not passing the CSMA soak test was cycled for 500 to 700 times without perceptible fading (unpublished data).

Gold- and platinum-rimmed or banded glassware is very susceptible to alkaline detergent and chelates and normally is not recommended for machine washing. Discoloration is said to be especially rapid when manganese is present in the water.

Plasticware pattern also requires some type of fading-evaluation test but the generally good durability, the shorter expected life, and lower cost of these wares has aroused less interest in this area.

Aluminum discoloration and weight loss is a real problem. In many cases, hard water alone will discolor aluminum; properly formulated detergent solutions usually will not discolor aluminum if present in sufficient concentration. However, low concentrations of detergent may fail to prevent discoloration. Aluminum strips may be soaked in detergent solutions at various concentrations and visual as well as weight changes may be reported [55]. For evaluation in machines, a test coupon or aluminum pie plate may be placed in the machine and observed for discoloration at the same time that other evaluations are made. Weight loss allows definite numbers to be assigned to the rate of corrosion but, since in most cases visual appearance is more critical than weight loss, some judgment has to be used and most tests are comparative rather than absolute. Aluminum discoloration is a complex phenomenon dependent on water hardness, trace minerals and other factors; thus a given detergent formulation may work well in some waters and conditions but may not work in other waters or conditions and there appears to be no single answer or solution.

Other corrosion tests relating to materials for construction of machine and pot and pan handles must also be considered. If a material with a known history of usage is used as a standard, an accelerated corrosion test at high temperature and long soak may be improvised for comparative data.

Most commercial machines of current manufacture are made of relatively corrosion-resistant materials such as porcelain, stainless steel, inconel, durable plastics, polypropylene, etc., so that corrosion is not a serious problem.

G. Finalizing on Test Methods for Evaluation

There is no single method which will take into consideration all the various conditions under which the detergent product may be used. Some of the more important variables that need to be considered in designing a suitable test method are discussed below.

1. Variables in the Machine Itself

Usually the most widely used machine is chosen. However, for assessing the effect of foaming, one which is most susceptible to foam, to loss in wash pressure, and to foaming out of the machine should be used. The high-pressure machine is usually more susceptible to foaming and, therefore, would be affected more readily by foaming surfactant or would react to an inadequacy of defoaming surfactant in the detergent product. Machines vary in detergent carryover from wash to rinse cycle. Although there appears to be some benefit to carryover, there is no completely clear-cut correlation between carryover and aluminum protection. If detergent cup-caking is to be evaluated, one should select a machine which is most susceptible to cup-caking. In matters of corrosion and inadequate cleaning, filming, etc., inadequate detergent concentration is often the most important cause. Therefore the detergent cup size is an important factor. Although the detergent charge may be controlled to some extent in laboratory evaluation, the volume fill of the detergent cup is the controlling factor in home operation; consequently, it is important to provide a detergent product with high bulk density. This in turn is dependent on the raw materials used and the processing technique. Differences in the cycle vary from machine to machine. For many evaluations, a separate manual control which would override the automatic timer is found to be advantageous for laboratory operations. The silverware-basket location varies from machine to machine as does the location of the spray arm(s). Some home machines have an upper spray head or arm for top rack and a conventional wash arm for the lower rack to provide greater washing efficiency. Some of the above comments apply also to commercial machines. Detergent dispensers for commercial machines are of an entirely different design, as discussed elsewhere.

Racking and spray pattern must be considered, as pointed out by Kimmel et al. [46] and Mallman et al. [20]. Kimmel et al. suggest finding locations which tend to film or spot in varying degrees, and marking and using these positions for locating the test pieces.

2. Wash and Rinse Temperatures

Wash and rinse temperatures affect cleaning performance and drying. For home machines the temperature may vary from as low as 115° F to as high as 160°F. The actual wash temperature will usually be 5 to 10°F below this supply temperature. Thus evaluations should be carried out at several temperatures ranging possibly from as low as 110°F to as high as 150°F. In order to maintain temperature uniformity, a thermostatted reservoir of fairly large capacity located near the dishwasher should be employed, rather than depending on the ordinary water heater which may show considerable temperature difference between "high" and "low." The actual wash temperature of the machine should be recorded always.

If a commercial machine is used in the evaluation, the wash temperature usually employed is 140 or 160°F and the rinse temperature is 180°F. The wash-tank temperature may be maintained with the heater thermostat supplied with the machine. A booster heater or a separate 180°F-rinse supply is needed. It should be kept in mind that rinse temperatures may run as high as 200° F in hospital kitchens and this may accentuate filming problems in waters of high hardness.

3. Water Hardness

Control of water hardness is an important factor in detergent evaluation. As hardness varies by geographical location as well as the source of water, it is important to evaluate detergents in at least two water hardness levels and in softened water. Many laboratories are located where a city water supply of 100 to 120 ppm hardness is available and this is a good level for the softer water. In addition, if well water of about 300 ppm hardness is available, this may be another satisfactory level of test. However, for testing at higher water hardness, the well water may be fortified with a calcium and magnesium chloride (2:1 ratio) and sodium bicarbonate [56], and made up to 500 ppm hardness. The use of alkaline earth chlorides and sodium bicarbonate does not completely duplicate natural hard water, since this will result in the presence of higher amounts of sodium chloride than is usually found in natural waters high in calcium bicarbonate. High-solids water such as that found in the Southwest must be duplicated by analyzing and fortifying water with the appropriate salts. Tests should also be carried out in softened water. Such effects as softened-water etch may require a large number of cycling before it becomes visible.

Water hardness aggravates filming as well as tea and coffee staining. Tea brewed in hard water has deeper color and stronger flavor and taste and results in a greater tea-staining problem [57]. This is especially true in England where tea is the favored beverage and where strong brew is the rule. Whichever water is used, the hardness should be determined and recorded once a day or at frequent intervals. Daily fluctuation in hardness is not usually large but large seasonal fluctuations may occur, especially if the source is river water.

4. Uniformity of Soiling and Curing

Soil-curing conditions will be found in the various references cited previously. It should be noted that high temperature and short time used by some workers may lead to variations in the degree of curing, as temperature may not equilibrate under these conditions. Spray or dipping conditions should be controlled carefully to assure uniform soil thickness. Uniformity of soil removability should always be checked against a standard reference detergent.

5. Selection of Test Pieces

Selection of test pieces may be important. Straight-sided tumblers are preferred to curved tumblers, and plain china is preferred to patterned ones. Kimmel et al. [46] suggest washing plate glass for at least four cycles in order to get improved reproducibility. Preconditioning of tumblers and all test pieces should be done in order to remove any traces of material (or polishing agents in the case of plate glass) which may be present. Flatware is difficult to clean, especially that which has egg soil residue; and flatware should be included in the tests. Spotting of flatware is also another common problem and one that is noticed by the public.

The CSMA Test Method is probably the most widely used test for evaluation of products for the home dishwashing machine. The procedure may be modified by a more precise measurement of film using the photometer and flat glass pieces as described by Kimmel et al. [46], or the tumbler photometer described elsewhere in this chapter. Foam measurements may also be reported by recording the revolutions per minute of the wash arm [44,45] and by various other refinements, as suggested elsewhere in this chapter. The CSMA test is primarily for soil deposition and china corrosion, but a more complete evaluation would include soil removal tests as well as corrosion of silver plate, aluminumware, etc. The china corrosion test mentioned in the CSMA procedure may be applicable to evaluation of detergents for the home machine where overglaze patterns are involved, but it is not usually important for evaluation of detergents for commercial machines where underglaze patterns are involved. The exposure of dishware to detergent solution in the commercial machine is much shorter than in the home machine, and this must be taken into account in the interpretation of the results.

IV. FUNCTIONS AND PROPERTIES OF VARIOUS DETERGENT COMPONENTS

The following section gives a brief description and properties of the various components used in dishwashing detergents. The references cited give more detail and complete information.

A. Polyphosphates

Polyphosphate is a generic name used to describe complex phosphates and, in the jargon of detergent chemistry, is usually limited to dimeric or higher linear polymers of the phosphate.

Unfortunately, the naming of phosphates in the early literature, when complete information was unavailable, led to some confusion which still exists today. For example, a product having an Na_2O/P_2O_5 ratio of about 1.0:1.0 obtained by melting monosodium orthophosphate has been called Graham's salt, and sodium hexametaphosphate. A product with an Na_2O/P_2O_5 ratio of 1.1:1.0 has also been called sodium hexametaphosphate. A glassy phosphate having an Na_2O/P_2O_5 ratio of 1.2:1.0 has been called polyphosphate by the trade.

Van Wazer [58] suggests that the proper way of naming phosphate glasses is to describe them in terms of either (1) the Na_2O/P_2O_5 mole ratio, (2) the number average or other specified chain length, or (3) the average composition. Following this convention, Table 2 lists common polyphosphates which are or have been in use during recent years. All are essentially linear polymeric phosphates except the trisodium phosphate which is monomeric. It should be noted that the average chain length is

TABLE 2

Na_2O/P_2O_5 Ratio of Common Phosphates Used by the Detergent Industry

Na_2O/P_2O_5 mole ratio	Average chain length	Approximate composition (% P_2O_5, (w/w)[a]	Compound	Usual form
1.0:1.0	60	69.6	Graham's salt	Glass
1.1:1.0	14	67.6	Sodium hexameta-phosphate	Glass
1.33:1.00	6	63.20	Sodium hexaphosphate	Glass
1.5:1.0	4	60.43	Sodium tetraphosphate	Glass
1.67:1.00	3	57.88	Sodium tripolyphosphate	Crystal
2.0:1.0	2	53.38	Sodium pyrophosphate	Crystal
3.0:1.0	1	43.3	Trisodium (ortho) phosphate	Crystal

[a]Phosphate glasses may contain a small amount of water which affects chain length.

somewhat misleading in that the distribution, particularly above the tripolyphosphate, may be quite wide and contain a range of chain lengths. The only ring phosphate of importance to the detergent industry today is the trimetaphosphate, which is readily converted to sodium tripolyphosphate by caustic soda or by other alkaline material such as soda ash, and higher silicate in the presence of water and at moderately high temperatures and high ionic strength.

The preparation and manufacture of various polyphosphates and the thermal transitions are beyond the scope of this chapter. They are discussed in detail by Van Wazer [58], Quimby [59], and Bell [60]. A summary of known thermal transitions in the systems Na_2O/P_2O_5 is reproduced in Fig. 6 [61]. Of particular interest are the thermal and cooling requirements to obtain the so-called form I and form II sodium tripolyphosphate. The importance of these two forms is in relation to their rates of hydration. Form I, also known as the "high temperature rise" is characterized by hydrating rapidly and exothermically; its temperature rise can be measured readily. The form II hydrates slowly showing a much smaller heat rise. The commercial sodium tripolyphosphate is a mixture of form I and form II. There is said to be more than 30 different grades of sodium tripolyphosphate in regard to the balance between form I and form II and as to bulk density,

Pyrophosphates

$Na_4P_2O_7\text{-III} \xrightleftharpoons{410 \pm 10^\circ C} Na_4P_2O_7\text{-II} \xrightleftharpoons{520 \pm 10^\circ C} Na_4P_2O_7\text{-I} \xrightleftharpoons{985^\circ C} \text{melt}$

Tripolyphosphates

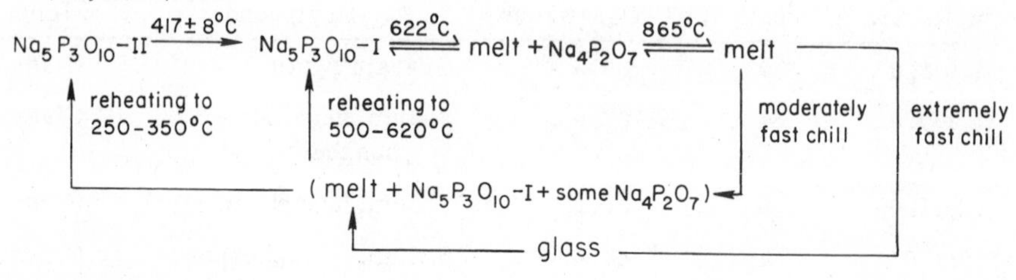

Metaphosphates

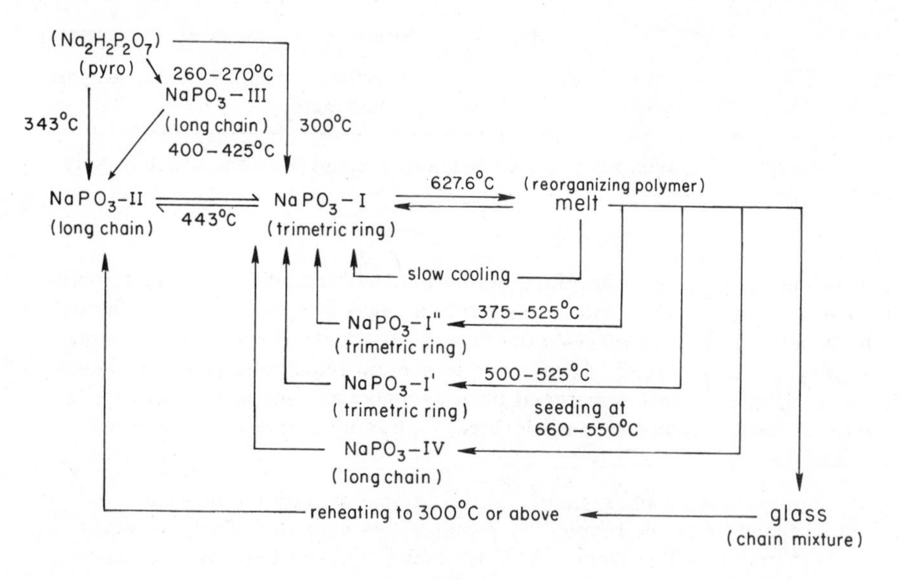

FIG. 6. Known thermal transitions in the system Na_2O-P_2O_5. From Ref. 61.

particle size, etc. Sodium tripolyphosphate designated form I is predominantly high temperature rise with small amounts of form II; and form II is predominantly low temperature rise and contains a small amount of form I. The hydration of form II may be suppressed by the addition of glycerine; thus the temperature rise may be used to estimate the amount of form I in a mixture [62].

The form I is of interest when simple mixing equipment is used and moisture is added such as in the form of liquid sodium silicate and when it is desired to have a rapid uptake of water and granule formation. Form II is used generally for processes involving detergent slurries where slower hydration is desired and granule formation is not desired, such as in spray drying. The rate of hydration is usually retarded by the presence of other dissolved solids; the processing method, order of mixing, temperature, and other factors all affect the overall results. Shen [63] discussed these various factors involved in crutcher and spray-tower operations.

1. General Properties of Polyphosphates

The sequestering properties of sodium polyphosphates are well known and discussed elsewhere. The polyphosphates are rather unique in exhibiting a threshold effect [59], i.e., they inhibit the formation of nuclei, which, in turn initiates precipitation of calcium carbonate in amounts far less (as low as 1 to 5 ppm) than the theoretical amount needed to sequester completely the calcium ion. Other chelates such as ethylene diamine tetraacetic acid (EDTA) or nitrilotriacetic acid (NTA) and their sodium salts do not exhibit this property. Although polyphosphates are added to detergent formulations primarily as calcium sequestering agents, there are many other benefits that make them truly unique builders. They show surface active properties in a sense that they are absorbed by the glass surface as well as by the oil globule affecting its electrokinetic properties [12, 64], aiding in emulsification and/or desorption of oil from glass surface, and reducing redeposition. They exhibit general deflocculating, peptizing, and dispersing properties [65], thus minimizing redeposition. There is good evidence that lipstick removal is enhanced by high polyphosphate content of the detergent product. They will solubilize casein (milk solids). Presumably this is due to an exchange process which results in the formation of a calcium polyphosphate complex and soluble sodium caseinate. They show antideposition properties [11] of fatty soils. Shen [63] has speculated that the synergistic effect between a nonionic and a polyphosphate may be due to a lowering of the critical micelle concentration of the nonionic. The marked reduction in filming of glassware in hard water is the most noticeable property of the polyphosphates.

2. Properties and Selection of Sequestering Phosphate

The particular polyphosphate to be used must be matched with the product processing and end use. Such factors as sequestration economy, hydrolytic stability, pH of end use, physical form of detergent, solubility, and functionality in the detergent must be considered. The polyphosphate content of the detergent is ideally from 30 to 50%; thus it is a major constituent in bulk as well as in cost.

3. Sequestration Economy

The data in Figs. 7 and 8 give the sequestration value of the various phosphates, EDTA, and NTA in the pH range of 5 to 12 at 25 and 65°C using oxalate indicator. The data indicate that in the pH range of alkaline detergent (pH 10 to 12) sodium tripolyphosphate outperforms the other polyphosphates, particularly the tetrasodium pyrophosphate. The widespread usage of sodium tripolyphosphate in dishwashing detergents is probably explained to a large extent by these values. In addition, the tripolyphosphate does not cake or tend to pick up moisture and form a sticky surface as do the glassy phosphates. The longer chain phosphate sequesters relatively well at the lower pH and is being used more widely today for water conditioning than for detergent use; they were once used quite extensively in detergents during the 1940s and early 1950s. The data indicate that sodium tripolyphosphate would be the most economical form for alkaline detergent use provided that hydrolytic stability is not a factor. Most processing, i.e., agglomeration, can be controlled so that hydrolytic cleavage is kept to a minimum; thus the important factor is whether or not cleavage occurs in their use.

4. Hydrolytic Stability in Aqueous Solutions

The dehydrated and condensed phosphates are dehydrated orthophosphate, i.e., molecularly dehydrated at high temperature, and may rehydrate in water solution. They are stable in the dry form. The tripolyphosphate forms a stable hexahydrate $Na_5P_3O_{10} \cdot 6H_2O$. The rate of hydrolytic cleavage, i.e., reversion to the orthophosphate, is a function of the temperature, time, and the particular polyphosphate, as well as the pH and other factors such as electrolyte concentration. Low pH will accelerate and high pH will slow down this hydrolytic cleavage.

Several investigators, such as Bell [66], Van Wazer [67], and Crowther and Westman [68], have studied the physical stability, hydrolytic cleavage, and reaction kinetics. The Graham's salt (Na_2O/P_2O_5 = 1.0:1.0) undergoes a complex cleavage and in neutral and alkaline solutions forms substantial amounts of trimetaphosphate (ring structure) and other ring phosphates which produce in turn tripolyphosphate, pyrophosphate, and orthophosphate. Since the tripoly- and pyrophosphate will further undergo hydrolysis, the ultimate species remaining on complete cleavage will be the orthophosphate.

The sodium tripolyphosphate will undergo cleavage into one molecule each of pyro- and orthophosphate, and the pyrophosphate formed here undergoes further cleavage ultimately giving three molecules of orthophosphate.

The sodium pyrophosphate can only split into two molecules of orthophosphate. The rate of cleavage varies with the different polyphosphates. The sodium pyrophosphate is considerably more stable than are the sodium tripolyphosphate or longer-chain phosphates.

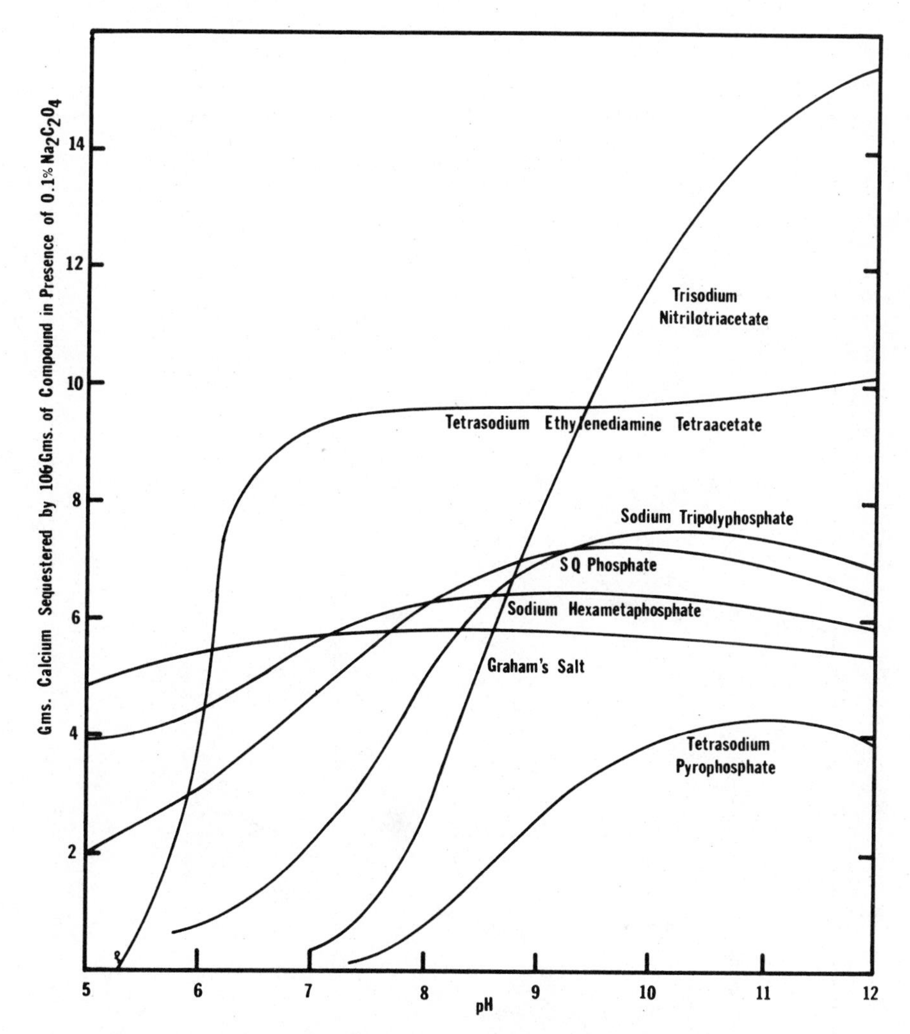

FIG. 7. Sequestration values for various chelating agents at 25°C. Titration with Ca/Mg solution in 2:1 ratio. From Ref. 91.

In dishwashing operations where the wash temperature is 160°F and where the solution life in a home-type machine is not over 5 to 10 min, and in commercial-type machines not over 4 to 5 hr, with continuous replacement of polyphosphate (detergent) this hydrolytic degradation is of little or no consequence. However, in boil-tank operations and soak-tank operations

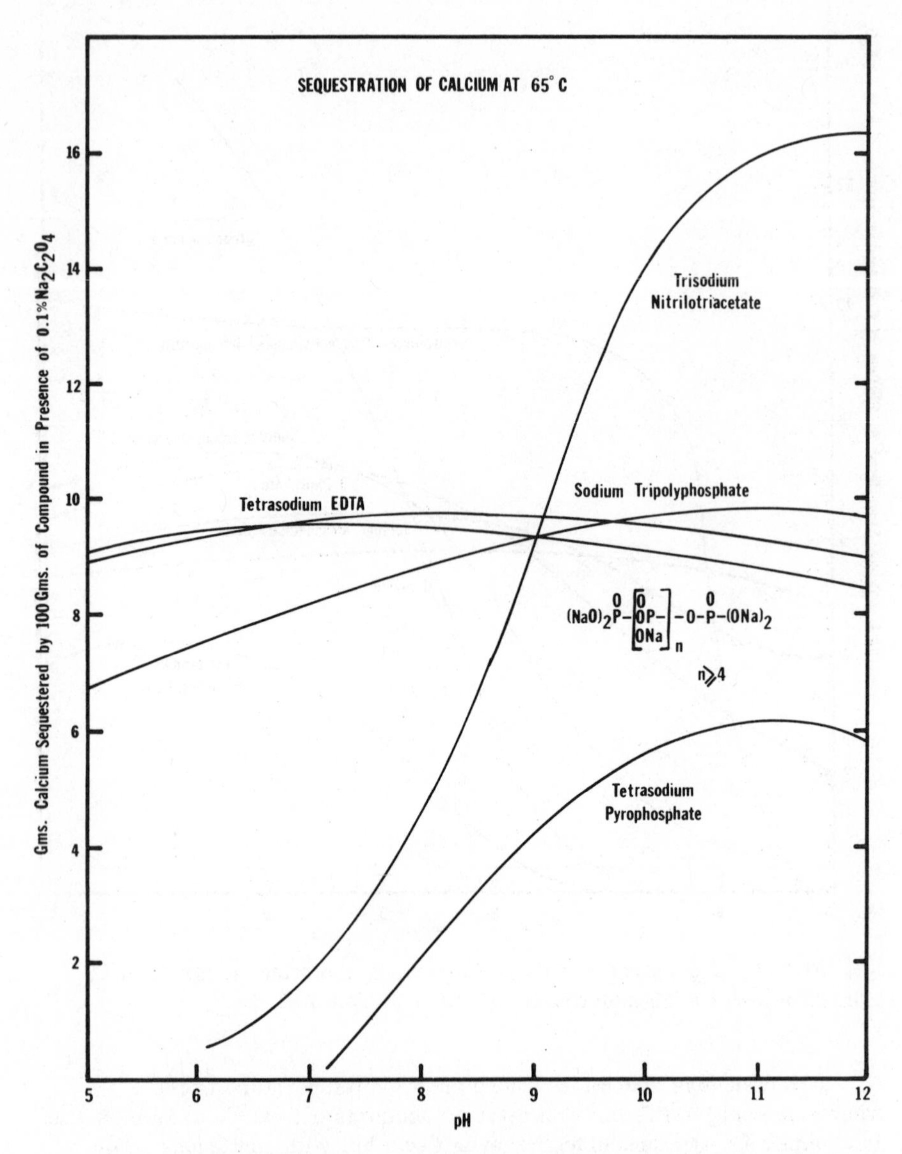

FIG. 8. Sequestration values for various chelating agents at 65°C. Titration with Ca/Mg solution in 2:1 ratio. From Ref. 91.

where solutions are reused over a long period of time at high temperatures, the hydrolytic cleavage must be considered. See Ref. 66 for rate of degradation at 70 and 100°C.

The shelf life of liquid detergent concentrates containing polyphosphate is governed by the hydrolytic degradation. Experience indicates that the shelf life of home products must be at least 1 year and for institutional products at least 6 months, assuming that the stock is well managed and rotated.

A convenient method for estimating the extent of degradation in terms of half-life of a liquid detergent containing polyphosphate, i.e., the time required for 50% degradation, is given in a nomograph reproduced from Monsanto Technical Bulletin. Figure 9 gives the half-life for sodium pyrophosphate and sodium tripolyphosphate as a function of pH and temperature. Correction factors for electrolyte concentration are also given. See also Ref. 67.

5. Solubility

The solubility of polyphosphate is important in several different ways. First, in liquid detergent concentrates high solubility means that a highly concentrated product may be prepared, resulting in economies in freight and container costs. Liquid detergents for home dishwashing machines are not yet on the market and would require a special dispenser linked to the cycle timer before wide acceptance is expected. In the commercial machine market small inroads have been made by the liquid detergent systems. The main advantages are convenience, safety, storage of drums away from the dishwashing area (thus enabling dispensing from a remote area in commercial installations), freedom from differential solubility problems, immediate solubility especially in short-time wash, and the possibility of varying the detergent composition to suit water conditions if a two-solution system is employed. Secondly, rate of solubility is important in the dispensing of powdered detergent in commercial machines. That is, if all components of a detergent do not dissolve at approximately the same rate, differential solubility may result in a nonconstant composition in the wash tank. In commercial machines one type of dispenser in use today is a hopper tank with a water-inlet line and overflow. The water inlet is controlled by a solenoid valve actuated by an electrical conductivity sensor in the wash tank. The powdered detergent is placed in the hopper, and water flows in and forms a saturated solution. If the sensor calls for detergent, the water flows in and the saturated solution overflows into the wash tank until the preset concentration is achieved. Differential solubility and caking of the detergent in the dispenser particularly, on overnight standing when the dispenser warms up during the day and cools off at night, is a problem because the supersaturated solution tends to cake (crystallize on cooling) making redissolving difficult.

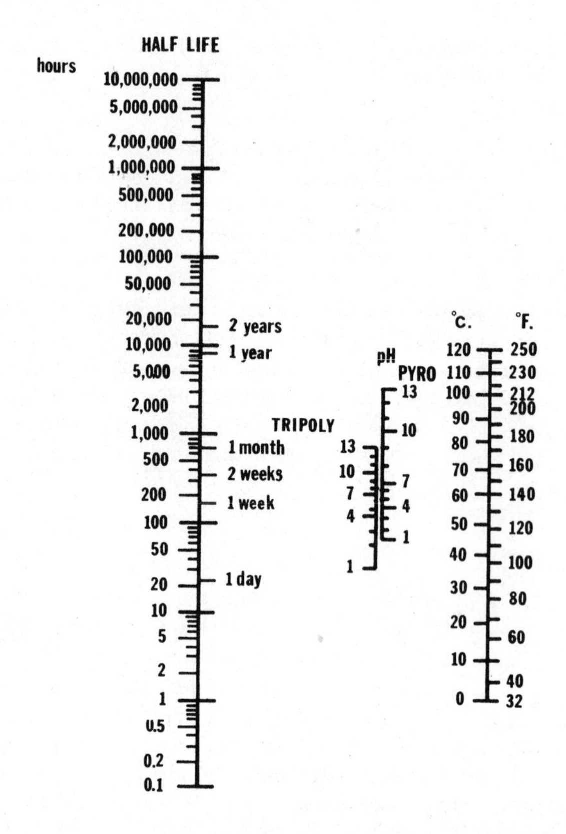

FIG. 9. A nomograph for the hydrolytic degradation of sodium pyrophosphate and sodium tripolyphosphate. This nomograph may be used for estimating, by an appropriate correction factor, the half-life of a system (liquid detergent) containing other salts: for sodium phosphate (pH less than 11) multiply by 0.7; for sodium phosphate plus 10% NaCl (pH 4 and greater) multiply by 0.5; for potassium phosphate (35% solids) multiply by 0.35. From Ref. 91.

In the home-type machine having a double wash cycle, the second detergent cup has a water shield which opens during the second wash cycle allowing the water to wash the detergent out of the second cup. Water vapor and leakage of water into the cup causes "cup-caking" or formation of a gummy mass which is removed with difficulty. This cup-caking is aggravated by any components which tend to stick together on hydration. Anhydrous tripolyphosphate aggravates cup-caking and hydrated tripolyphosphate [69] minimizes it.

Thus, not only solubility but rate of solubility and hydration, etc., and other components of the detergent, must be selected to minimize these problems.

The solubilities of the various detergent polyphosphates and the equilibrium crystalline phases as given by Shen [63] are reproduced in Fig. 10. These and other data indicate that tetrapotassium pyrophosphate has a high solubility in the range of 60% solution. Because of this high solubility and good hydrolytic stability, tetrapotassium pyrophosphate is usually used in liquid detergent concentrate. Glassy phosphate whose Na_2O/P_2O_5 mole ratio is 1.42:1.00 also has good apparent solubility. Although such glasses

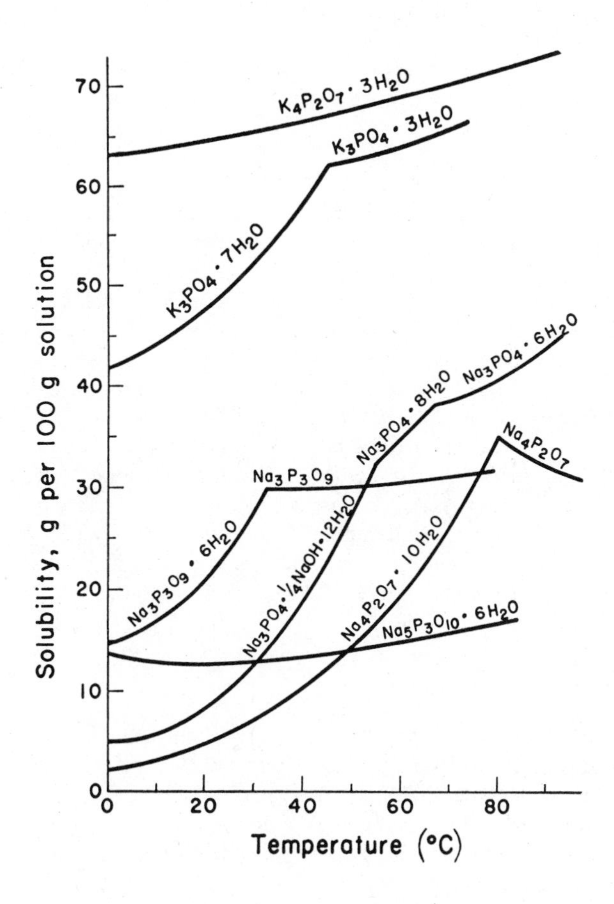

FIG. 10. Solubility of various detergent phosphates and the equilibrium crystalline phases as a function of temperature. From Ref. 63.

contain some pyro- and tripolyphosphate ions, in extremely high concentrations, the high viscosity of such a solution inhibits the rate of formation of the crystal nuclei of the pyro- and tripolyphosphate so that it may appear clear for extended periods of time. Figure 11 from Van Wazer indicates that solutions containing less than 48% glass did not precipitate in 3 years.

The maximum solubility (not equilibrium solubility) before a precipitate is immediately formed versus the chain length is shown in Fig. 12 from Van Wazer. This graph indicates that as the $\bar{n}$ average chain length changes approximately from 3.0 to 4.5, the immediate solubility increases from about 15 to 50%. Thus, from a visual appearance standpoint, it might be possible to formulate a semistable liquid concentrate from a glassy phosphate whose Na_2O/P_2O_5 ratio is 1.5:1.0 to 1.4:1.0 of such viscosity as to delay or inhibit the rate of formation of the crystal nuclei and prevent the clouding phenomenon in such solutions. The addition of electrolyte and additives could adversely affect this apparent stability.

6. Processing

The use of polyphosphates in dishwashing detergents is so well established that it need not be documented. However, special uses involving the polyphosphate, particularly in processing, are of interest. Milenkevich and Henjum [69] teach the art of producing agglomerated detergent from sodium tripolyphosphate, liquid silicate, and chlorinated trisodium phosphate, hydrating the tripolyphosphate in situ and obtaining a product free of cup-caking. Oberle's [70] patent teaches the use of tripolyphosphate and liquid tetrapotassium pyrophosphate to "encapsulate" liquid organic defoamers for use in powdered detergent to minimize the interaction between alkali metal dichloroisocyanurate and the defoamer. Kaneko et al. [71] describe a method for preparing a free-flowing, chlorine-stable dishwashing product using a low-foaming nonionic surfactant, tetrasodium pyrophosphate, sodium metasilicate pentahydrate, and chlorinated trisodium phosphate. In all of these processes, the order of mixing and the method of processing is rather important.

7. Trimetaphosphate

Sodium trimetaphosphate ($Na_3P_3O_9$), the trimeric ring form, is commercially available. Its preparation and its structure are described by Bell [60] and Van Wazer [72]. Its greatest potential probably lies in processing as in crutching and spray drying or agglomeration, where it is converted to the sodium tripolyphosphate by caustic soda or other strongly alkaline materials. The conversion of the sodium trimetaphosphate to the sodium tripolyphosphate is said to occur in two steps:

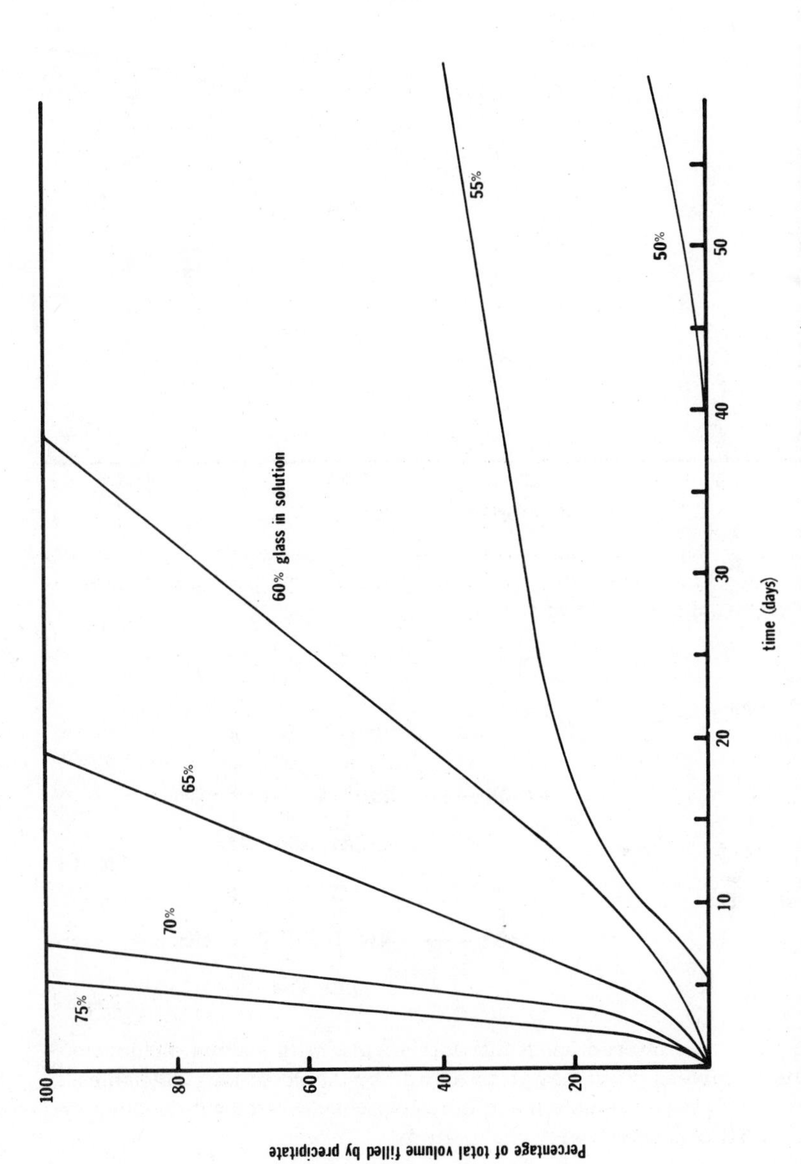

FIG. 11. Precipitation in aqueous solutions of sodium phosphate glass (Na_2O/P_2O_5 = 1.42) on storage at room temperature. The family of curves represents various initial concentrations of the glass. Solutions containing less than 48% glass did not give a precipitate in 3 years. From Ref. 67.

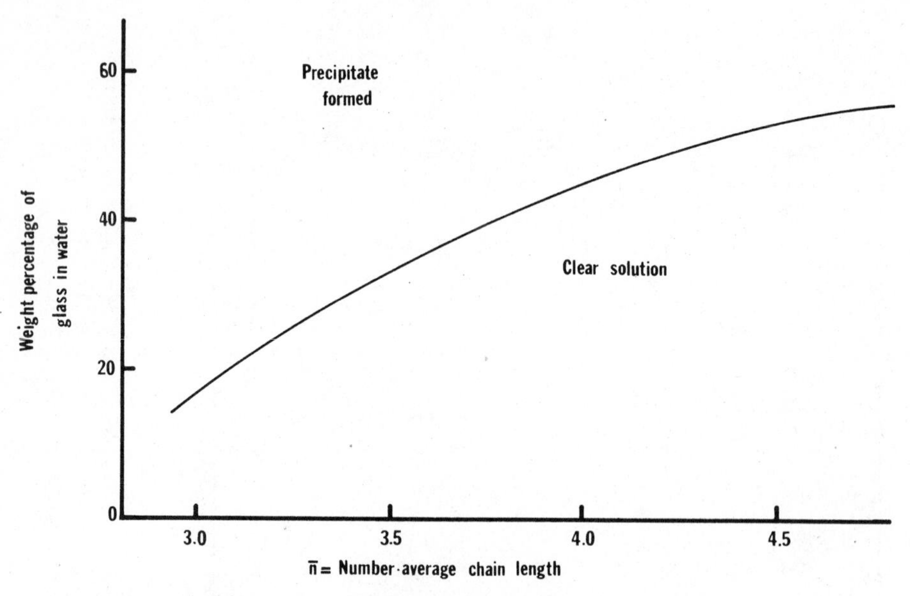

FIG. 12. Region of composition (number average chain length) and concentration at which sodium polyphosphate glasses form precipitate immediately after dissolution. From Ref. 67.

```
        O    ONa
         \\  /
          P
         / \
        O   O
        |   |
      O=P   P=O
       / \ / \
    NaO   O   ONa

                                O    O    O
                                ||   ||   ||
    + NaOH  ----->      HO-P-O-P-O-P-ONa
                                |    |    |
                               ONa  ONa  ONa        (Step 1)

                                O    O    O
                                ||   ||   ||
    + NaOH  ----->      NaO-P-O-P-O-P-ONa
                                |    |    |
                               ONa  ONa  ONa        (Step 2)
```

FIG. 13. The interaction of nitrilotriacetate with sodium dichloroisocyanurate at various pH values as measured by the remaining available chlorine. The pH values shown are initial values adjusted with sodium hydroxide. NTA concentration = 0.2 g/liter.

The first step is slow; therefore, it is the rate-controlling step. The rate of step 1 is increased tenfold by a 25°C rise in temperature. The rate increases approximately exponentially with the ionic strength of the aqueous medium. The conversion is said to be essentially complete in a few minutes under proper process conditions with essentially all of the sodium trimetaphosphate converted to sodium tripolyphosphate and with little or no side products. Sodium sulfate has a catalytic effect on the conversion rate. Thus these rate-controlling factors may be used to advantage in processing.

The high heat of reaction offers the potential of lower cost of heat and equipment necessary for drying the finished product. The properties of trimetaphosphate, once converted to the tripolyphosphate would be essentially the same as for tripolyphosphate in regard to chelation, pH optimum, etc. Although sodium trimetaphosphate has been known for some years, it has only been available as a commercial item since about 1965. The sodium trimetaphosphate would be of interest where a "processed" detergent is desired. The technical information on conversion and processing is given in a technical bulletin on sodium trimetaphosphate [73].

B. Chelates other than Polyphosphates

The organic sequestrants such as sodium ethylene diamine tetraacetate and sodium nitrilo triacetate have been known and available for some years [74, 75]. More recently, the organo phosphonates have been studied for use as detergent builders as indicated by patents issued to Quimby [76-79]. While the organo phosphonates are excellent sequestering agents, the current interest is in nonphosphorous detergents because of concern for phosphate pollution of lakes and rivers.

Thus interest has been renewed in the possible use of nitrilotriacetate, citric acid, and polycarboxylate compounds [80-83] and other nonphosphorous sequestering agents in laundry detergents and water-conditioning uses. While nitrilotriacetate at one time was considered seriously as a replacement for tripolyphosphate in laundry detergents, its use has been shelved temporarily pending further investigation of possible toxicological implication in combinations of cadmium or mercury.

Machine dishwashing compositions containing (1) polyacrylate, polyacrylamide, and polyphosphates, (2) polyacrylate and nitrilotriacetate, and (3) sodium citrate and polycarboxylate polymers, have been disclosed in the patent literature [84-88]. Alkali metal salts of citric acid as a replacement for polyphosphate has also been suggested in the literature [89, 90].

The above systems are largely in the experimental or early stages of development and there appears to be no real universal replacement for the polyphosphate for machine dishwashing detergent as yet. While nitrilotriacetate is an excellent chelating agent, the majority of the machine dishwashing formulations contain active chlorine which will react, unfortunately, with nitrilotriacetate, resulting in a substantial loss of chlorine. Figure 13 shows that the loss of available chlorine (using sodium dichloroisocyanurate as the source of hypochlorite) is quite rapid in dilute solution when sodium

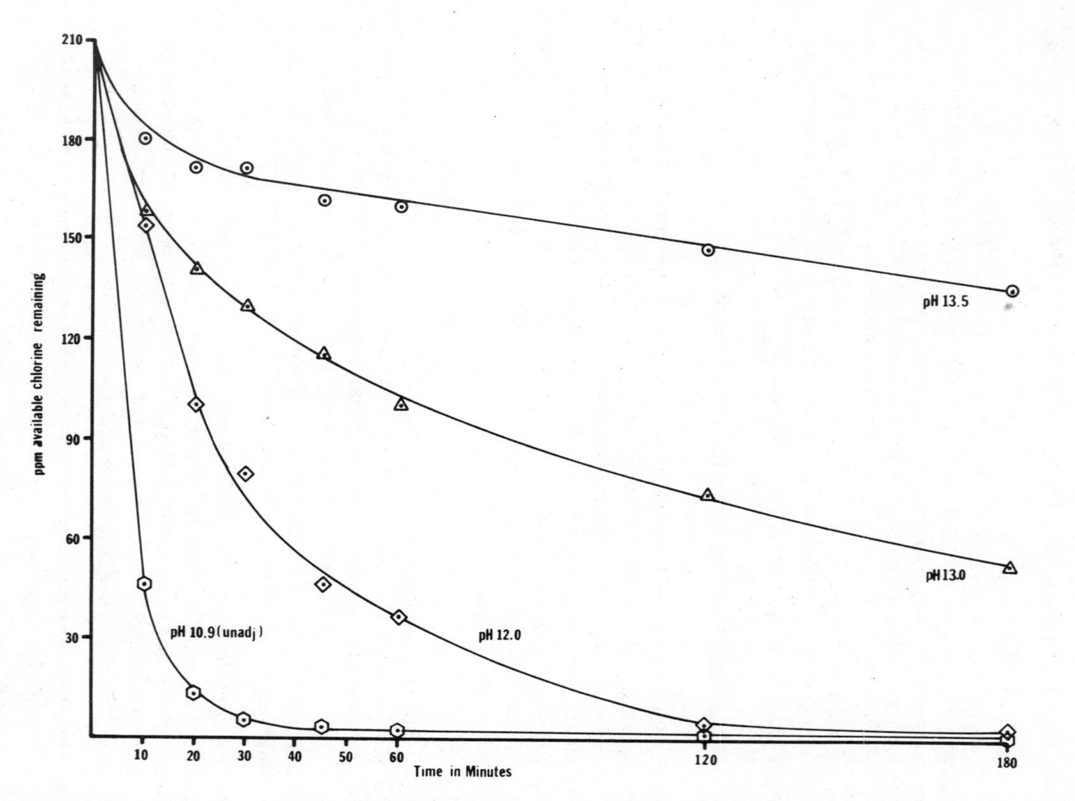

FIG. 13. The interaction of nitrilotriacetate with sodium dichloroisocyanurate at various pH values as measured by the remaining available chlorine. The pH values shown are initial values adjusted with sodium hydroxide. NTA concentration = 0.2 g/liter.

nitrilotriacetate is present. Similar curves were obtained with sodium hypochlorite. Moreover, the use of nitrilotriacetate in machine dishwashing detergent is limited by their lack of threshold properties; aluminum discoloration is another problem. These problems must be solved to an acceptable degree.

In selecting a sequestering agent for detergents, the prime considerations are performance, cost, and safety. Performance is dependent on calcium and magnesium binding power and capacity as well as threshold effect and other properties associated with a particular chelate. Table 3 is a summary of the binding (complex stability constant) power, binding capacity, and molecular weight of several calcium-binding compounds.

The figures given in Table 3 may be used to estimate the relative cost of sequestration, but it is not rational to compare the cost of commodity and specialty chemicals because cost will usually drop substantially if a specialty chemical becomes a commodity chemical. It should be noted that the figures in Table 3 were from several sources and, as shown in the key, different methods were used to arrive at the values.

Although the data are not shown in Table 3, a technical bulletin [91] indicates that sodium pyrophosphate sequesters significantly more magnesium than calcium, whereas the reverse is true for sodium tripolyphosphate. In general, the Ca/Mg ratio in natural waters is said to be 2:1; thus the values given in Figs. 7 and 8, where the titrant was a mixture of Ca and Mg in this ratio, would be valid for most water situations.

C. Silicates

Sodium and potassium silicates, like phosphates, may be defined in terms of the silicon dioxide/alkali metal oxide ratio. Lange [92], Vail [93], Kramer [94], and others have described the various alkali metal silicates and their properties. Table 4 shows the nomenclature of the various silicates and their composition.

The ortho-, sesqui-, and metasilicates are normally available as crystalline, solid material. They are available in anhydrous and hydrated forms. The silicates of higher SiO_2 content than the metasilicate are available in solid as well as in liquid form. The liquid forms of these higher silicates are usually used, as the rate of solubility of dry powder forms which are high in SiO_2 is poor and they require heat or steam for rapid dissolution. These liquid silicates are usually used in liquid detergent formulation or used to spray on solids and form agglomerated detergents as described in the Milenkevich and Henjum patent [69] and the Monsanto Technical Bulletin [73]. The water from the liquid silicate is taken up, for example, by anhydrous sodium tripolyphosphate which forms the solid

TABLE 3

Calcium Sequestering Capacities[a] and Stabilities at 25°C of Various Compounds

Compound	Molecular weight	Calcium sequestering capacities (g Ca/100 g)	Stability constants calcium (Log K_e)
Graham's salt[b]	—	5.3	—
Tetrasodium pyrophosphate[b]	266	3.9	—
Sodium tripolyphosphate[c]	368	10.9	5.2
Nitrilotriacetate $Na_3 \cdot H_2O$[c]	275	14.6	6.0
Ethylene diamine tetraacetate $Na_4 \cdot 2H_2O$[d]	416	9.6	10.7
Citric acid[e]	192	—	3.15
Aminoethylidene diphosphonic acid[c]	271	23.6	6.0
Hydroxethylidene diphosphonic acid[f]	206	31.6	7.09

[a]Some slight discrepancy in sequestration value will be noted in the various references, probably due to differences in purity of material tested and indicator and amount of indicator used, pH, temperature, test method, etc.

[b]At pH 10.5, oxalate indicator, 25°C.

[c]At pH 10, 25° C (Calcium electrode).

[d]At pH 11.

[e]Ref. 81.

[f]At pH 10, $CaCO_3$ indicator, 25°C.

TABLE 4

Nomenclature of Various Commercial Alkali Metal Silicates and their Composition

Compound	Formula	M_2O/ SiO_2	pH 0.1% Soln.	Liquid silicate Wt. (%) M_2O	SiO_2
Powdered silicate					
Sodium orthosilicate[a]	Na_4SiO_4	2/1	12.0		
Sodium sesquisilicate[a]	$Na_3SiO_4 \cdot 5H_2O$	1.5/1	11.8		
Sodium metasilicate[a]	Na_2SiO_3	1/1	11.8		
Liquid silicate[b]					
B-W silicate[c]	Na_2O/SiO_2	1/1.6		19.5	31.2
C silicate[c]	Na_2O/SiO_2	1/2.0		18.0	36.0
RU silicate[c]	Na_2O/SiO_2	1/2.4		13.8	33.1
K silicate[c]	Na_2O/SiO_2	1/2.9		11.0	31.9
N silicate[c]	Na_2O/SiO_2	1/3.22		8.9	28.7
Potassium silicate[c]	K_2O/SiO_2	1/3.34		12.6	26.8

[a]Ref. 94.

[b]Trade name, Philadelphia Quartz Company.

[c]Philadelphia Quartz Company Bulletin.

hexahydrate resulting in a free-flowing detergent. A free-flowing granular blend of liquid silicate sprayed onto sodium tripolyphosphate and known as "silicated tripoly" is also known to the trade.

Potassium silicates are usually used in liquid detergent concentrate along with tetrapotassium pyrophosphate because of the greater solubility of the potassium salts, such as the potassium pyrophosphate. (See Fig. 10 for the difference in solubilities of sodium and potassium pyrophosphates.)

The main functions of the silicates in dishwashing detergent are twofold. They prevent or minimize corrosion of metals and ceramic dish surface and/or pattern due to other detergent ingredients present, such as polyphosphate and high alkalinity. Aluminum discoloration and/or corrosion due to other detergent ingredients, or sometimes by hard water alone, is minimized by the presence of a silicate. As a general rule, corrosion is more severe at the higher pH, and requires greater concentration of the silicate as the pH of the detergent rises. The type and amount of silicate to be employed will depend on the amount of sodium tripolyphosphate, alkalinity of the other ingredients present, the concentration used, and the degree of protection desired versus the cleaning ability required of the detergent formulation. In general, this will have to be determined by suitable corrosion tests. The cause and prevention of glassware corrosion is discussed by Rutkowski [95].

For commercial-dishwashing-machine detergent, the short-wash duration requires higher alkalinity and usually sodium metasilicate is used; but for home-type dishwashing-machine detergents and for maximum corrosion protection, sodium silicate with an SiO_2/Na_2O ratio of 2:1 or higher is recommended [92]. The protective action of the silicate is attributed to sorption of the silicate and formation of, for example, alumino-silicate on aluminum surface which protects it from hydroxyl and chelate ions [92]. Baker [96] has reported on weight loss of strips of tin, zinc, and aluminum immersed in solutions of trisodium phosphate, caustic soda, sodium carbonate, and various silicates at 60°C for varying periods of time and at different concentrations, and concludes that with one exception silicates impaired (damaged) the surface less than other alkalies tested. Recently, hydrous sodium polysilicate, in the form of powder or granules, of reduced alkalinity has been made available for wet or dry blending and is claimed to be suitable for use in low-phosphate detergents [97].

The second most important property of the silicate is its contribution to alkalinity and buffering. Silicates are selected with the final alkalinity of detergent in mind. Alkalinity is an important factor in solvation, hydrolysis, salt formation, etc. Sodium metasilicate is usually used in dry-blended detergent and where limited corrosion may be tolerated. Silicates with an SiO_2/Na_2O ratio of 2:1 or higher are used where greater corrosion protection is desired and where an agglomeration process is used.

When silicate of a higher ratio (SiO_2/Na_2O) than 1.0:1.0 is used in a detergent formulation, suitable precaution must be taken to protect it from free moisture and carbon dioxide in order to prevent reversion of the silicate to the insoluble silicon dioxide. Such insolubles are troublesome in that they will deposit as specks on the dishes and glasses and on the interior of the dishwasher. (In addition, the softened-water etch mentioned previously is attributed to use of silicates high in SiO_2.)

Another property ascribed by Lange [92] is the colloidal properties of the silicate whose SiO_2/Na_2O ratio is greater than 2:1. Lange presents data wherein both the molecular weight and intrinsic viscosity is plotted against the SiO_2/Na_2O ratio. The slope of these curves changes abruptly at a ratio of 2:1 suggesting polymerization at this point. The polymerization is said to be reversible, thus the pH of the final solution will determine whether such polymers exist. This colloidal nature is said to promote soil suspension and corrosion inhibition.

D. Alkaline Builders

Alkaline builders used primarily for alkalinity and without other obvious tangible benefits will be described here. Alkalinity will aid dissolving of many soils. Perhaps the best example of this important function is the use of caustic soda in bottle washing. Milk and other soils are dissolved by soluble salt formation and/or hydrolysis. Many organic soils may be hydrolyzed by sufficient alkalinity, time and temperature, as is well known in destructive hydrolysis in isolation of proteins and amino acids. In protein and carbohydrate hydrolysis, acid is more often the favored agent because it is less destructive and the end-product may be isolated in a less degraded form; but for detergent use speed and complete solubilization is the more important factor.

The pH of some common alkalis at 20°C and various concentrations is shown in Table 5 [98]. Caustic soda, sodium carbonate, trisodium phosphate, and sodium metasilicate are some of the common alkalis most frequently used in dishwashing detergent. It should be noted that pH is a logarithmic function so that, for example, the pH of 5% NaOH of 13.8 represents over a hundredfold increase in the $[OH^-]$ over 5% Na_2CO_3 which has a pH of 11.6.

1. Caustic Soda

Caustic soda is derived from sodium chloride by electrolysis using either the mercury amalgam cell or the diaphragm cell. The mercury cell produces a higher quality product having less iron contaminant, and may be preferred for some uses.

Caustic soda is by far the most alkaline agent used. It is not used in dishwashing products for home use because of its corrosive action on aluminum, overglaze pattern, and soft metals. Furthermore, there may be a possibility of damaging painted and varnished surfaces around the kitchen if it is accidentally spilled and it may cause burns on contact with the skin. However, in commercial dishwashing, where wash times are short, corrosiveness is less of a problem, usage is closely controlled, and where, due to the conditions of use, its superior cleaning power is more obvious,

TABLE 5

pH Values of Some Common Alkalis

Concentration Wt. (%)	pH at 20°C					
	NaOH	Na_2CO_3	Na_2SiO_3	Na_3PO_4	$Na_2B_4O_7$	$NaBO_2$
0.1	11.9	10.7	11.3	11.0	9.26	10.70
0.5	12.7	11.3	12.1	11.8	9.23	11.04
1.0	13.1	11.4	12.3	11.9	9.24	11.18
2.0	13.3	11.5	12.7	12.2	9.24	11.36
5.0	13.8	11.6	13.1	12.4	9.32	11.72

Source: From Ref. 98.

caustic soda has found a fairly widespread acceptance. Reference to its use in dishwashing detergent formulations is cited in several patents [8, 99] and articles [100, 101].

Because caustic soda is hygroscopic and is also very reactive with some components of the detergent system, certain precautions must be used in formulating a caustic-soda-containing detergent. Caustic soda forms various hydrates and may hold up to approximately 40% water and still remain a solid at room temperature [102]. However, when caustic soda flakes are exposed to moisture, the moisture tends to stay at the surface and form a slurry, and this slurry tends to react readily with surfactants and chlorine donor compounds. Caustic-stable, "capped" nonionic surfactant such as alkyl phenoxy polyethenoxy benzyl ether [103] may be used, or the surfactant may be "encapsulated" or isolated as indicated in the Oberle patent [70]. The use of a good water scavenger such as anhydrous sodium tripolyphosphate is also helpful in keeping the moisture away from the caustic soda, thus minimizing interaction and/or caking. The use of hydrated silicate and hydrated trisodium phosphate with caustic soda should be avoided because of potential caking problems during storage of the product.

2. Trisodium Phosphate

Trisodium phosphate is obtainable as anhydrous (Na_3PO_4), trisodium phosphate monohydrate ($Na_3PO_4 \cdot H_2O$), trisodium phosphate nonohydrate ($Na_3PO_4 \cdot 9H_2O$), and trisodium phosphate dodecahydrate ($Na_3PO_4 \cdot 12H_2O \cdot 1/4\ NaOH$). A number of other orthophosphate salts are described in the literature [104]. Trisodium phosphate, like sodium carbonate, softens water by precipitation of the calcium salt. Trisodium phosphate is much less of an offender in filming of glasses than are soda ash or sodium metasilicate, but not as good as the complex phosphate as shown by Wilson and Mendenhall [35], Madden [36], and unpublished work [37]. The calcium orthophosphate apparently is much less tenaciously absorbed by a glass surface than are calcium carbonate or calcium silicate.

Trisodium phosphate as such is not extensively used in dishwashing machine detergents today, although the chlorinated trisodium phosphate (discussed under active chlorine donor in Sec. IV. E) is still used in high tonnage. Early detergents, prior to the introduction of polyphosphates and for some time thereafter, did contain a substantial amount of trisodium phosphate.

Trisodium phosphate ranks after caustic soda and sodium metasilicate in alkalinity (see Table 5), and is used in hard-surface cleaners (wall and paint cleaners), and is a very useful and efficient electrolyte for electrolytic detarnishing of silverware using a sacrificial metal such as aluminum. The higher hydrates have low melting points and tend to promote caking of the product during storage.

3. Soda Ash

Soda ash or sodium carbonate is still in common use in dishwashing detergents despite the filming tendency, because of its alkalinity, availability, and low cost. It is available in several forms. Light ash is a commercial-grade soda ash in fine granular form, having a relatively low-bulk density, hence the name light ash. The bulk density of light ash is in the range of 34 $lb/ft.^3$ Dense ash of approximately the same composition is a larger, heavier particle-size material having bulk density of approximately 61 $lb/ft.^3$ Bulk density may vary with different suppliers. Light ash alone, when dissolved in water, must be added with good agitation as, like other salts with large heat of hydration and rapid hydration, it tends to fuse together and form cake which does not dissolve readily. Dense ash dissolves much more slowly and does not exhibit this tendency to cake. The common sal soda or "washing soda," the 10-mole hydrate, is not used in dishwashing detergent because of its tendency to cake on storage.

Another form of anhydrous soda ash was at one time available under the trade name of "Flozan" [105]. It is a granular material having a very porous structure (large specific surface) and has a very unusual capacity for taking up large amounts of water yet not forming a slurry. It may hold up to about 25% of its weight in water and is suggested for use where water or liquid is to be sprayed on solids to form a free-flowing product without a drying step.

Other forms of carbonate are sodium bicarbonate and sodium sesquicarbonate ($Na_3H(CO_3)_2 \cdot 2H_2O$). Neither of these is used to any extent in dishwasher detergent. A special noncaking form of potassium or sodium bicarbonate treated with silicone is available but not used in detergents; its use is primarily in dry, powder-type fire extinguishers.

Dense and light soda ash used alone or in excessive amounts or with inadequate sequestering promotes filming due to the formation of insoluble $CaCO_3$. Underground waters invariably pick up calcium bicarbonate due to dissolved carbon dioxide and trickling over limestone deposits. This hardness is in the form of the relatively soluble $Ca(HCO_3)_2$. When such waters are made alkaline, the following reaction takes place, resulting in the precipitation of the $CaCO_3$:

$$Ca^{2+} + 2HCO_3^- + 2OH^- \rightarrow CaCO_3 + 2H_2O + CO_3^{2-}$$

If a detergent containing a large amount of soda ash and an insufficient amount of sequestering agent is dissolved in hard water, there is a common ion effect and a pronounced tendency to precipitate out the calcium as calcium carbonate. This may be illustrated by considering the following. The dissociation for calcium carbonate is given as:

$$CaCO_3 \rightleftarrows Ca^{2+} + CO_3^{2-}$$

If the reaction is reversible, and equilibrium is attained instantaneously, the mass-action law may be applied giving the equation: (It is recognized that this situation is not ideal in that the forward reaction may be slow but the reverse may be faster; the ideal situation probably exists only for complex ion formation.)

$$\frac{[Ca^{2+}][CO_3^{2-}]}{[CaCO_3]} = K$$

Under these assumptions, since K is a true constant, increasing the $[CO_3^{2-}]$ must decrease the $[Ca^{2+}]$ in solution, and this can only happen with the precipitation of $CaCO_3$. Thus it is seen that the addition of Na_2CO_3 to hard water will result in the precipitation $CaCO_3$.

Further, in an ideal situation, where Ca^{2+} and CO_3^{2-} are present and in the absence of other outside influence, this common-ion effect may be calculated by substituting the K value for solubility product of $CaCO_3$ of 1.2×10^{-8}. When this is done and solubility of $CaCO_3$ is plotted against the $-\log[Ca^{2+}]$ and $-\log[CO_3^{2-}]$, the plot shown in Fig. 14 is obtained. The plot shows that Ca^{2+} may be precipitated quantitatively in the presence of excess CO_3^{2-}.

Actually, $Ca(HCO_3)_2$ rather than $CaCO_3$ is present in hard waters and the pH, the presence of other salts, and the presence of chelates will affect

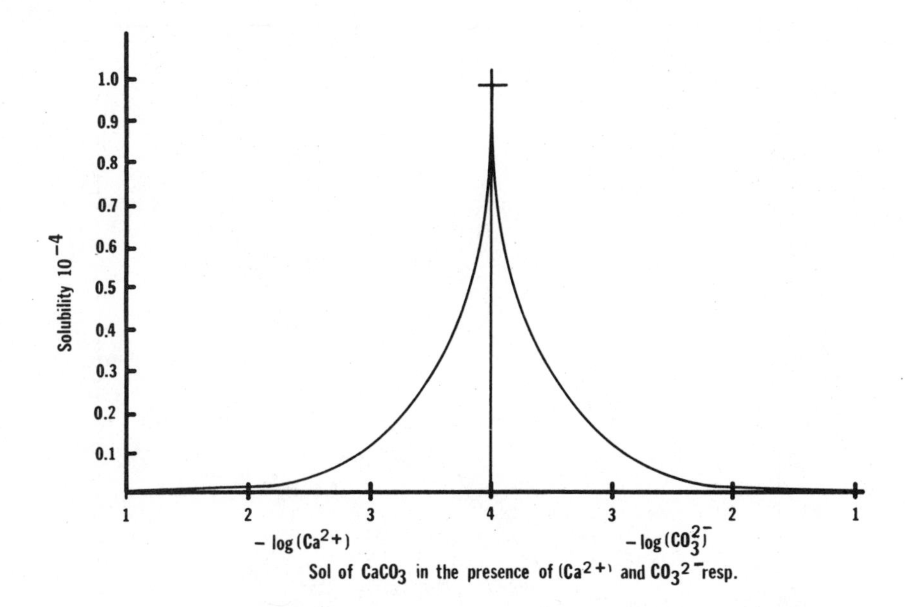

FIG. 14. Common ion effect: the solubility of $CaCO_3$ in the presence of excess Ca^{2+} and CO_3^{2-}.

the equilibrium; thus the above equation is at best theoretical and qualitative but does indicate the necessity for judicious use of soda ash in detergents, especially in extremely hard waters.

The selective-ion electrode might find application in studies of this type where the free Ca^{2+} may be determined, independent of other factors in a complex system.

4. Borax

Borax has been known for a number of years but has not been used in mechanical dishwasher detergent to any important extent. The sodium borate $Na_2B_4O_7$ is commercially available as the anhydrous, penta, and decahydrate forms. Borax decahydrate has a low rate of solubility and a relatively low equilibrium solubility. At room temperature it is soluble to the extent of about 2.5% by weight, and at 50°C it is soluble to the extent of about 10%. For other temperatures see Fig. 15.

Anderson and Wegst [4] mention the use of borax and sodium borate along with other detergent ingredients. A sodium-borate-containing dishwashing formulation with a destaining property is suggested by Cohen [106].

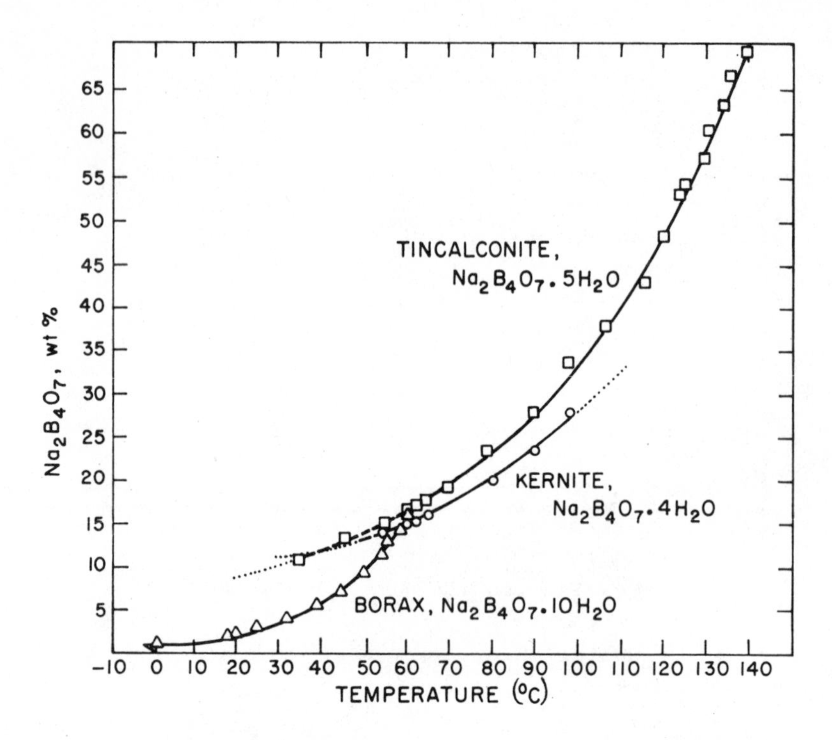

FIG. 15. Solubility/temperature curves of the various hydrates of sodium borate. From Ref. 158.

E. Active-Chlorine Compounds

Active chlorine compounds refer to such compounds that possess or are able in solution to provide OCl^- or MOCl. The active chlorine compounds are used in detergents primarily to remove tea, coffee, and other food stains and to minimize filming or spotting of glasses. Detergents for home machines usually contain from about 0.50 to 0.80% available chlorine, while detergents for commercial machines contain about 1.50 to 2.00% available chlorine.

1. General

The active chlorine compounds are inorganic alkali metal or alkaline earth metal hypochlorite (such as NaOCl, $Ca(OCl)_2$, LiOCl) and N-chloro compounds usually containing an organic radical. N-chloro compound is usually characterized by a double bond on the atom adjacent to the trivalent nitrogen and the chlorine (Cl^+) attached to the nitrogen is readily exchanged with H^+ or M^+ (where M^+ is a common metal ion such as Na^+, K^+, etc.) so as to release HOCl or OCl^- on hydrolysis.

TABLE 6

Percentage of Total Chlorine in Hypochlorite Solutions Present as Hypochlorous Acid at Various pH Levels

pH at 25°C	As HOCl (%)	pH at 25°C	As HOCl (%)
5.0	99.7	9.0	3.1
5.5	99.1	9.5	0.99
6.0	96.9	10.0	0.31
6.5	91.0	10.5	0.10
7.0	76.0	11.0	0.03
7.5	50.0	11.5	0.01
8.5	9.1		

Source: From Ref. 28.

As shown in Table 6, in alkaline detergent solutions in the range of pH 10 or higher, the free chlorine exists predominantly (>99%) in the OCl^- form.

The solid sodium hypochlorite is unstable, decomposing explosively and spontaneously [107]; thus the common form of sodium hypochlorite is the 5% NaOCl household liquid bleach. However, a solid form in mixture or complex-salt form is possible as in the chlorinated trisodium phosphate or a chlorinated caustic-tripolyphosphate, as indicated by a U. S. patent [99]. The calcium hypochlorite is very stable by itself, but has not been used in detergent formulations because it adds to the water hardness, thus limiting its usefulness. Calcium hypochlorite is described in a U. S. patent [108]. Lithium hypochlorite has been a more recent entry among the inorganic hypochlorites, but it has not made much of an inroad so far as its use in dishwashing detergents is concerned.

The organic active-chlorine compounds have been investigated by a number of workers such as Muskat and Chenicek [109] on chloromelamine and related products, Guiteras and Schmelkes [110], Schmelkes and Horning [111] on α,α'Azo bis[chloroformamidine](azochloramide), Mallman and others on sodium p-toluene sulfonchloramide (Chloramine-T) [43]. These earlier investigators studied the chlorine compounds more for their sanitary (antibacterial) values than for destaining or dishwashing aspects. More recently, several patents have been issued to Corliss et al. [6], Mizuno and Oberle [8], Oberle [70] on the use of various forms of dichlorocyanurates. Keast and co-workers [100, 101] disclose the use of dichloroisocyanurates. A patent on the use of a complex salt of potassium dichloroisocyanurate and trichloroisocyanuric acid in detergent formulations has been granted to Symes [7]. Hardy [112] teaches the use of trichloroisocyanuric acid in detergent compositions.

Schmolka et al. [30] have made a comprehensive study in which attempts were made to chlorinate 80 potential chlorine carriers. Only 27 active chlorine compounds were obtained which could be tested for bleaching activity. A number of references to U. S. patents and articles were cited by Schmolka. Schmolka further reported that N, N'-dichloro-N, N'-dimethyl-oxamide, N, N'-dichloro-N, N'-dimethyl-α, α-dichloromalonamide, and dichloroisocyanuric acid are superior to dichlorodimethyl hydantion in cotton bleaching at 140°F. It is believed that these bleaching studies might show some correlation to usefulness as an active chlorine compound in dishwashing. A more recent entry into this market is a dichloro derivative of glycoluril, produced under the trade name of Nobechlor-70 (Centerchem Products, Inc., New York). The usefulness of the various N-chloro compounds is dependent on stability, compatibility with other detergent ingredients, solubility in water, and the availability of the hypochlorite, which is dependent primarily on the hydrolysis constant. The hydrolysis constant for various N-chloro compounds is given in Table 7. It would appear that a hydrolysis constant of approximately 3×10^{-4} or larger (as well as good solubility) is needed to get effective bleaching or performance in a dishwashing compound as indicated by various studies [30, 113]. It is to be noted that compounds with a relatively small hydrolysis constant like Chloramine-T ($K = 4.9 \times 10^{-8}$) are ineffective for use as a bleach in dishwashing compounds. The very small K for albuminoid chloramines possibly explains why many proteinaceous materials very rapidly dissipate the available chlorine in wash solutions.

Other compounds of possible interest are oxidizing agents such as sodium perborate and potassium monopersulfate. The latter at a pH range of 9.0 is able to convert Cl^- into OCl^-. Other strong oxidizing agents such as sodium chlorite have potential bleaching properties.

2. Chlorinated Trisodium Phosphate

Chlorinated trisodium phosphate is a crystalline hydrated double salt of trisodium phosphate and sodium hypochlorite prepared by crystallizing it from a blend of aqueous sodium hypochlorite, caustic soda, trisodium phosphate, and disodium phosphate. The resulting product is assigned the formula $4(Na_3PO_4 \cdot 11H_2O)\ NaOCl$ by Bell [104] when made by the Mathias method [114] but contains NaCl when made by the Adler process [5]. The commercial chlorinated trisodium phosphate available today is made by the Adler process and conforms to the following specifications: sodium hypochlorite, over 3.25%; sodium phosphate expressed as $Na_3PO_4 \cdot 12H_2O$,

TABLE 7

Hydrolysis Constants, (HOCl) (RR'NH)/(RR'NCl), of Various Chloramines

Chloramine	Constant
Trichloroisocanuric acid	6.71×10^{-4}
Dichloroisocyanuric acid[a]	3×10^{-4}
1,3-Dichloro-5-dimethylhydantoin, 1st chlorine	2.54×10^{-4}
Monochloroisocyanuric acid[a]	1.9×10^{-4}
1-Chloro-5,5-dimethylhydantoin	1.14×10^{-4}
N-Chlorosuccinimide	6.6×10^{-5}
N-Chlorosulfamate[a]	$? \times 10^{-6}$
N-Chloro-p-nitroacetanilide	3.24×10^{-5}
N-Chloro-o-nitroacetanilide	1.12×10^{-5}
N-Chloro-m-nitroacetanilide	9.9×10^{-6}
N-m-Dichloroacetanilide	2.3×10^{-6}
N-p-Dichloroacetanilide	1.64×10^{-6}
Dichloramine-T	8×10^{-7}
N-Chloropropionanilide	7.5×10^{-7}
N-Chlorobutyranilide	7.0×10^{-7}
N-Chloroacetanilide	6.70×10^{-7}
N-o-Dichloroacetanilide	6.4×10^{-7}
N-Chloro-p-acetotoluide	3.48×10^{-7}
N-Chloro-m-acetotoluide	2.61×10^{-7}
N-Chloroformanilide	1.26×10^{-7}
N-Chloro-o-acetotoluide	1.11×10^{-7}
Chloramine-T	4.9×10^{-8}
Ammonia monochloramine	2.8×10^{-10}
Albuminoid chloramines[a]	up to 10^{-20}

[a]Estimate.

Source: From Ref. 29.

over 91.75%; sodium chloride, under 5.00%. The latter is a by-product in the solution at low temperature and retained by the chlorinated trisodium phosphate. The Adler method [5] is a unique process wherein the product is prepared by direct crystallization in its own molten mass. The Stamm patent [115] teaches the method of stabilizing chlorinated trisodium phosphate by the addition of a small amount of sodium tripolyphosphate. The Anderson and Wegst patent [4] describes the use of chlorinated trisodium phosphate in a dishwashing-detergent formulation using a simple dry-blending operation and describes its property of bleaching cups (tea and coffee stains). The Milenkevich and Henjum patent [69] teaches the art of obtaining a stable blend consisting of combining chlorinated trisodium phosphate, anhydrous sodium tripolyphosphate, and liquid sodium silicate, ending up with an agglomerated, chlorine-stable, product which is free of cupcaking.

Chlorinated trisodium phosphate has several limitations which may be overcome partially by proper formulation. In order to maintain a long shelf life, the chlorinated trisodium-phosphate hydrate must be kept intact. That is, other detergent salts present in the formulation must not break down this hydrate by taking away its water of hydration; thus, hydrated salts such as sodium metasilicate pentahydrate are used [4]. In other cases, the amount of moisture added to the blend is controlled to form approximately the sodium tripolyphosphate hexahydrate [69] to provide proper moisture balance. Because of the relatively low melting point of chlorinated trisodium phosphate, products containing it must not be subjected to temperatures much above room temperature. Exposures to temperatures of 100 to 110°F may cause a large loss in available chlorine within several weeks to several months time. The presence of caustic soda in a formulation containing chlorinated trisodium phosphate is known to cause a rapid loss of the available chlorine, and physical instability.

3. Chloroisocyanurates

There are various forms of the chloroisocyanuric acids and their salts. They are all crystalline products or crystal agglomerates with varying solubilities and all have relatively high melting points >200°C. Table 8 is a summary of their structure and some of their physical and chemical properties. The sodium and potassium salts and the trichloroisocyanuric acid-potassium dichloroisocyanurate complex are used extensively in dishwashing detergent, primarily because of their good stability in a detergent formulation and good bleaching quality (large hydrolysis constant). These products are available in a "granular" form made apparently in a large-diameter-tower drying process wherein hard, solid spheres are formed. Preferably, they are sized to approximately 60% through U. S. no. 50 sieve, and essentially 100% on U. S. no. 100 sieve. In this form where the particles are not broken down by attrition during mixing and handling, and

TABLE 8

Typical Physical and Chemical Properties of the Chloroisocyanuric Acids and their Salts

Chemical Nomenclature	Trichloroiso-cyanuric acid	Dichloroiso-cyanuric acid	Sodium dichloro-isocyanurate	Potassium dichloroiso-cyanurate	Trichloro-potassium dichloroisocynaurate complex
Structure	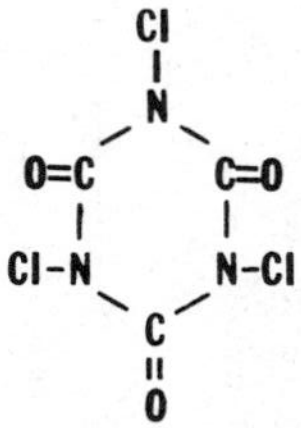	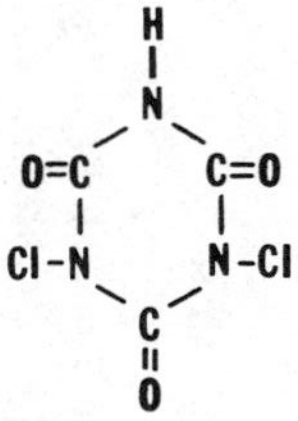	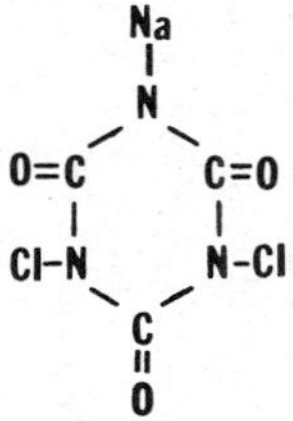	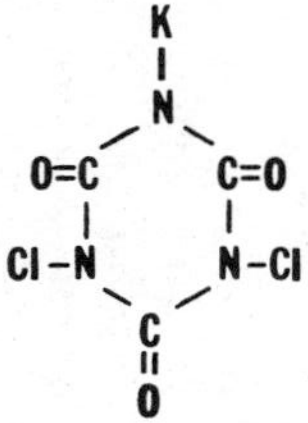	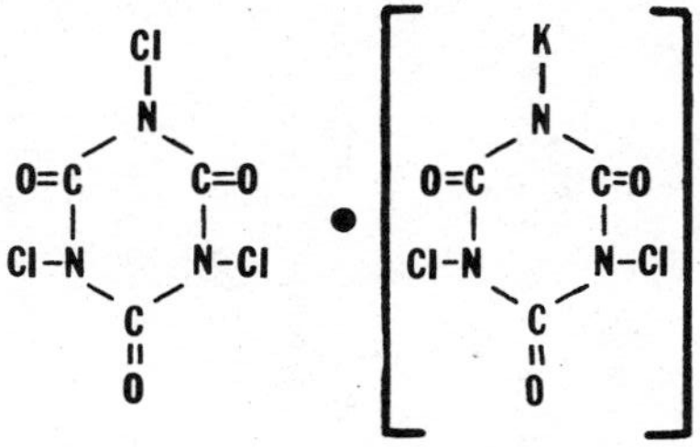
Molecular weight	232.44	197.99	219.98	236.08	1176.76
Avail. chlorine					
Theoretical	91.54%	71.66%	64.5%	60.08%	66.0%
Typical	90.0%	70.6%	61.0%	59.6%	66.0–67.0%
Physical form	Cryst. Solid	Cryst. Solid	Cryst. Solid	Cryst. Solid	Cryst. Solid
Color (pure sample)	White	White	White	White	White
pH 1% solution	2.7–2.9	2.7–2.9	5.9	5.9	—
Solubility in water	1.2	0.8	25	9	2.5
Grams solute/100 g solvent					
Melting Point (° C)	225–230	230–235	240–250	240–250	260–275

where contact with other detergent components is minimized, they show good chlorine stability.

The use of chloroisocyanurates is well established in dishwasher detergents for both home and commercial machines. The principal advantages of the use of chloroisocyanurate are that: (1) a dry, stable product may be formulated even incorporating caustic soda [8]; (2) the chloroisocyanurate has a relatively large hydrolysis constant compared to other N-chloro compounds as shown in Table 7, and the chlorine is nearly as reactive as sodium hypochlorite in its bleaching action [113]; (3) the chloroisocyanurate has a very high available chlorine content, thus minimizing the cost of the carrier and, in addition, providing flexibility in the formulation; and (4) the chloroisocyanurates are relatively safe to handle and do not present fire hazards as do some other chlorine compounds.

A number of patents have been granted for the manufacture and use of chloroisocyanuric acid and its salts. There are various routes for making the chloroisocyanurates. The first step is pyrolysis of urea to form the cyanuric acid by one of several methods [116-118]. The cyanuric acid must be fairly pure and free from ammonia values. The cyanuric acid is then converted to the sodium salt by neutralization in an aqueous solution corresponding to the number of chlorine substitution(s) desired. Direct chlorination of the hydrogen in cyanuric acid is difficult while the sodium is readily replaced by Cl^+ (OCl^- or HOCl). The sodium salt (i.e., disodium) is chlorinated at low temperature to form the sparingly soluble dichloroisocyanuric acid which may be recovered and washed free of the sodium chloride, and then coverted to the potassium or sodium salt [7, 119-122].

F. Neutral Salts and Urea

Various neutral salts have been used in dishwasher detergent formulations. Neutral salts such as sodium chloride and sodium sulfate have been tried but are not in wide use today. The effect of neutral salts may be ascribed to the attractive forces of the ions of the added salt and the oppositely charged ions of the substrate, thus causing an increased solubility due to complex ion formation.

Neutral salts may also be linked to the salting-in and salting-out phenomenon such as observed with globulin proteins [123]. That is, neutral salts may have a profound solvent action on globulins (salting-in). Proteins also exhibit a salting-out effect which is usually ascribed to dehydration of the protein by adding a very large amount of salt. The various cations show different abilities to salt out proteins and may be arranged in a series known as the Hoffmeister or lyotropic series. A typical Hoffmeister series in order of decreasing tendency to salt out is: $MgCl_2$, $CaCl_2$, LiCl, NaI, NaBr, $NaNO_3$, NaClO, NaCl, KCl, sodium acetate. Thus, by a series of manipulations it is possible to fractionate proteins by a salting-in and salting-out process. In mixed soils such as are usually encountered in dishwashing, it is difficult to show any significant benefits of the common neutral salts such as NaCl or Na_2SO_4 and, for the most part, they must be regarded as fillers or as aids to processing in agglomerated or spray-dried detergents.

Urea (not a salt in the usual sense) has also been suggested as an ingredient in detergent formulations. Urea has been advocated as a solvent for cellulose but no clear-cut evidence of direct benefit has been shown. Urea will react with hypochlorite, and thus should not be used with formulations containing active chlorine.

G. Corrosion Inhibitors

Sodium silicate is the most widely used corrosion inhibitor. As mentioned previously, the silicates high in SiO_2 are more effective members. However, the amount of SiO_2 and the pH of the use solution are the controlling factors. For a product with a use solution of about pH 10, sodium silicate content of about 10% ($SiO_2/Na_2O > 2:1$) will give good protection to china and china patterns, but some discoloration of aluminum may occur even under these conditions in a relatively few exposures. For a higher alkalinity product, about 25 to 30% of sodium metasilicate would be required. The degree of protection versus the ability of the detergent to clean, freedom from insoluble silica, and the tendency to form "permanent etching" on glasses (in softened water) are the factors that have to be "fine tuned" and related to the average use conditions.

The sorption of silicate on aluminum was studied by McCune [124] who determined the amount of SiO_2 absorbed by aluminum granules. From this and other evidence, he concludes that a reasonable mechanism of inhibition of aluminum corrosion by silicate might be the hydrous aluminum oxide reacting with monomeric or low-molecular-weight silicic acid, perhaps slightly ionized, depending on the pH to produce an aluminum silicate inert layer.

Wegst et al. [125] has been granted a patent on the use of an alkali-soluble zinc compound (ZnO equivalent of 1.5 to 7.6%) to inhibit corrosion of glass and vitreous ceramic surfaces. Green [126] has been granted a patent on the use of aluminate, berylate, and zincate for ceramic corrosion inhibition; Lintner [127] was granted a patent on the use of basic aluminum acetate and aluminum formate as a corrosion inhibitor against china pattern fading. Knapp and Thompson [128] were granted a patent on low-foaming detergent containing alkali aluminate and zincate as corrosion inhibitors. Rubin [129] was granted a patent on a corrosion-inhibitor system for platinum and other precious-metal pattern based on the use of gluconates with cerium salt and gluconate with ammonium dipersulfate in a chlorine-containing composition.

The aluminate- and zincate-type inhibitors have a tendency toward filming glasses under certain conditions and must be formulated carefully. At the current time these inhibitors are not known to be used widely, probably for this reason.

Another type of corrosion which occurs with moderate frequency, is the leaching of materials from the glass surface as pointed out by Vance [130] and Rutkowski [95]. Glass is considered to be a solid solution and lacks the molecular structure pattern of crystalloids. Moreover, being a mixture, there may be surface areas richer in certain alkaline-earth or alkali-metal elements. These may be leached from the surface and, by the very definition of the term glass, there may be some migration of elements from the subsurface to the surface. Laboratory studies (unpublished) have indicated that in some cases at least, the alkaline earth as well as the alkali metals are involved. The leaching effect was visually more pronounced with a detergent product high in sodium tripolyphosphate with a pH of about 10 than with a product low in sodium tripolyphosphate but high in pH. This would lead one to believe that the alkaline earth metal is involved. The leaching was carried out at 140°F for 150 hr. The visual criteria for leaching is the appearance of scratchlike lines at the surface (stress lines). Single glasses were so arranged that the surface could not have touched any other surface; thus it is not a result of mechanical damage. This same type of scratch line is observable in old glasses washed under home conditions.

For further and detailed information, the reader is referred to "The Structure of Glass" [131] and to articles by Douglas and El-Shamy [132] on the reaction of glass with aqueous solutions, by Weyl and Marboe [133] on various aspects of the glass, and by Morey [134] on composition and properties of glass.

Douglas and El-Shamy [132] studied the kinetics of the reactions of various glasses with water and aqueous solutions with a pH from 1 to 13. Most of the work was conducted on a simple binary glass composed of K_2O or Na_2O and SiO_2. His work indicates that higher temperatures and longer exposure times gave proportionally greater increases in leaching; K_2O extraction was markedly decreased at a pH above 9, whereas silica extraction increased rapidly above pH 9.0. The effect of other glass components (alkaline earth metal oxides) on the one hand, and the presence of chelating agents in detergent solution on the other hand, must be considered in this leaching phenomenon.

The glass tumbler is an important item in dishwashing and the one utensil most noticed by the consuming public in regard to film, etch, scratch, soil deposition, etc.; such factors must be weighed heavily in detergent formulation and evaluation.

H. Low-Foaming Surfactants and Defoamers

The majority of dishwashing detergents contain either a low-foaming or a defoaming surfactant (about 1 to 3% by weight). There is no clear-cut differentiation between them except as measured by their performance under appropriate conditions. Crecelius [135] describes the use of defoamers in detergents and the dependence on temperature for defoaming. Most defoamers tend to foam slightly at temperatures below about 110°F, but are nonfoamers at temperatures greater than about 120° F. Fineman et al. [45] describe the technique for evaluating low-foaming surfactants and differentiate between several commercial foaming and low-foaming surfactants. The Pluronic Grid [136] is a useful guide to selecting low-foaming surfactants. Reich et al. [137] describe a dynamic foam meter which may be used to study foam buildup with time as well as to determine the effect of temperature on the foaming characteristic of low-foaming surfactants. With suitable manipulation, it may be used to measure defoaming properties, i.e., by adding the foaming agent (such as egg) followed by the addition of defoamers. Kelly and Borza [138] describe a foam-test method in which intermittent foaming action is used so as to allow a buildup of steady-state foam level. Regardless of the method used, it is necessary to have a complete clean out of the system between different surfactants to obtain reproducible results. These laboratory methods are convenient for screening surfactants and defoamers, but there is no substitute for tests in foam-sensitive machines for "foam out" and clocking the revolutions per minute [45] of the wash arm or measuring the manifold pressure under use, or simulated use, conditions.

Foaming in dishwasher and loss in washing efficiency were discussed in Sec. II. E. Foaming due to protein soils such as egg and milk were pointed out by Martin and Temple [9], Podas and Crecelius [10], and Crecelius [135]. Recognition of the problem, and the importance of proteins as foaming agents led to much investigation and the eventual correction of this problem by the development of defoaming agents.

There is insufficient information to draw any conclusions about the structure-function relationship that could be considered rules to predict whether a compound is a defoamer or a low-foaming surfactant (except for a homologous series). Both classes of compounts have hydrophilic and hydrophobic portions of the molecule, which makes them a surfactant but the molecular weights, the hydrophile-hydrophobe balance, and the terminating groups all affect their properties as surfactants. However, it may be worthwhile noting several observations.

The low-foaming surfactants such as the Pluronic series are three-block polymers which terminate with hydrophilic-ethanoxy ethanol groups at both ends. It is hypothesized that they may form a horseshoe-type structure in which the hydrophobic end is the rounded part of the horseshoe and the hydrophilic ends are the legs of the horseshoe. Compounds such as alkyl phenoxy polyethenoxy benzyl ether [139], alkyl polyethenoxy benzyl ether [140], that described by the Martin and Temple patent [9] which is a five-block polymer terminating at both ends with polypropenoxy propanols, and the benzyl ether of this species [141] are all defoamers. It thus appears that defoaming in these cases is favored by hydrophobic terminating groups, although this is certainly not the sole criterion involved.

The caustic stability definitely appears to be favored by capped-end polymers in which the hydroxyl end groups of the alkalene oxide are capped with benzyl ether groups [139-141]. The general characteristics of the products used as low-foaming and defoaming surfactants are described in Sec. 5. Where surfactants are blended with caustic, it is important to use capped polymers to prevent interaction between the surfactant and the caustic, as caustic may degrade noncapped polymers into products which may foam significantly.

Phosphate ester surfactants have been described some time ago [142]. Their synthesis and use was described by Nunn and Hesse [143,144] and Dupre and Fordyce [145]. More recently, certain phosphate esters have been found to be excellent defoamers and are used in minor amounts with low-foaming nonionics (1/5 to 1/20 of the amount of nonionic) for defoaming proteinaceous food soils such as egg.

Groves [146] described the use of an acid or neutral phosphate ester produced from the reaction of concentrated phosphoric acid and 1 mol ethylene oxide adduct of a linear C_{16} primary alcohol and butylene-oxide-capped nonionic.

Simmons [147] described the use of a fatty alcohol ethoxylate and alkyl or alkenyl acid phosphate and salts such as sodium monostearyl acid phosphate.

Schmolka and co-workers [148, 149] have described the use of stearyl and oleyl acid phosphate esters with a number of different types of low-foaming nonionics containing oxyalkylene groups. The combinations studied by Schmolka and co-workers were reported to give excellent defoaming on egg and milk soils from temperatures of 100 to 160°F. Several levels of soil concentration and water hardness from 0 to 560 ppm $CaCO_3$ were studied and reported. Of importance is that these phosphate ester defoamers have the ability to suppress foam even at fairly low temperatures (below 110 to 115°F) where the ordinary defoamers of low-foaming nonionics fail to maintain their defoaming or low-foaming properties. These phosphate esters are used in combination with other low-foaming nonionic surfactants.

Where chlorine and surfactants are used in the same blend, it is important to prevent the interaction of these two components by encapsulation such as that mentioned in the Oberle [70] patent, or by other suitable means. Here the loss of available chlorine is probably of greater importance than is the degradation of the surfactant by the chlorine, although evidence for both possibilities definitely exists.

I. Enzymes

Currently, there is a great deal of interest in the use of enzymes in laundry detergents. Enzymes have been used during the past 50 years for desizing in the textile industry and as clarifiers in food products. Although some laboratory and commercial applications were studied at least as early as 1944 for removal of starchy soils from dishes and from 1952 in laundry products [150], the recent use of the enzymes in laundry detergents undoubtedly had its start in Europe. The home laundering operations in Europe conventionally use a long presoak and are more adapted to the use of enzymes than is the U. S. laundry practice of short soak or no soak.

Enzymes are derived from a number of sources such as papaya, pineapple, and fig, and animal organs such as stomach, pancreas, etc. However, the current interest is in the bacterial enzymes, especially from certain species of _Bacillus subtilis_. The poundage of enzymes available from plant or animal sources may be limited, but the poundage from a microbial source is almost unlimited and production may be more readily geared up to demand.

Liss and Langguth [151] have shown the benefit of the addition of alkaline protease (low in amylase) and an alkaline protease containing neutral protease and a high level of α-amylase in two laundry detergent formulations using the Terg-O-Tometer for evaluation. EMPA Cloth no. 116 (fabric stained with blood, milk, and Japanese ink) and EMPA Cloth no. 112 (fabric stained with cocoa, milk, and sugar) were used and the studies showed the effect of pH, temperature, and anionic- versus nonionic-based detergent. They also studied the shelf-life stability of enzyme-based detergent.

Cayle [152] made similar studies and presented data which indicated that EMPA Cloth no. 116 is suitable for establishing the effectiveness of detergents containing proteases, whereas EMPA Cloth no. 112 responds to protease as well as to amylase. They developed an experimental test fabric uniformly stained with starch and lampblack which responded more specifically to the amylase. They also reported on compatibility between enzyme and various detergent builders.

Soil-removal results on mixed soils need to be interpreted carefully as mixed soil sometimes shows peculiarity of an unexpected nature. A case in point is a study made by food technologists to obtain dehydrated products which may be reconstituted more readily and to resemble natural fresh product by the addition of some other component prior to dehydration. For example, a study by Schultz et al. [153] indicated that, in dried egg yolk, the removal of water irreversibly changes the structure of the low-density lipoproteins and releases foam-inhibiting free lipid from these lipoproteins. Thus, when the yolk is reconstituted, it loses its ability to form a stable foam. However, when yolk is codried with 15% added carbohydrates, the carbohydrates partially protect the lipoproteins from this irreversible structural change and result in a product which may foam nearly as much as fresh egg yolk. Thus it is important to determine soil removal on individual soils as well as on mixtures, as the mixture may not be representative of a normal soiling-desoiling situation due to interactions between components in a mixed soil and as it affects binding between soil and substrate.

Enzymes are biocatalysts that perform a wide variety of functions, but each enzyme is quite specific. In general, they are classified as lipase or fat splitting, protease or protein splitting, and amylase or starch splitting enzymes. While enzymes are unique in many ways, they also have some definite limitations. For example, they are inactivated by high temperature, the maximum temperature of use being about 140 to 160°F. (Reaction rate versus inactivation rate must be considered.) The maximum pH is about 9.5 [154,155]. They are destroyed by strong oxidizing agents such as sodium hypochlorite. Although the amount of substrate that a unit of enzyme may turn over is quite large (turnover number may be several million or more), the time required to do the job is usually long. For example, in clothes washing, soaking from 30 min to overnight is recommended for an enzyme-detergent soak.

Very little information is available at this time on the use of enzymes in dishwasher detergent. However, much of the information gained in the laundry studies of pH, temperature and time optima and maxima, and sensitivity to strong oxidizing agents, etc., might be expected to hold in formulating a dishwashing product based on enzymes.

If enzymes can be incorporated successfully into dishwashing detergents, it would appear that they ought to be less corrosive to dishware and metal. However, to obtain maximum utilization of the enzyme, a longer wash cycle may be required.

It would appear that enzymes would have greatest potential in the removal of such soils as oatmeal, rice, cream of wheat, egg, and casserole dishes, etc., in which a combination of long soak and specific enzymes might do the job. At this time it would appear that enzymes are in a developmental stage so far as use in dishwashing detergent is concerned, and more developmental work needs to be done in order to utilize fully the potentials offered by the enzymes.

V. RINSING AND RINSE ADDITIVES

Waters with hardness of 200 to 300 ppm $CaCO_3$ or higher, and waters with high solids, will often give filming and spotting problems that are particularly noticeable on glasses.

Laboratory studies indicate that the final rinse has a major effect on filming and spotting. For example, if sufficient distilled or deionized water is used for the final rinse, there would be minimum spotting and filming, even though they are washed in hard water. Unfortunately, such large amounts of water would be required that this has only been practical in laboratory glassware washing (tissue culture laboratories) where such high costs may be justified. In commercial machines, the spent rinse water normally is collected and fed into the recirculated (power) rinse tank, and then successively into the wash tank (part may be diverted to the prescrap) to renew these solutions so that the rinse water represents nearly the total water usage of the machine.

Some studies have been made to assess the practicality of recirculated final deionized water rinse which would be passed through a deionizer to remove continually solids from the rinse water. The cost of such a system appears to be prohibitive at the present time. Ion-exchange systems where the calcium ion is exchanged for the sodium ion are in fairly wide use. This system appears to help in many cases but is not a completely satisfactory answer to the hard-water problem. The softened-water etch problem as pointed out by Madden [36] is one of the problems. The treatment does not reduce the total solids in high-solids water. Therefore, there have been many attempts to treat the rinse water either to reduce the precipitation of the hardness or to minimize the amount of water which clings to the utensils.

A. Polyphosphates

The addition of threshold amounts of polyphosphates to the rinse water has been tried. As already mentioned, there is a pronounced tendency for the calcium bicarbonate to decarbonate at high temperature (180°F and higher) and form calcium carbonate. Polyphosphates may delay such precipitation; but the results are not consistent and the process has not been accepted widely.

B. Surfactants

The first surfactant-based rinse additives to appear on the market were based on alkyl phenoxy polyethenoxy ethanol and were used as silverware dips [156]. The silverware was sent through the dishwashing machine and dipped for a few seconds in a thermostatted bath containing the additive and was drained dry.

However, the surfactant was a relatively high-foaming material and could not be used directly in the rinse line as the rinse water eventually ends up in the wash tank where it would create a foaming problem.

The first true rinse additives were made possible with the introduction of block polymers of ethylene and propylene oxide [136, 157], which could be tailor-made as to average molecular weight and hydrophobe-hydrophyl balance. These were among the first true low-foaming surfactants available. The alkyl phenoxy polyethenoxy benzyl ethers [139] and the alkyl polyethenoxy benzyl ethers [140] came later and these types of material are also in common use as rinse additives. These surfactants are characterized by having a relatively large molecular weight, low cloud points in the range of 24 to 28°C in 1% distilled water solution, low-foaming or defoaming nature (Ross-Miles initial foam height not greater than about 50 mm and one of not greater than about 15 mm after 5 min), and surface tension in the range of 37 to 40 dynes/cm at 20°C. The values will vary somewhat with the structure, and the figures are characteristic of products which have been found useful and are not intended as limiting specifications.

These products are of sufficiently low foam that they could be added directly to the rinse line without foaming problems, and some have definite defoaming properties. When used properly (50 to 75 ppm active), these products minimize filming and spotting and, in institutional operations, eliminate towel drying which is a real step forward in sanitation as well as in labor saving.

Evaluations of performance of rinse additives, aside from the characteristics listed above, are subjective. Filming and spotting (CSMA) evaluation as mentioned previously may be used. Filming and spotting are important criteria in establishing performance as well as in determining the optimum concentration range of the rinse additive, as excessive concentration will result in filming even with the most effective additive. Another method is to measure the drying times under controlled conditions. The effective rinse additives promote thin water films that evaporate quickly and evenly as opposed to poor additives that may wet completely but form thick films. Drying time is a realistic measure since it is not only a functional criteria but also a measure of film thickness when other factors are controlled. A common method of evaluation consists of using a front-load home machine, replacing the door with a Plexiglas front with a small hole drilled in the door to permit injection of the rinse additive with a hypodermic syringe. The "sheeting" may be observed after incremental addition of the additive and the minimum amount of additive needed to "sheet" may be used as a basis for comparing their effectiveness. (The machine is stopped momentarily to make these observations.) At the same time this allows a visual observation on the film thickness. Thin films are characterized by rapid and uniform film "breaks" due to draining and/or evaporation of water, and they appear as fine bubbles which erupt over a wide area of the test surface. Because of the wide differences in the nature of the surfaces involved, such tests must be conducted with all of the commonly used materials such as glass, vitrified china, aluminum, and stainless steel. Other methods of evaluation have been tried such as attaching two parallel electrode strips to a glass plate and measuring conductivity (or loss of conductivity as the film evaporates).

Foaming properties of these low-foaming surfactants may be measured by one of several methods [44,45,137]. The temperature has a very pronounced effect on foaming and, characteristically, a surfactant may be a low foamer at, say, 140°F or higher but may show undesirably high foaming at 110 to 115°F or lower. Thus it is important to measure foaming properties at two or more different temperatures.

C. Acids

The use of acid additives either by themselves or with surfactants has been explored. The theory is to add sufficient acid to convert the bicarbonate ion to water and carbon dioxide and to form a soluble, nonfilming calcium salt which would be noncumulative with respect to filming from one wash cycle to another. Although this appears to work in some operations, results are not completely consistent. This method has one drawback of not reducing solids and another of being corrosive to the rinse plumbing. Some acids are chelates such as, for example, the hydroxy carboxylic acids, and they may perform a double function. Acid rinse additives are very popular in Europe where adverse water conditions are encountered more frequently.

D. Iron Staining

Another problem which is commonly associated with the water is iron staining. Iron staining of chinaware is of common occurrence where iron in water is about 0.25 ppm and higher. The iron-stain problem occurs more frequently in commercial dishwashing where the turnover of dishes may be much higher than in the home. The iron is in the colorless ferrous form in intact water, but is air-oxidized readily and is oxidized when the water is chlorinated into the reddish brown ferric hydroxide. The water may pick up iron from piping as well as be present in the groundwater.

Iron in water is also suspected of intensifying tea staining by the formation of a ferro-tannin complex. Iron removal units may be helpful in overcoming the iron-staining problem.

VI. SUMMARY

This chapter covers briefly the developmental aspects of dishwashing, some theories on detergency, evaluation of detergent, and functions and properties of the various detergent components.

An attempt has been made to draw together samplings of information available and pertinent to dishwashing; the references and patents cited will give more detailed information on the various topics and many more references on the subject.

As is often true, there are many unanswered questions and further work needs to be done in many areas. For example, the recent work reported by Loeb and Shuck [16] and the proposed definition of a detergent as "any substance that either alone or in a mixture, increases the efficiency with which mechanical energy is utilized in a cleaning process [pp. 299]" is a very unique concept and a substantial contribution in the philosophical, chemical, and physical approach to the study of detergents. More fundamental, quantitative studies of this nature would undoubtedly help in understanding how a detergent works. This concept gives a new perspective to the role of detergents.

There is a need for more effective corrosion inhibitors, particularly in high pH products, without the tendency to form film on glassware. Aluminumware is very susceptible to attack by chelates and alkalinity and to discoloration by minerals present in many waters. The corrosion may take the form of etching or unsightly discoloration.

The problem of attack by detergent of gold and platinum banding on glassware and dishes is far from being solved. Part of the problem here, of course, is the thinness of the metal foil as well as the water constituents.

Finding a suitable substitute for polyphosphate is one of the greatest challenges facing the detergent industry. Polyphosphate has become such a low-cost commodity and possesses so many good attributes that it is very difficult to find a replacement. Ideally, a replacement for sodium tripolyphosphate for use in dishwashing detergent ought to possess the following properties: it should not contain phosphorus or nitrogen; it should be reasonable on cost of usage; it should have sequestering capacity and complex constant for calcium and magnesium at least as good as sodium tripolyphosphate, good stability in shelf life and in solution, not reactive with hypochlorite; it should possess some of the "colloidal/surface active" properties attributed to the polyphosphate; possess threshold effect; it should not be hygroscopic; it should have some benefits which would lend themselves for use in agglomeration or processing; lastly, but importantly, it should minimize filming on glasses and dishes at least as well as the polyphosphates.

The use of enzymes has not been fully explored. Though the current generation of enzymes have some limitations, if the pH optimum or stability to chlorine can be increased even further, it might be possible to formulate an enzyme product which would have all of the properties of chlorinated alkaline detergent plus the enzyme activity. Or if the potency of the enzyme can be improved and the substrate spectra widened, it might be possible to formulate a noncorrosive detergent safe for washing aluminum, gold- and platinum-banded ware, etc.

Some agent to replace chlorine which would be more effective but less corrosive to metals, particularly silverware, and still destain and minimize spotting, would be very useful.

Another important area which was hardly touched on is water treatment or management. Water quality is very fundamental to cleaning results and particularly to filming and spotting. Any treatment to upgrade the water quality would go a long way toward improving results.

Along the same vein, since water quality is said to be deteriorating rapidly with the increasing population and industrialization, there is a great need to anticipate what, if any, changes need to be made in dishwashing detergent formulations to cope with this problem.

Most of the surface-tension data on low-foaming and defoaming surfactants is reported at 25°C. Since the use temperature is 55 to 70°C, there is need for surface tension data at these temperatures.

Newer improved instruments such as the FT (Fourier Transform) high resolution nuclear magnetic resonance involving different nuclei and high resolution scanning electron microscopes are available for a more complete structure analysis of surfactants, following the degradation of polyphosphates and for studying corrosion of surfaces, film formation, and the likes.

Finally, one of the greatest challenges facing the industry today is energy conservation; to maintain in the future, when reduced energy levels might be mandated by necessity and/or high cost, the same high degree of cleanliness and sanitation that was achieved when energy availability was taken for granted.

ACKNOWLEDGMENT

The author wishes to thank Economics Laboratory, Inc., and Mr. William M. Podas, Senior Vice-President and Director of Research, for making this chapter possible. My thanks also to Dr. John L. Wilson, Dr. Thomas E. Brunelle, Vice-President, Mr. Thomas M. Oberle, Vice-President, Messrs. Edward D. Berglund and Donald E. Wal for many helpful suggestions, and to Mrs. Lois Jackels for her assistance.

REFERENCES

1. United States Potters Association Teaching Guide, United States Potters Association, East Liverpool, Ohio, 1956.
2. D. A. Meeker, The Story of The Hobart Manufacturing Co., Newcomer Publications in North America, Princeton University Press, Princeton, N. J., 1960.
3. D. H. Morrish, History of Dishwashers, General Electric Co., Louisville, Ky., 1967.
4. D. E. Anderson and W. F. Wegst (to Wyandotte Chemical Corp.). U.S. Patent 2,689,225, Sept. 14, 1954.
5. H. Adler (to Victor Chemical Works). U. S. Patent 1,965,304, July 3 1934.
6. D. S. Corliss, R. R. Keast, and J. S. Thompson (to FMC Corp.). U.S. Patent 3,352,785, Nov. 14, 1967.
7. W. F. Symes (to Monsanto Company). U.S. Patent 3,272,813, Sept. 13, 1966.
8. W. G. Mizuno and T. M. Oberle (to Economics Laboratory, Inc.). U.S. Patent 3,166,513, Jan. 19, 1965.
9. A. T. Martin and N. S. Temple (to Economics Laboratory, Inc.). U.S. Patent 3,048,548, Aug. 7, 1962.
10. W. M. Podas and S. B. Crecelius, J. Milk Food Technol., 30, no. 2, 39 (1967).
11. R. M. Anderson, J. Satanek, and J. C. Harris, J. Am. Oil Chem. Soc., 37, 119 (1960).
12. J. C. Harris and J. Satanek, J. Am. Oil Chem. Soc., 38, 169 (1961).
13. W. G. Jennings, S. Whitaker, and W. C. Hamilton, J. Am. Oil Chem. Soc., 43, 130 (1966).
14. W. R. Kelly, J. Am. Oil Chem. Soc., 43, 358 (1966).
15. A. M. Schwartz, J. Am. Oil Chem. Soc., 48, 566 (1971).
16. L. Loeb and R. O. Shuck. J. Am. Oil Chem. Soc., 46, 299 (1969).
17. L. Loeb and R. O. Shuck, J. Am. Oil Chem. Soc., 48, 25 (1971).
18. I. M. Koltoff and E. B. Sandell, Textbook of Quantitative Chemistry, 3d ed., Macmillian, New York, 1956, p. 60.
19. U.S. Department of Health, Education and Welfare, Public Health Service Publication no. 934, 1962.
20. W. L. Mallman, L. Zaikowski, and D. Kahler, A Study of Mechanical Dishwashing (National Sanitation Foundation Research Bulletin, no. 1), 1947.

21. A. M. Scalzo, R. W. Dickerson, Jr., and R. B. Read, Jr., J. Milk Food Technol., 32(1), 20 (1969).

22. I. M. Koltoff and E. B. Sandell, Textbook of Quantitative Chemistry, 3d ed., Macmillian, New York, 1956, p. 62.

23. J. R. Van Wazer and M. E. Tuvell, J. Am. Oil Chem. Soc., 35, 552 (1958).

24. S. Ross and G. J. Young, Ind. Eng. Chem., 43, 2520 (1951).

25. S. Ross and T. H. Bramfitt, J. Phy. Chem., 61, 1261 (1957).

26. S. Ross and R. M. Haak, J. Phys. Chem., 62, 1260 (1958).

27. M. M. Fineman, private communication, 1956.

28. R. E. Kirk and D. F. Othmer, Ency. Chem. Tech., 4, 911 (1964).

29. R. E. Kirk and D. F. Othmer, Ency. Chem. Tech., 4, 913 (1964).

30. I. R. Schmolka, M. Cenker, M. Kokorudz, and W. K. Longdon, Paper presented at 37th Annual Fall Meeting of AOCS, Minneapolis, Sept. 30, 1963.

31. D. Briggs, private communication, 1960.

32. W. M. Podas and C. S. Bloomberg, Paper presented at the Klenzade Seminar, Chicago, March 1962.

33. R. W. R. Baker, Biochem. J., 41, 337 (1947).

34. L. W. Magnuson and K. Mizuno, unpublished data, (Feb. 7, 1963).

35. J. L. Wilson and E. E. Mendenhall, Ind. Eng. Chem., 16, 251 (1944).

36. R. E. Madden, Soap Chem. Spec., p. 45 (April 1967).

37. Midwest Research Institute Project No. 2231-C, Progress Report nos. 2-4.

38. R. F. Vance, J. Am. Oil Chem. Soc., 46, 639 (1969).

39. E. H. Mann and C. C. Ruchhoff, Public Health Rep., 61(24), 877 (1946).

40. G. J. Hucker, Progress Report on Detergent Evaluation Investigation, April 1942—March 1943, N. Y. Agricultural Experiment Station, New York.

41. L. O. Leenerts, J. F. Pietz, and J. Elliot, J. Am. Oil Chem. Soc., 33, 119 (1956).

42. W. W. Benton, A. T. Martin, C. H. Shiflett, and J. L. Wilson, Proc. Minnesota Acad. Sci., 24, 681 (1956).

43. W. L. Mallman and D. Kahler, Studies on Dishwashing (National Sanitation Foundation Research Bulletin, no. 2), 1949.

44. Detergent and Cleaning Compounds Test Methods Compendium, First Edition, CSMA, Washington, D.C. (Dec., 1978).

45. M. N. Fineman, H. L. Greenwald, and C. G. Gebelein, Soap Chem. Spec., (August, 1955).

46. A. L. Kimmel, H. M. Gadberry, and D. O. Darby, Soap Chem. Spec., (April, 1961).

47. E. Armbruster and G. Ridenour, Paper presented at the 38th Midyear Meeting, CSMA, Boston, June 9, 1952.

48. C. H. Francis, Unpublished data, Sept. 9, 1960.

49. F. W. Gilcreas and J. E. Obrien, Am. J. Public Health, 31, 143 (1941).

50. Anonymous, Consum. Rep., (1949) (December, 1959).

51. E. Domingo, Q. Bull. Assoc. Food Drug Off. U.S., 13(3), 100-107 34, 12.

52. E. H. Armbruster and G. M. Ridenour, Sanitarian, 23(2), 103 (1960).

53. F. Feigl, Qualitative Analysis by Spot Tests. Vol. 1:Inorganic Applications; Vol. 2: Organic Applications. Elsevier, New York, 1954.

54. A. K. Pryor, Unpublished report, (Oct. 5, 1958).

55. Aluminum Company of America, Private Communication, 1951.

56. Official Methods of Analysis of The Association of Official Agricultural Chemists, 9 Ed., p. 71, Washington, D.C., 1960.

57. J. C. L. Resuggan, Private communication, 1969.

58. J. R. Van Wazer, Phosphorus and its Compounds, Vol. I: Chemistry, Interscience, New York, 1958.

59. O. Quimby, Chem. Rev., 40(1), 158 (1947).

60. R. N. Bell, Inorganic Synthesis, Vol. III (L. F. Audreieth, ed.). McGraw-Hill, New York, 1950, p. 85.

61. R. E. Kirk and D. F. Othmer, Ency. Chem. Tech., 15, 245 (1964).

62. J. D. McGilvery, ASTM Bull., 191, 45 (July, 1953).

63. C. Y. Shen, J. Am. Oil Chem. Soc., 45, 510 (1968).

64. M. E. Ginn, R. M. Anderson, and J. C. Harris, J. Am. Oil Chem. Soc., 41, 112 (1964).

65. G. B. Hatch and O. Rice, Ind. Eng. Chem., 31, 51 (1939).

66. R. N. Bell, Ind. Eng. Chem., 39(2), 136 (1947).

67. J. R. Van Wazer, Phosphorus and its Compounds, Vol. I: Chemistry, Interscience, New York, 1958, pp. 454, 784, 785, and 788.

68. J. P. Crowther and A. E. R. Westman, Can. J. Chem., 32, 42 (1954).

69. J. A. Milenkevich and J. E. Henjum (to the Procter & Gamble Co.), U.S. Patent 2,895,916, July 21, 1959.

70. T. M. Oberle (to Economics Laboratory, Inc.), U. S. Patent 3,306,858, Feb. 28, 1967.

71. T. M. Kaneko, I. R. Schmolka, and J. W. Compton, J. Am. Oil Chem. Soc., 45, 855 (1968).

72. J. R. Van Wazer, Phosphorus and its Compounds, Vol. I: Chemistry, Interscience, New York, 1958, p. 683.

73. Monsanto Technical Bulletin on Sodium Trimetaphosphate, Report No. 5685, (April 10, 1964).

74. F. C. Bersworth (to the Martin Dennis Co.). U.S. Patent 2,387,735, Oct. 30, 1945.

75. J. J. Singer, Jr., and M. Weisberg (to Hampshire Chemical Corp.). U.S. Patent 3,061,628, Oct. 30, 1962.

76. O. T. Quimby and J. B. Prentice (to the Procter & Gamble Co.). U. S. Patent 3,400,151, Sept. 3, 1968.

77. O. T. Quimby (to Procter & Gamble Co.). U. S. Patent 3,400,148, Sept. 3, 1968.

78. O. T. Quimby (to Procter & Gamble Co.). U. S. Patent 3,400,176, Sept. 3, 1968.

79. O. T. Quimby (to Procter & Gamble Co.). U.S. Patent 3,400,149, September 3, 1968.

80. R. R. Pollard, Soap Chem. Spec., p. 58 (Sept., 1966).

81. T. A. Downey, Soap Chem. Spec., p. 52 (February, 1966).

82. R. G. Pristine, Jr., and A. J. Stirton, Paper presented at AOCS Meeting, Minneapolis, Oct. 5, 1969.

83. F. L. Diehl (to Procter & Gamble Co.). U. S. Patent 3,308,067, March 7, 1967.

84. P. M. Sabatelli, E. R. Loder, C. A. Brungs, and C. R. Sarge (to Chemed Corporation). U. S. Patent 3,623,991, November 30, 1971.

85. P. M. Sabatelli and C. A. Brungs (to W. R. Grace & Co.). U. S. Patent 3,579,455, May 18, 1971.

86. P. M. Sabatelli and C. A. Brungs (to W. R. Grace & Co.). U.S. Patent 3,535,258, October 20, 1970.

87. P. M. Sabatelli, C. A. Brungs, E. R. Loder, and C. R. Sarge (to Chemed Corp.). U. S. Patent 3,627,686, (Dec. 14, 1971).

88. W. G. Mizuno, J. L. Copeland, and A. E. Scholze (to Economics Laboratory, Inc.). U. S. Patent 3,700,599, October 24, 1972.

89. Procter & Gamble Co., W. German patent 2,212,619, October 5, 1972.

90. Pfizer, Inc., Soap Cosmet. Chem. Spec., 49(5), 98 (1973).

91. Monsanto Technical Bulletin, Special Report no. 4666 and attachment; R. R. Irani and C. F. Callis, J. Phys. Chem., 64, 1398 (1960).

92. K. R. Lange, J. Am. Oil Chem. Soc., 45, 487 (1968).

93. J. G. Vail, Soluble Silicates, Vols. I and II (ACS Monograph Series, no. 116), Reinhold, New York, 1952.

94. M. G. Kramer, J. Am. Oil Chem. Soc., 29, p. 529 (Nov. 1952).

95. B. J. Rutkowski, J. Am. Oil Chem. Soc., 48, 166 (1971).

96. C. L. Baker, Ind. Eng. Chem., 27, 1358 (1935).

97. H. Weldes, Soap Cosmet. Chem. Spec., 48, 72 (1972).

98. R. E. Kirk and D. F. Othmer, Ency. Chem. Tech., 3, 624 (1964).

99. W. G. Mizuno (to Economics Laboratory, Inc.). U. S. Patent 3,166,512, Jan. 19, 1965.

100. R. R. Keast, E. S. Roth, and J. S. Thompson, Soap Chem. Spec., p. 56 (June, 1968).

101. J. S. Thompson, Soap Chem. Spec., p. 45 (June, 1964).

102. R. E. Kirk and D. F. Othmer, Ency. Chem. Tech., 1, 741 (1964).

103. Rohm and Haas Bulletin on Triton, Low Foam Detergents: Physical Properties, Performance Characteristics. SAN 238-6. Rohm and Haas Co., Philadelphia (1963).

104. R. N. Bell, Ind. Eng. Chem., 41(12), 2901 (1949).

105. Diamond Alkali Company (Diamond Shamrock), Report no. 1055 (June 9, 1961).

106. B. Cohen, Soap Chem. Spec., 31 (April, 1970).

107. R. E. Kirk and D. F. Othmer, Ency. Chem. Tech., 5, 15 (1963).

108. R. D. Gleichert (to Pittsburgh Plate Glass Co.). U. S. Patent 3,134,641, May 26, 1964.

109. I. E. Muskat and A. G. Chenicek (to Pittsburgh Plate Glass Co.). U. S. Patent 2,184,886, Dec. 26, 1939.

110. A. F. Guiteras and F. C. Schmelkes, J. Biol. Chem., 107(1), 235 (1934).

111. F. C. Schmelkes and E. S. Horning, J. Bacteriol., 29(3), 323 (1935).

112. E. E. Hardy (to Monsanto Chemical Co.). U. S. Patent 2,607,738, Aug. 19, 1952.

113. Monsanto Technical Bulletin, I-177 (Sept. 1960).

114. L. D. Mathias (to Victor Chemical Works). U. S. Patent 1,555,474, Sept. 29, 1925.

115. J. K. Stamm (to W. R. Grace & Co.). U. S. Patent 3,364,147, Jan. 16, 1968.

116. R. H. Westfall, U. S. Patent 2,943,088, June 28, 1960.

117. E. C. Sobocinski and W. F. Symes (to Monsanto Co.). U.S. Patent 3,357,979, Dec. 12, 1967.

118. W. P. Moore and C. B. R. Fitz-William, Jr. (to Allied Chemical Corp.). U.S. Patent 3,3½8,887, May 9, 1967.

119. R. J. Fuchs and R. A. Olson (to FMC Corp.). U.S. Patent 3,336,228, Aug. 15, 1967.

120. S. J. Kovalsky and R. A. Olson (to FMC Corp.). U. S. Patent 3,270,017, Aug. 30, 1966.

121. W. P. Moore (to Allied Chemical Corp.). U. S. Patent 3,295,916, Nov. 15, 1966.

122. K. Merkel, H. U. Werner, and A. Palm (to Badische Anilin). U. S. Patent 3,073,823, Jan. 15, 1963.

123. E. J. Cohn and J. T. Edsall, Protein, Amino Acids and Peptides (ACS Monograph Series, no. 90), Reinhold, New York, 1943, pp. 609 and 621.

124. H. W. McCune, J. Electrochem. Soc., 106(1), (1959).

125. W. F. Wegst, L. R. Bacon, and R. H. Vaughn (to Wyandotte Chemicals Corp.). U. S. Patent 2,447,297, Aug. 17, 1948.

126. R. L. Green (to FMC Corp.). U.S. Patent 3,350,318, Oct. 31, 1967.

127. A. E. Lintner (to Calgon Corp.). U. S. Patent 3,128,250, April 7, 1964.

128. K. W. Knapp and J. S. Thompson (to FMC Corp.). U. S. Patent 3,255,117, June 7, 1966.

129. F. K. Rubin (to Lever Brothers. U. S. Patent 3,303,104, Feb.7, 1967.

130. R. Vance, Personal communication, 1964.

131. Proceedings of the Third All-Union Conference on the Glassy State, Leningrad, Nov. 16-20, 1959 [English translation]. Consultant Bureau, New York, 1960.

132. R. W. Douglas and T. M. M. El-Shamy, J. Am. Ceram. Soc., 50(1), 1 (1967).

133. W. A. Weyl and E. C. Marboe, The Constitution of Glasses: A Dynamic Interpretation, Vols. I and II, Parts I and II, Interscience, New York, 1962, 1964 and 1967.

134. G. W. Morey, Properties of Glass (ACS Monograph, no. 124), Rheinhold, New York, 1954.

135. S. B. Crecelius, J. Milk Food Technol., 28(9), 291 (1965).

136. Wyandotte Chemical Pluronic Grid, Bulletin, form 189-61.

137. H. E. Reich, J. T. Patton, and C. V. Francis, Soap Chem. Spec., p. 55 (April, 1961).

138. W. R. Kelly and P. F. Borza, J. Am. Oil Chem. Soc., 43, 364 (1966).

139. W. D. Niederhauser and E. J. Smialkowski (to Rohm and Haas Co.). U.S. Patent 2,856,434, Oct. 14, 1958.

140. L. M. Rue, T. E. Brunelle, and W. G. Mizuno (to Economics Laboratory, Inc.). U.S. Patent 3,444,242, May 13, 1969.

141. T. E. Brunelle, L. M. Rue, and S. B. Crecelius (to Economics Laboratory, Inc.). U.S. Patent 3,334,147, Aug. 1, 1967.

142. M. V. Shelanski and M. W. Winicov (to West Laboratories, Inc.). U.S. Patent 2,710,227, June 7, 1955.

143. L. G. Nunn, Jr., and S. H. Hesse (to General Aniline and Film Corp.). U.S. Patent 3,004,056, October 10, 1961.

144. L. G. Nunn, Jr. (to General Aniline and Film Corp.). U.S. Patent 3,004,057, October 10, 1961.

145. J. Dupre and D. B. Fordyce (to Rohm and Haas Company). U.S. Patent 3,294,693, December 27, 1966.

146. W. L. Groves (to Continental Oil Company). U.S. Patent 3,595,968, July 27, 1971.

147. J. K. Simmons and E. A. Kitchen (to Procter & Gamble Company). U.S. Patent 3,630,923, December 28, 1971.

148. I. R. Schmolka and T. M. Kaneko, J. Am. Oil Chem. Soc., 45, 563 (1968).

149. I. R. Schmolka and M. H. Earing (to Wyandotte Chemicals Corp.). U.S. Patent 3,314,891, April 18, 1967.

150. John L. Wilson and A. T. Martin, unpublished data, Jan. 26, 1944 and Oct. 1, 1952.

151. R. L. Liss and R. P. Langguth, J. Am. Oil Chem. Soc., 46(10), 507 (1969).

152. T. Cayle, J. Am. Oil Chem. Soc., 46(10), 515 (1969).

153. J. R. Schultz, H. E. Snyder, and R. H. Forsythe, J. Food Sci., 33, 507 (1968).

154. J. C. Hoogerheide, Enzymes as Additives to Laundry Detergents. AOCS Meeting, Chicago, Oct. 1967.

155. H. E. Worne, Deterg. Age, 5(9), 19 (Sept., 1968).

156. J. L. Wilson, W. G. Minzuno, and S. B. Crecelius, Soap Chem. Spec., 34(2), 48 (Feb., 1958).

157. L. G. Lunsted (to Wyandotte Chemicals Corp.). U. S. Patent 2,674,619, April 6, 1954.

158. R. E. Kirk and D. F. Othmer, Ency. Chem. Tech., 3, 624 (1963).

159. Anonymous, Merchandising, 3(3), 34 (March, 1978).

Chapter 22

DETERGENTS AND OUR ENVIRONMENT

Mark W. Tenney

TenEch Environmental Consultants, Inc.
South Bend, Indiana

Judith B. Carberry

Department of Civil Engineering,
University of Delaware, Newark, Delaware

I. INTRODUCTION 900

II. ENVIRONMENTAL ASPECTS 903

A. Ecological Considerations 903
B. Hydrological Considerations 904
C. Pollution of the Aqueous Environment 907

III. POLLUTIONARY CHARACTERISTICS OF SYNTHETIC DETERGENTS 923

A. Components 923
B. Elemental Impacts. 924

IV. WASTEWATER TREATMENT 940

A. General 940
B. Primary Treatment 943
C. Secondary Treatment 944
D. Tertiary and Advanced Wastewater Treatment 953
E. Sludge and Brine Handling 965

V. SUMMARY 980

GLOSSARY 986

REFERENCES 988

I. INTRODUCTION

Synthetic detergents have been so totally accepted by modern consumers that their use currently must be assumed to be a necessity of life. During the period of their initial adoption in the United States (ca. 1942-50), little if any attention was given to their environmental impact. In fact, it has been only during the last 15 years that any appreciable consideration has been given to the relationship of synthetic detergents to the environment. This recent concern has resulted in modifications both in the detergent formulation and in wastewater-treatment practice. The last few years, for example, have seen the change from biodegradable surface active agents, legal restraints in several states on the maximum percentage of phosphatic builders, and phosphate removal standards imposed upon wastewater effluent discharges within certain drainage basins.

Extensive research and development on the part of the detergent manufacturers has lead to the manufacture of specific, patented detergent formulations which have been developed to provide optimum cleaning efficiency. Since it is technically feasible and currently demanded by the consumer literally to tailor detergent formulations to the type of soil and water anticipated, the suitability of possibly precluding the consumer now from achieving maximum detergent utilization because of environmental considerations poses complex considerations.

The imposition of usage or formulation constraints on products such as synthetic detergents because of their potential environmental impact places the utmost importance on achieving an accurate assessment of their environmental impact. This consideration has now reached such importance in the United States that the manufacturer currently must consider it with the same degree of regard as the other aspects of business, e.g., market analysis, technical competence, economic aspects, etc.

In order to make an assessment of the impact of synthetic detergents on the environment, an accurate determination of each of the following considerations is paramount:

1. Specification of the quality of the environment desired and elucidation of the fundamental biological, chemical, and biogeochemical interrelationships inherent to the desired environmental quality
2. Identification of the potential pollutionary constituents in synthetic detergents and determination of how they would affect the desired environmental quality if discharged
3. Evaluation of the current and future technology of wastewater treatment relative to the potential pollutionary elements

It is difficult, however, to obtain much more than an instantaneous picture of any of the above elements as each one can change both with time and geographical location. Such an analysis can best be performed by the use of computer simulation and modeling techniques.

Figure 1 depicts the normal flow pattern of a synthetic detergent from its manufacture through to its ultimate discharge to the environment and, as such, serves as the basic model for obtaining an assessment of the impact of synthetic detergents on the environment. The interrelationships of each of the considerations mentioned above is clearly illustrated in this figure. The dynamic nature of the system is apparent immediately when one considers that at any given time detergent formulations are being improved, consumer demands are changing, the state of the art of wastewater treatment is improving, and the desired level of environmental quality is changing. The tendency of this system is to reach a steady-state condition wherein the contributions characteristic to each of the first three components of Fig. 1 (detergent manufacture, consumer demands, and wastewater treatment capability) for that instant are maintained in balance with the instantaneous level of environmental quality desired (or tolerable).

As indicated in Fig. 1, the final decision as to the acceptable or tolerable level of environmental quality will be established primarily by the citizens themselves (i.e., the consumers) after considering both the current characteristics of synthetic detergents and the state of the art of wastewater collection and treatment. In establishing or seeking an acceptable level, typical considerations would be to determine how great a financial sacrifice

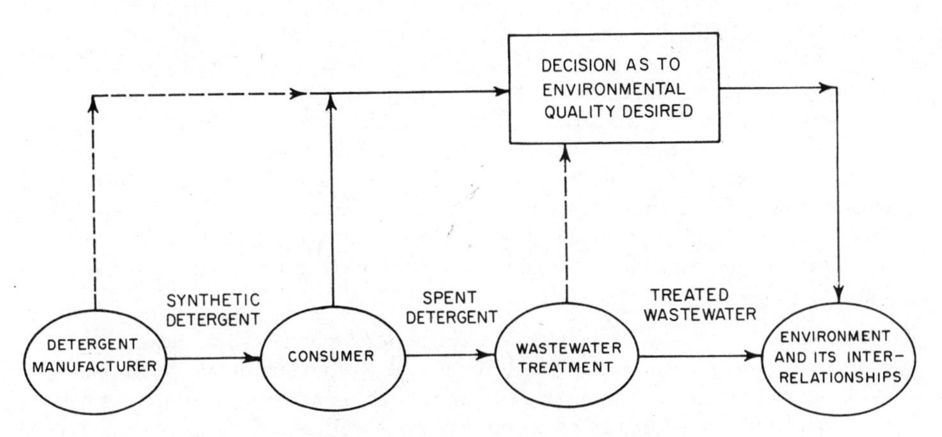

FIG. 1. Summary flow diagram for a synthetic detergent from its manufacture to its ultimate discharge into the environment.

must be made in order to obtain an environment relatively safe with respect to public health and, perhaps, to obtain an environment satisfactory for public use and enjoyment. Economic sacrifices in order to obtain benefits much in excess of these would probably not be acceptable at this time. On the other hand, the sacrifice of eliminating the use of detergents in order to ensure a pristine environment clearly would be equally unacceptable to the populace as a whole.

A more rational approach currently would be to anticipate that the model of Fig. 1 could be controlled in a steady-state condition wherein consumers would, no doubt, desire to maintain their demand for the efficient synthetic detergent formulations; at the same time, they would be willing to bear at least the cost of partial wastewater collection and treatment. Since they would expect to discharge some pollution to the environment, ultimately they would be accepting a somewhat reduced level of overall environmental quality. The exact relationships within this model would vary with both time and geographical location and thus no definitive statements could be developed which would be applicable to all cases.

Historically, however, environmental problems have not been solved in this manner and various short-term solutions have been presented. For example, in a 1971 hearing before the Federal Trade Commission relative to the potential pollutionary impact of phosphate in synthetic detergents, Okun [1] cited statistics indicating that 30% of the country's population released domestic wastes directly as ground discharges in septic tanks or similar devices and that 55% of the country's population reside in cities whose municipal wastewaters are discharged into the oceans or major river systems that flow to the oceans. Only the remaining 15% of the population thus discharged their wastewater into freshwater systems capable of eutrophication (see Sec. II. C. 2 for a detailed discussion of eutrophication). The implication of this observation would appear to be that if only 15% of the nation's population contribute to the eutrophication of freshwater bodies by detergent phosphate fertilization, why should the unaffected 85% be forced to accept possibly inferior detergent performance as a result of having the phosphate builder removed? The City of Chicago in 1972 banned the sale of detergents containing phosphate builders, even though they would fall in the 85% of the population which release their wastewaters to noneutrophiable waters.

On the other hand, if an analysis of the pertinent technical information were to be made and combined with the specific environmental conditions existing at a given time and location, an accurate assessment of the alternative steady-state conditions could be obtained intelligently by use of a viable model such as that depicted in Fig. 1. Subsequent ranking of the various alternatives in terms of the desired priorities ultimately will provide the means for the establishment of the requisite decision. In this regard, the remainder of this chapter describes those significant factors for the effective evaluation of the impact of synthetic detergents on the environment.

II. ENVIRONMENTAL ASPECTS

A. Ecological Considerations

The natural life of the biosphere exists in land, air, and water. Each of these physical phases is interrelated to the other, so that an examination of one phase must not be considered as an isolated system, but rather as part of the whole biosphere. Each physical phase receives physical and chemical inputs to which its living and nonliving forms must respond by absorbing or expending energy. Each physical phase exists in a dynamic state, always receiving inputs and giving off outputs to the other physical phases of the ecosystem. If the dynamic forces upon a particular phase are in balance with its response, then a steady-state condition exists and there is no strain on any particular phase or combination of phases. If an excessive pollution load or a continuously increasing pollution load is exerted on the system, the strain upon the natural balance of the system will cause an abnormal response and the natural steady-state condition of the system will no longer exist.

Because of their dynamic natures, each phase of the biosphere has an inherent absorptive capacity. The maintenance of some elements in the biosphere, in fact, is necessary to sustain the natural chemical and biological cycles. The most significant consideration when evaluating potential pollutionary releases into the environment is to ensure that none of the phases of the biosphere becomes polluted beyond its absorptive capacity. In this regard, it is necessary to establish accurately those steady-state levels appropriate to a given environmental quality and minimize or ideally eliminate any pollutionary inputs which would cause an imbalance from the desired level. The assessment of the environmental impact of synthetic detergents, as with any product of technological developments, would be liable to this consideration.

Since they are utilized in water, synthetic detergents are generally discharged into wastewater collection systems and, as such, their main pollutionary impact would be to the nation's waters. Because the degree of treatment of detergent-containing wastewaters determines the extent of their impact on the environment, both consumer and manufacturer must accept the fact that the greater the degree of wastewater treatment, the smaller the impact the discharge water will have on the receiving water. The environmental impact of synthetic detergents, however, is not limited strictly to waters. For example, the greater the degree of wastewater treatment, the greater generally will be the quantity of waste sludge which ultimately will require disposal into the air or onto the land. Because environmental effects of residual sludge disposal cannot be eliminated (see Sec. IV.E), wastewater treatment alone cannot be considered the ultimate solution for preventing undesirable impacts on the environment. Also, it should be noted that in those areas where no sewage collection system exists, synthetic detergents may cause pollutional inputs to the land (e.g., from septic systems, direct discharges, etc.).

B. Hydrological Considerations

The total volume of water associated with our planet is fixed and exists in various phases and locations which are collectively referred to as the hydrological or water cycle. The world's supply of freshwater, for example, is obtained almost entirely as precipitation resulting from the evaporation of seawater. The major stages of the water cycle are: precipitation, percolation, runoff, and evaporation. A schematic diagram of the water

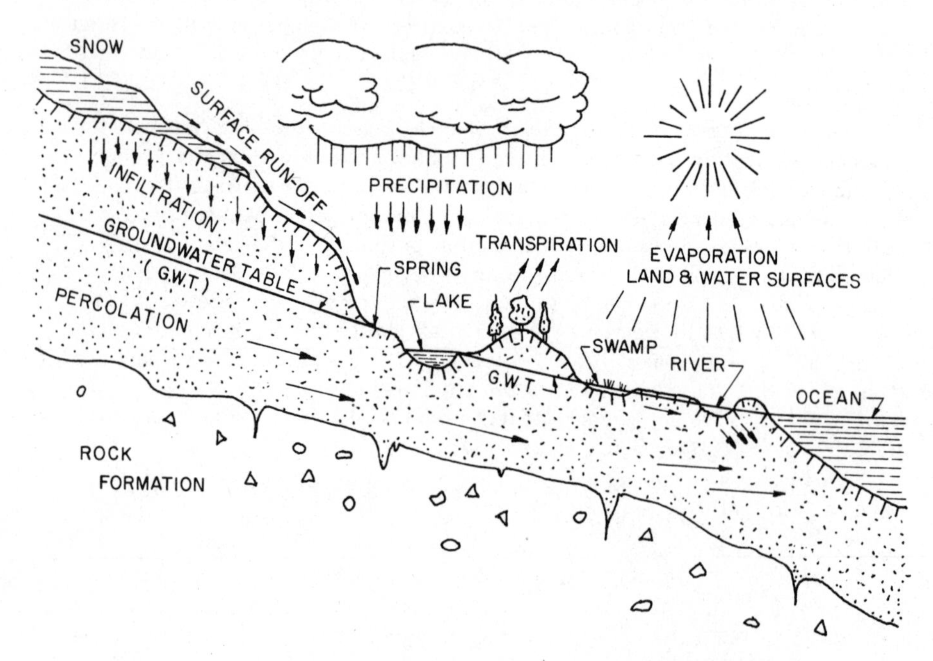

FIG. 2. Schematic diagram of the water cycle. From Ref. 2.

cycle is shown in Fig. 2. As can be observed from this figure, the major fraction of the water which falls to the earth as precipitation falls directly upon water surfaces and is returned directly to the atmosphere by evaporation. Of the minor fraction which falls onto land masses, part is lost to the atmosphere by evaporation and transpiration from vegetation, part flows overland to receiving waters or surface runoff, and part enters the soil. The rainwater infiltrating the soil flows downward under the influence of gravity until it reaches the groundwater table to join the subterranean reservoir within the earth's crust. Most of the groundwater eventually discharges at the surface of the earth through springs and seepage outcrops or it passes, at or below the water level into streams and standing bodies of water, including the oceans.

Synthetic detergents can be discharged to any of the surface or groundwater locations within to the water cycle and, as such, can represent pollution input to that component. The extent of pollution in each case will depend on such factors as: the extent of treatment which the wastewater receives prior to its discharge; the composition and concentration of the pollutants remaining in the wastewater; and the characteristics of the water body receiving the pollution.

In assessing the potential impact of synthetic detergents to surface waters, one of the initial considerations must be to determine the relative quantity of water in the various categories of the hydrological cycle. Table 1, for example, indicates the water budget of the world. As indicated in this table, the total amount of water on this planet amounts to 326,000 cubic miles, but only a dramatically small fraction is freshwater. Approximately 97.3% is in the oceans of the world and approximately 2.1% is in ice caps and glaciers; only about 0.01% is in freshwater lakes and rivers and only 0.61% in groundwater.

The distribution of water in the conterminous U.S. is shown in Table 2. Of particular significance are the relative volumes of water in freshwater lakes, stream channels, and groundwater compared to the total volumes of water in oceans shown in Table 1. The ultimate pollution potential of synthetic detergents will depend to a large degree on the nature of the receiving water.

The world's oceans possess the greatest capacity to dilute and, therefore, absorb pollution impacts. Moving rivers and streams possess a moderate capacity to absorb pollutional effects and yet maintain an acceptable level of water quality. However, standing streams and freshwater lakes have demonstrated the least capacity to absorb pollution impacts without diminishing the quality of water for aesthetic or usage purposes. Yet the latter two more fragile categories comprise the entire freshwater resources of the nation.

TABLE 1

World Water Supply and Budget

Water item	Volume (thousands)		Total water (%)
	Cubic miles	Cubic kilometers	
Water in land areas			
Freshwater lakes	30	125	0.009
Saline lakes and inland seas	25	104	0.008
Rivers	0.3	1.25	0.0001
Soil moisture and vadose water	16	67	0.005
Groundwater to depth of 4,000 m (~ 13,100 ft)	2,000	8,350	0.61
Icecaps and glaciers	7,000	29,200	2.14
Total in land area	9,100	37,800	2.80
Atmosphere	3.1	13	0.001
World ocean	317,000	1,320,000	97.3
Total	326,000	1,360,000	100
Annual evaporation			
From world ocean	85	350	0.026
From land areas	17	70	0.005
Total	102	420	0.031
Annual precipitation			
On world ocean	78	320	0.024
On land areas	24	100	0.007
Total	102	420	0.031
Annual runoff to oceans from rivers and icecaps	9	38	0.003
Groundwater outflow to oceans	0.4	1.6	0.0001
Total	9.4	39.6	0.0031

Source: From Ref. 3, p. 62.

TABLE 2

Distribution of Water in the Conterminous United States

Water category	Area (square miles)	Volume (cubic miles)	Annual Circulation (million acre-feet per year)	Detention period (years)
Frozen water				
Glaciers	200	16	1.3	40
Ground ice		(seasonal only)		
Liquid water				
Freshwater lakes[a]	61,000	4,500	150	100
Salt lakes	2,600	14	4.6	10
Average in stream channels	—	12	1,500	0.03
Groundwater				
Shallow	3,000,000	15,000	250	200
Deep	3,000,000	15,000	5	10,000
Soil moisture (3-ft root zone)	3,000,000	150	2,500	0.2
Gaseous water				
Atmosphere	3,000,000	45	5,000	0.03

Source: From Ref. 3, p. 61.

[a]United States part of Great Lakes only.

C. Pollution of the Aqueous Environment

Because so many different items can represent pollution, from discrete elements to heterogeneous mixtures of complex organic material, the measurement of pollution must be capable of adequately measuring or representing each situation. One of the more common classifications of pollutants is by the use of physical, chemical, and biological categories. Table 3 illustrates such a breakdown with some of the more significant pollutant considerations shown for each category. Typically, when one wishes to determine the pollutional strength of a wastewater, analytical determinations are performed for items such as those shown in Table 3.

TABLE 3

Common Measurements of Water Quality[a, b]

A. Physical Measurements
   1. Temperature
   2. Color
   3. Turbidity
   4. Suspended materials

B. Chemical Measurements
   1. Organic
      a. Biochemical oxygen demand
      b. Chemical oxygen demand
      c. Total organic carbon
   2. Inorganic
      a. Fertilizing elements
         i. Nitrogenous
         ii. Phosphatic

C. Biological Measurements
   1. Coliform
   2. Fecal coliform
   3. Standard plate count

[a]See Glossary for definitions of the measurements outlined in this table.

[b]The listing is far from complete. "Standard Methods for the Examination of Water and Wastewater" [59] should be consulted for a more detailed listing and description of the available analytical techniques.

If one wishes to determine the impact of pollution on the environment in a broad sense, one can regroup all pollutants into one of two categories: (1) organic pollutants which contain hydrocarbon fragments as a major portion of their composition, and (2) inorganic pollutants or those compounds which do not have hydrocarbon fragments (e.g., dissolved minerals and salts which would exist in crystalline form in their pure states).

This broad distinction is convenient from an environmental perspective because there are two types of indigenous organisms in nature, one group which is stimulated primarily by organic materials (i.e., animals), and another group which is stimulated primarily by inorganic constituents (i.e., plants). In a biological sense, this distinction between plants and animals exists down to the simplest unicellular level. At the unicellar level, algae represent the plant kingdom and bacteria represent the animal kingdom. It is at this level of the ecosystem that any undesirable pollution has its most direct and significant impact. The distinction between animal and plant activity throughout the ecosystem is known as heterotrophic metabolism and autotrophic metabolism, respectively.

Heterotrophic organisms require organic material as an energy source for growth, whereas autotrophic organisms primarily utilize solar energy for growth. In the case of heterotrophic activity, complex organic forms of the elemental components (i.e., mainly carbon, nitrogen, phosphorus, and sulfur, as well as any inorganic trace elements) are utilized by the microorganisms to gain energy in order to synthesize new cellular protoplasm. In general, these microorganisms respire; that is, they utilize oxygen in their metabolism. The overall biochemical process is shown in Fig. 3.

In contrast, autotrophic activity forms the basis for the production of new organic material. In this case, inorganic material (carbon dioxide, nitrogen, phosphorus), is utilized by autotrophic microorganisms via the photosynthetic process in which solar energy is captured. This reaction is presented in Fig. 4.

The autotrophic process is important in many ways, for not only does it harness solar energy but also it helps to remove the products of combustion and respiration, i.e., carbon dioxide and water. Also, in turn, it renews the oxygen supply to those organisms which respire and provide new organic matter for the whole food chain of respiring organisms, including man.

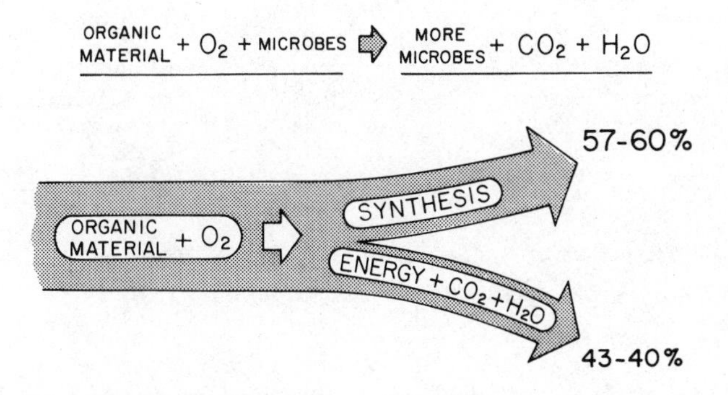

FIG. 3. Schematic representation of heterotrophic microbial activity under aerobic conditions [Eq. 1].

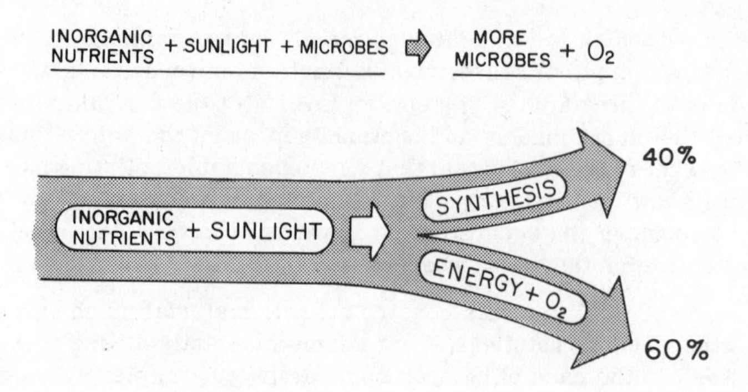

FIG. 4. Schematic representation of autotrophic microbial activity [Eq. 2].

The autotrophic and heterotrophic systems coexist with each other in a closed cycle or feedback loop. The end products from one type of metabolism provide input to the other type of metabolism, and the flow from one position within the loop to another is controlled by the quantity of material in the storage locations and the environmental conditions described in Fig. 5.

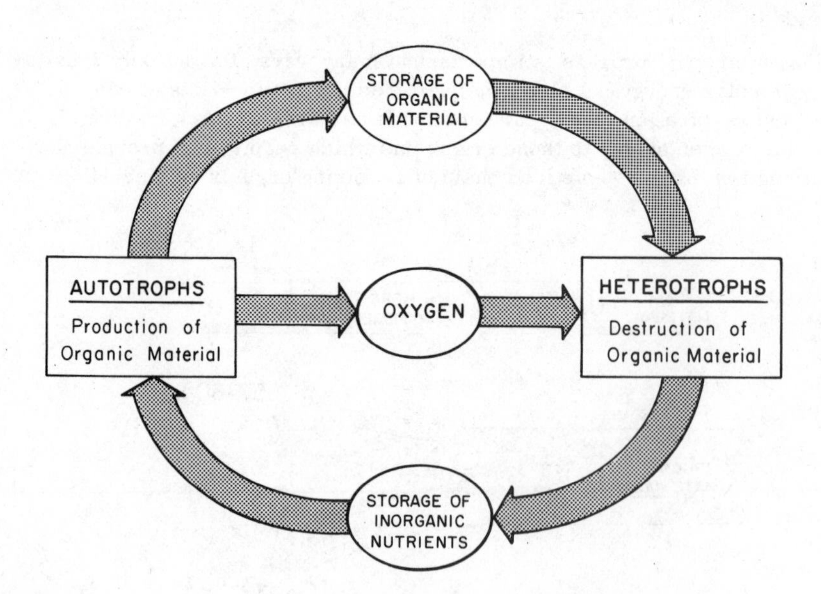

FIG. 5. Heterotrophic/autotrophic balance between microorganisms.

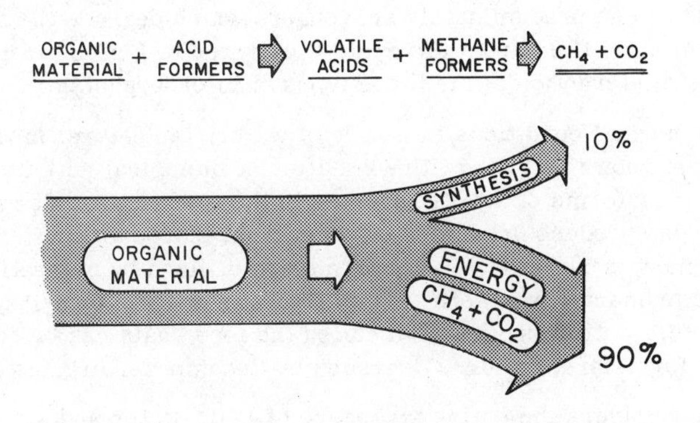

FIG. 6. Schematic representation of heterotrophic microbial activity under anaerobic conditions [Eq. 3].

In the integrated ecological system of the biosphere a steady-state between production and destruction of organic material as well as between production and consumption of oxygen seems to be maintained, thus providing a constant surplus of oxygen in the atmosphere. In a smaller ecological system, however, the balance between production and destruction may become disturbed easily. The introduction of either excessive organic or excessive inorganic nutrients can cause pollutional effects, i. e., an imbalance between heterotrophic and autotrophic functions. If excessive organic materials are introduced into the system, for example, heterotrophic activity will be in excess of photosynthetic oxygen release and oxygen depletion will result. If, on the other hand, excessive inorganic nutrients are introduced into this system, autotrophic production rates will become larger than the rates of destruction and mass blooms of algae will occur. These two phenomenona represent the two primary effects of pollutional discharges to the environment and form the basis for evaluating the impact or disturbance due to pollution.

A distinction must be made in the case of heterotrophic activity when oxygen is absent. A specific class of heterotrophic organism exists which utilizes organic matter and produces simple organic gases as end products. These organisms are called decomposers and their metabolic processes are illustrated in Fig. 6.

These microorganisms coexist with the autotrophs and aerobic heterotrophs. They are predominately scavengers which degrade the spent organic fauna and flora from the cycle illustrated in Fig. 5. Figure 7 indicates the interaction of the three types of microorganism.

Under normal conditions in a body of water, the natural input to such a cycle as that shown in Fig. 7 allows sufficient biological activity to provide food for higher forms of aquatic species in the food chain. The autotrophs, since they can produce organic life forms from inorganic nutrients, are called primary producers. These forms are utilized by heterotrophic cells serving as primary consumers. The latter are utilized by secondary consumers, etc., until a progression called the food chain can be traced. These life forms are, in turn, degraded by decomposer organisms.

If one considers the entire biosphere of land, water and air phases, this food chain must be described as a three-dimensional food web. Each ecological niche is occupied by many species which are competitive with each other and dependent upon the next lower niche for survival. The food web is delicately balanced from the highest niche of humans down to the primary producer or autotrophic microorganisms. The food web thus, for all practical purposes, can be considered as a pyramid with each level dependent upon the next lower level for sustenance.

Any adverse effect upon the lower levels of the pyramid, where pollution has its most direct effect, will have a rippling effect upwards to each higher level. An appreciable loss of energy occurs, however, at each

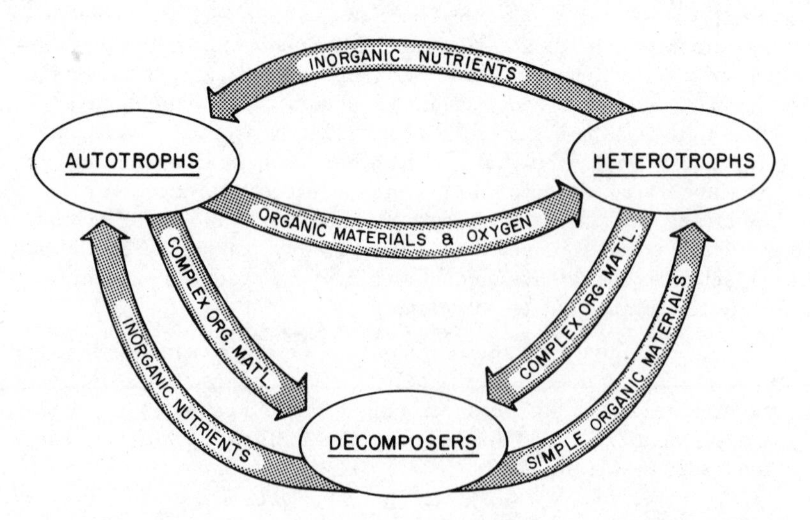

FIG. 7. Interrelationship between heterotrophic/autotrophic activity and decomposition.

trophic level of energy transformation, due to the fact that each biological energy conversion is inherently extremely inefficient. Therefore, the adverse effects, felt most dramatically at the lower trophic levels, will be dampened successively at each higher level. By the time these adverse effects reach the human niche, their magnitude might be so diminished that humans could ill-advisedly ignore them as being short-term developments or rationalize them away as being unimportant.

At the lowest unicellular level, the chemical composition of microbial cells can be considered to be approximately $C_{106}H_{180}O_{45}N_{16}P$. Other elements are also present, but only in trace amounts. The cells utilize various chemical compounds or substrates in this ratio as they synthesize new cellular protoplasm of this exact chemical formulation, thus producing identical new daughter cells. If the requisite chemical elements are not present in this ratio, one or more elements will be in excess and one or more elements will be present at a minimum concentration. The latter elements are considered as the limiting substrates and their concentration will determine the amount of microbial growth in a given system. If either the inorganic or the organic materials, for some reason, become depleted to a level which cannot sustain either the autotrophic or the heterotrophic organisms, then the population of one or both diminishes to a level which the loop can sustain continuously. Conversely, large pollutionary inputs to this cycle from human activities may stimulate an inordinately large proportion of one type of biological activity with respect to that of other types and adversely affect or completely destroy previously established ecological balances. Two primary environmental effects are observed normally in the aquatic environment as a result of pollutionary discharges: (1) oxygen depletion due to organic pollutants, and (2) fertilization by inorganic nutritional pollutants.

1. Oxygen Depletion Due to Organic Pollutants (Oxygen Sag Relationships)

Introduction of organic material into a receiving water stimulates heterotrophic biological activity and thus results in a depletion of the dissolved oxygen concentration of the water (see Fig. 3). Classically, the strength of organic pollutants consequently has been evaluated in terms of the potential amount of oxygen that would be required to convert biologically the organic material to carbon dioxide and water—biochemical oxygen demand (BOD)—and the impact of organic wastes or flowing waters has been determined historically by a consideration of dissolved oxygen relations in the stream.

The dissolved oxygen relationships inherent to the development of one of the classical pollutional expressions (i.e., the oxygen sag curve) and its mathematical formulation are shown in Fig. 8. As described above, the introduction of organic material results in the lowering of the dissolved

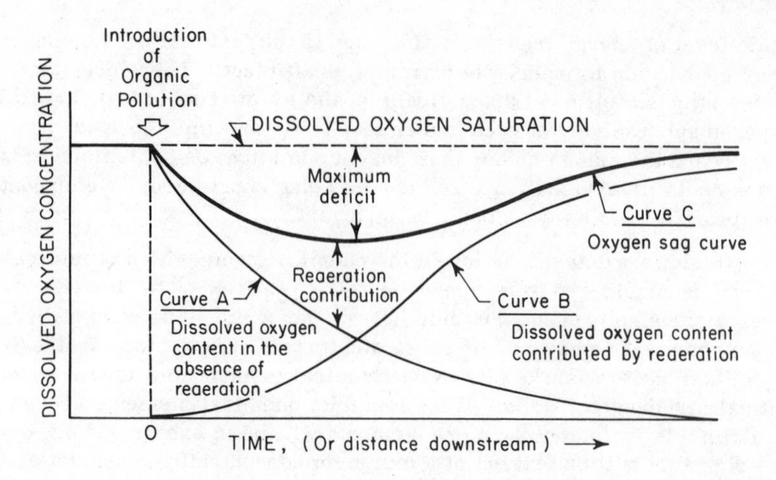

FIG. 8. Oxygen sag curve.

oxygen concentration in the water and results in an oxygen depletion curve as shown by curve A. This deficit of oxygen (i. e., the amount by which the actual dissolved oxygen concentration is below the saturation concentration) in turn initiates a reaeration reaction at the air-water interface of the flowing water; the greater the value of this deficit, the greater will be the rate of reaeration as described by the oxygen reaeration curve (curve B).

Summation of these two relationships as the water flows downstream from the point of organic discharge results in the oxygen sag curve depicted in curve C.

The mathematical formulation of this concept was developed in 1925 by Streeter and Phelps [4]. Their development was based on a combination and integration of the following two relationships:

1. Oxygen Depletion. The rate of change of organic material (L) (expressed in equivalent BOD oxygen units) with respect to time (t) was assumed to be equal to a constant of proportionality (K_1, the BOD constant) times the instaneous concentration of organic material, or

$$-\frac{dL}{dt} = K_1 L \tag{4}$$

2. Reaeration. The rate of reaeration (dD/dt) was assumed to be equal to a constant of proportionality (K_2, the reaeration

coefficient) times the dissolved oxygen deficit, D (i.e., D.O.$_{saturated}$ - D.O.$_{actual}$), or

$$\frac{dD}{dt} = K_2 D \qquad (5)$$

which yielded the following expression for the dissolved oxygen deficit (D) at time (t) (or distance) downstream from the source of pollution:

$$D = \frac{K_1 L_a}{K_2 - K_1} \left[10^{-K_1 t} - 10^{-K_2 t} \right] + D_a \, 10^{-K_2 t} \qquad (6)$$

with terms as defined above (L_a is the amount of organic pollution as BOD introduced at t = 0 and D_a is the dissolved oxygen deficit present at t = 0.

Equation 4 has been used to predict the depletion of dissolved oxygen caused by a measured quantity of organic pollution when released to a flowing water and for many years it served as the major yardstick for estimating aquatic pollutional impacts. Inspection of Eq. (6) indicates that the time (or distance) downstream where the greatest impact of a pollutional discharge will be felt (in terms of the dissolved oxygen deficit) will depend on the initial conditions of the river, the amount of organic material discharged (L_a), and the rate constants K_1 and K_2. Figure 9 demonstrates the fact that these conditions determine whether a reach of the river will become completely void of dissolved oxygen (i.e., anaerobic) or only partially deficient in dissolved oxygen. This figure also shows that regardless of the amount of oxygen deficiency, given sufficient time, and if no additional organic pollution is discharged, a zone of recovery will be observed in the river and the dissolved oxygen concentration ultimately will approach the saturation concentration once again.

In an evaluation of this nature, one of the most critical considerations is to determine the effect of this reduced dissolved oxygen level on the aquatic life. Fish are generally the most highly valued aquatic communities considered in this regard, with their species predomination depending directly on the available concentrations of dissolved oxygen. Fish populations characteristically will change from game fish to rough fish to undesireable species to ultimate extinction, as the dissolved oxygen level is reduced from saturation to zero. The Streeter-Phelps equation, then, can be used to calculate the amount of organic pollution which can be released into the environment or the amount of organic pollution which must first be

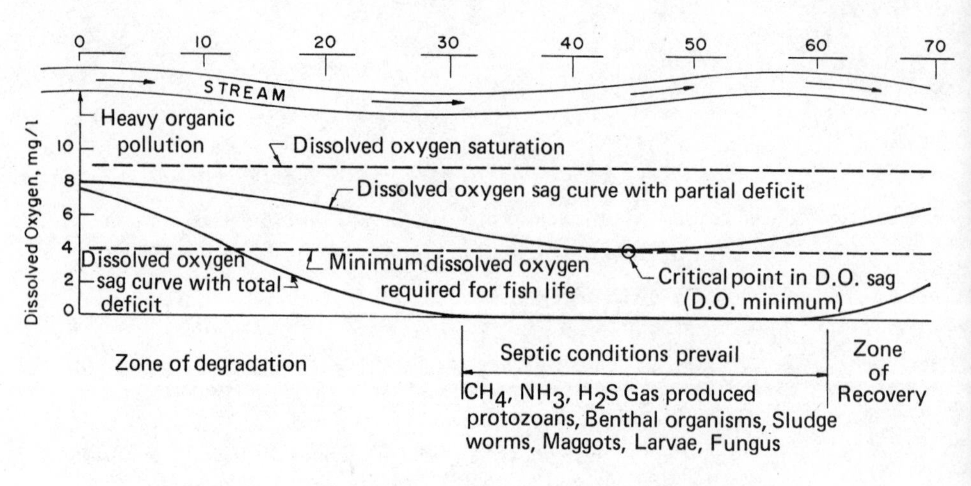

FIG. 9. Various oxygen sag curves and description.

removed from the discharge before it is released into the environment if a given dissolved oxygen concentration (or aquatic population) must be maintained.

The oxygen sag equation has many limitations (e.g., it is applicable only to channels of uniform cross section, it does not account for photosynthetic activity, the impact of bottom muds are not considered, the formulation for oxygen depletion is incomplete, etc.). Numerous refinements and modifications of this equation, consequently, have been developed since its initial publication in 1925, but the equation nevertheless stands as one of the first examples of an attempt to assess accurately the impact of a given pollutant on a receiving water. The refinement of such analyses, in terms of mathematical modeling concepts, is of importance in accurately assessing the impact of pollution. Such a consideration will serve not only to assess accurately impacts but also to enhance greatly the intelligent management of the environment. It is axiomatic that equitable decisions related to environmental management can be made only by utilizing such analytical techniques; in making decisions of this magnitude one must employ methods which will provide for a valid evaluation of all the pertinent factors and their interrelationships. Any assessment as to the effect of synthetic detergents on the environment in a given locality must be performed in a similar fashion if it is going to be accurate.

2. Fertilization by Inorganic Nutritional Pollutants (Eutrophication)

Introduction of inorganic nutritional pollutants (principally nitrogenous and phosphatic compounds) into a receiving water stimulates autotrophic biological activity as shown in Fig. 4 and thus results in an increase in algal growth. Commonly, these population increases are quite visible and are referred to as algal blooms. The blooms begin in late spring and continue through summer; they extend as deeply into the water as the light necessary for photosynthetic activity can penetrate. The algae reverse their metabolic role at night when they cannot photosynthesize and they respire in accordance with Eq. (1). Hence the dissolved oxygen concentration in a water body can vary markedly throughout the 24-hr day, particularly in highly productive waters. A typical diurnal variation in dissolved oxygen under these conditions is shown in Fig. 10.

The introduction of inorganic nutritional pollutants into bodies of water with long hydraulic detention times, such as lakes, reservoirs, impoundments, etc., can have startling effects—particularly if these water bodies are deep enough to stratify. Stratification occurs in those water bodies whose depths are sufficient to preclude the energy of the wind from mixing the entire contents of the body of water. Under these conditions the lake will stratify during the summer months in a temperate climate and will result in the warm, lightweight water rising to the surface and the cold dense water sinking to the bottom. The top warm waters will tend to be reasonably uniform in temperature due to the mixing action of the wind, but below this depth a sharp decrease in temperature occurs. This zone is known as the thermocline and the water temperature changes approximately

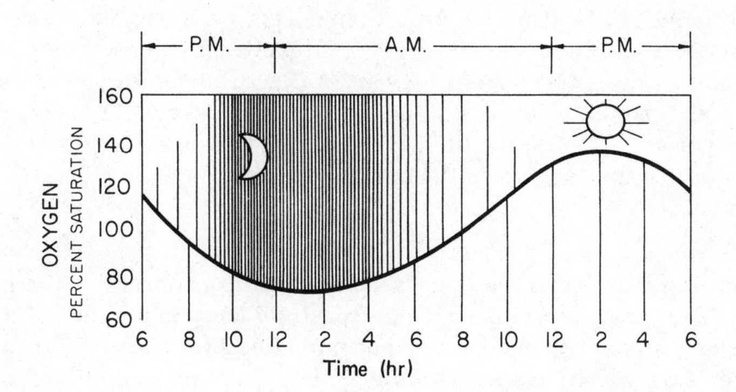

FIG. 10. Diurnal variations in dissolved oxygen because of photosynthetic activity.

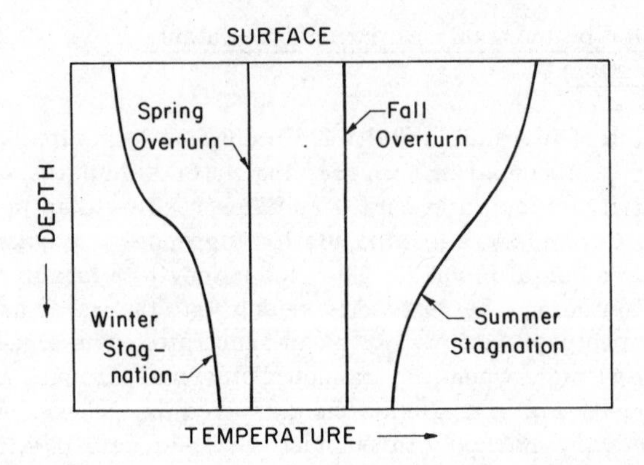

FIG. 11. Seasonal variations in temperature profiles in lakes in which stratification occurs.

1° C per meter of depth until the bottom water layer is reached. No mixing of waters occurs through the thermocline; under these conditions the zone of water above the thermocline is referred to as the epilimnion and the zone of water below the thermocline is referred to as the hypolimnion. A typical summer temperature profile, for conditions as described above, is shown in Fig. 11. In lakes which are stratified, there are only two time periods when the top and bottom waters mix and these are termed the fall and spring overturns. The fall overturn, for example, occurs as the prevailing atmospheric temperatures begin to drop, resulting in a cooling effect on the surface water layers; as the water cools it becomes denser and sinks, serving to displace warmer and less dense water to the surface where it is cooled. Figure 12 illustrates this process. This cycle continues until the water reaches 4° C, when water has its greatest density; progressive cooling reduces the water temperature toward 0°C and, since water becomes less dense and lighter below 4° C, the lake stratifies again. The inverted winter temperature profile for this condition is also shown in Fig. 11. A similar lake overturn occurs in the spring when the ice melts and the water warms and sinks as the water once again passes through its point of greatest density at 4° C.

During summer stratification, tremendous changes occur in the lake as a result of pollution. The algal blooms resulting from inorganic nutrient discharges cause the upper layers to become alternately supersaturated with oxygen during daily photosynthetic periods and totally devoid of oxygen during the night. The only oxygen replenishment available at night is through diffusion from the atmosphere at the air-water interface. The diffusional forces operating within the aqueous phase tend to distribute uniformly the resulting dissolved oxygen from the atmosphere downward

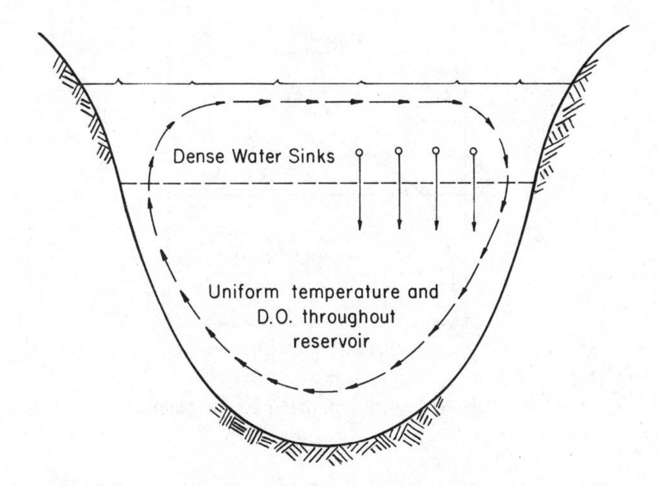

FIG. 12. Diagram of lake turnover relationships.

throughout the water mass. The respiring microbial species within the upper layers, however, utilize the dissolved oxygen from the atmosphere as fast as it can diffuse into the aqueous phase. During the daytime, a considerable amount of the profusion of oxygen produced by the photosynthetic microorganisms is utilized by the symbiotic respiring bacteria and the remainder diffuses downward as far as it can be transported.

During periods of heightened metabolic activity, the top stratification layer absorbs all the solar energy and contains any dissolved atmospheric or metabolically produced oxygen. Because the presence of the thermocline prevents mixing of the top and bottom layers, no replenishment of oxygen can reach the hypolimnetic waters and respiring organisms in this region will reduce the dissolved oxygen concentration or completely exhaust it—depending on the amount of organic pollution present. A dissolved oxygen profile typical of a highly polluted lake is shown in Fig. 13.

Throughout the periods of heightened biological activity in the summer, approximately 4% of the autotrophic microorganisms (algae) and the heterotrophic microorganisms (bacteria) produced in the epilimnetic waters die and settle through the thermocline each day to the bottom muds where they serve as substrate for the bottom heterotrophic decomposer organisms. Again, the degree of pollution present and resulting oxygen depletion will dictate whether this will be an aerobic or anaerobic reaction. Under polluted conditions, the bottom water generally becomes devoid of oxygen rather quickly and the anaerobic decomposition reaction of Fig. 6 generally occurs.

A cycle between the autotrophic and aerobic heterotrophic activity in the epilimnetic waters and the anaerobic decomposing microorganisms in the

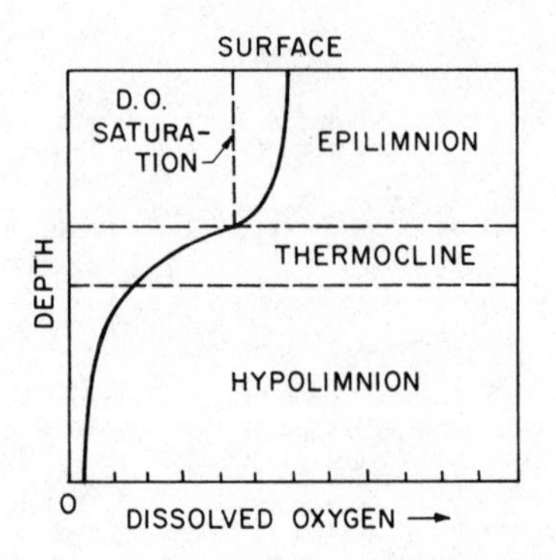

FIG. 13. Typical dissolved oxygen profile for a stratified eutrophic lake.

hypolimnion is thus established in the lake identical to that described in Fig. 7. Figure 14 shows a summary of these relationships and their effect on water quality.

The cycling of pollutants between the water and the biological species in a closed system is shown in more detail in Fig. 15. This cycling of pollutants is known as a biogeochemical cycle and is particularly significant in the case of a nutritional element such as phosphorus. As can be seen, such an element is conserved in the cycle and does not leave the system, except for that concentration which is being slowly washed out as a result of hydrological inflows and outflows.

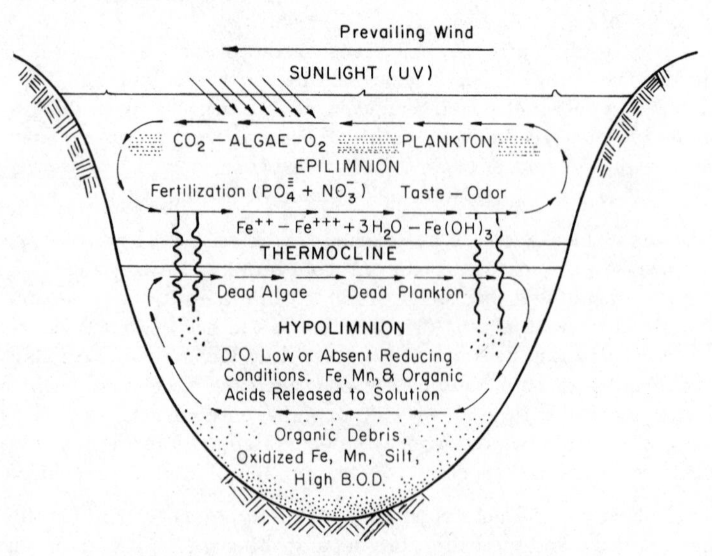

FIG. 14. Microbial and chemical activities characteristic of a stratified eutrophic lake and their effect on water quality.

During the winter stratification period, the bottom hypolimnetic waters seldom become devoid of oxygen. Therefore, any higher forms of aquatic life, such as fish, must exist in the epilimnion during the summer and the hypolimnion during the winter. Anaerobic decomposition of the previous summer's crop of dead microorganisms also occurs, but the rate of decomposition is decreased at lower temperatures. If pollution inputs to a water body are excessively large, rarely can a year's accumulation of biologically produced organic matter be decomposed by the decomposer microorganisms. The majority of residual organic matter accumulates in the bottom sediments, although some may be removed from the top layer of sediment during the turbulence of the succeeding spring overturn. The surplus dead organic material and inorganic metabolic products (i.e., phosphorus and nitrogen) are recirculated to the top epilimnetic waters during the overturn and are then present to begin again the annual symbiotic blooms of algae and bacteria.

This cycle continues year after year and its magnitude will be increased proportionately with increasing nutritional inputs to the body of water. Unfortunately, its magnitude will never be decreased unless the nutritional materials are physically removed from the cycle by some treatment process.

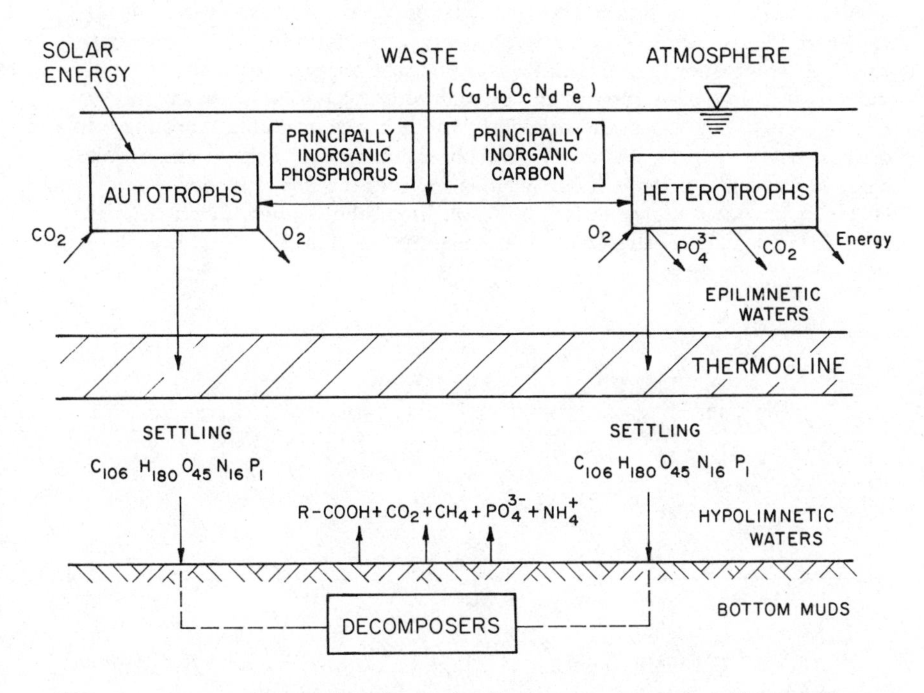

FIG. 15. Biogeochemical cycling of pollutionary elements within a lake.

In order to interrupt the biogeochemical cycle, such treatment schemes as water replacement with nutrient poor water, removal or covering of the bottom muds, harvesting of aquatic plants, etc., have been proposed. If the polluted water body is left untreated, however, the water quality deteriorates, the biological activity increases, the bottom muds become deeper and deeper, and the lake ultimately becomes a swamp and is completely lost as a resource. This process of lake aging is a natural one and is known as eutrophication. This word stems from the Greek word eutrophos meaning "well nourished" and refers to the increased productivity which occurs in lakes as they become richer in nutritional species. The three characteristics of water quality, biological activity, and depth of bottom muds can be used to classify a lake as to its degree of eutrophication. Extinction proceeds through three limnological classifications: oligotrophic (or youthful, with low nutrient content, little biological activity, and a clean bottom); mesotrophic (or middle aged, with moderate amounts of the three parameters); and eutrophic (or aged, with large amounts of nutrients, excessive biological activity and increasing volumes of organic sediments on the bottom). Figure 16 shows these stages and the magnitude of their corresponding biological cycles.

Although eutrophication is a natural process in nature, it takes tens of thousands of years to accomplish. Human activities, however, have so served to accelerate this process that lakes currently have been observed to become eutrophic in a matter of a few years due to the addition of nutrients. A graph showing the relationship between natural eutrophication and the acceleration of eutrophication by man's activities is shown in Fig. 17. In this figure, the extent of eutrophication is measured as the productivity of the surface. If the impact of eutrophication on the vertical axis is considered in terms of the number of uses for a water body, then when the lake is very young and unproductive the water would be of such high quality that it could be used directly for human consumption; as the lake grows older or

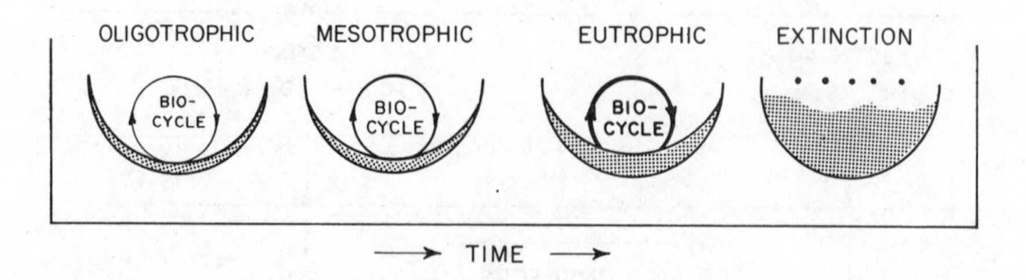

FIG. 16. Natural transition of a lake through various stages of productivity eventually resulting in extinction. From Ref. 5.

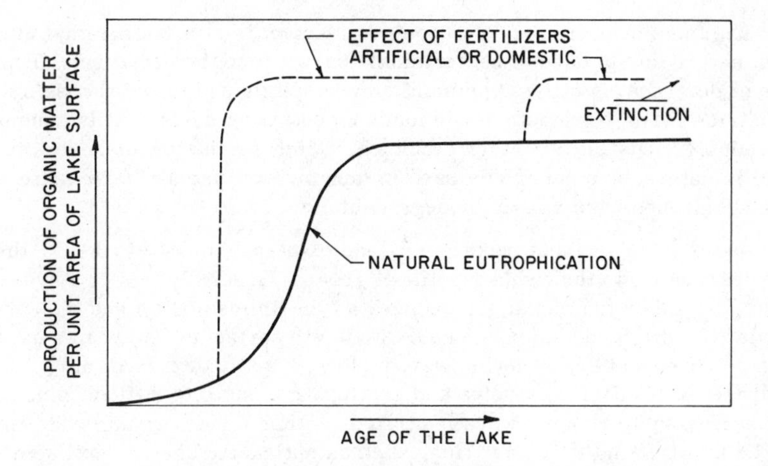

FIG. 17. The relationship of natural lake eutrophication and productivity and the impact of human activities on eutrophication.

more eutrophic, however, the water quality deteriorates so that first, one will no longer be able to drink it, or to swim in it, and then the fishing resource will be lost, until ultimately all the uses will be gone and the lake will be lost.

III. POLLUTIONARY CHARACTERISTICS OF SYNTHETIC DETERGENTS

A. Components

Great variations can occur in the formulation of different detergents from one brand to another to the extent that one type of component such as phosphate builders may be eliminated from the formulation altogether. Nonetheless, all the components may be grouped from an environmental impact consideration into two groups: organic and inorganic. This is a convenient distinction because, in general, the organic and inorganic components create differing impacts upon the environment and are treated differently for removal from wastewaters.

The predominate organic component of a synthetic detergent is the surfactant which is generally composed of a nonpolar hydrocarbon fraction and a polar fraction containing a cationic, nonionic, or anionic functional group. The anionic surfactants usually contain sulfates or sulfonates as their functional group and are totally biodegradable if their hydrocarbon side chain is linear.

The functional group of a cationic surfactant is of a tetramethyl ammonium nature and imparts a bactericidal character to the detergent. This type of detergent is utilized primarily by hospitals and food-processing industries. The tetramethyl-ammonium-containing surfactant is nonbiodegradable. Disteryldimethyl cationics, which can be low or nonbactericidal in nature, are commonly used in laundry practice as fabric softeners; these compounds are mostly biodegradable.

Nonionic surfactants contain ethylene oxide polymers attached to the hydrocarbon chain through a functional group. Generally this is a hydoxide group, but other functional groups such as the amides (NH_2) and the carboxyls (COOH) have also been used. Nonionic surfactant molecules have been reported to be both nonbiodegradable and easily degraded [6, 7], but comparison of different studies and conflicting results is difficult since the test surfactants are not always identified. Other minor organic components can be included in the formulation, such as anticaking agents, soil solubilizers, alcohol solvents, and carboxymethylcellulose dyes (both colored and fluorescent whitening agents).

The principal inorganic detergent fraction is due to the sequestering agents which are added to the detergent formulations in order to eliminate the effects of hard waters. The most commonly used sequestering agent is sodium tripolyphosphate (STP) used alone or in conjunction with higher polymers of phosphate. Other inorganic components of detergent formulations such as sodium sulfate, sodium borate and/or perborate, sodium chloride, sodium silicate, and sodium carbonate are used as builders in addition to or as a substitute for phosphates.

Table 4 presents typical detergent formulations of the anionic, nonionic and so-called "nonpolluting" types of detergent. Each component is grouped in the organic or inorganic class for future reference.

B. Elemental Impacts

One or more of the following immediate reactions may occur as a result of releasing synthetic detergents into the aquatic environment:

1. The components may react chemically with other dissolved or suspended ions or compounds.
2. The components may be taken up biologically and therefore stimulate metabolic activity of any indigenous organisms.
3. The components may remain in solution, thus increasing the concentration of total dissolved species in the system.

TABLE 4

Typical Types of Detergent Formulation

Type	Organic		Inorganic	
Anionic	15-20%	LAS surfactant	30-50%	STP
	0.5-1.5%	CMC	10-15%	Sodium sulfate
	2-4%	Foam stabilizer	5-10%	Sodium silicate
	0.1-0.2%	Perfume	0-5%	Sodium perborate
	0.5-1%	Anticake	0.3-0.6%	Fluorescent whitener
Nonionic	6-10%	Nonionic surfactant	30-50%	STP
	0.5-1.5%	CMC	0-10%	Sodium sulfate
	0.3-0.6%	Fluorescent whitener	5-10%	Sodium silicate
	0.1-0.2%	Perfume	0-20%	Sodium carbonate
	0.5-1%	Anticake	0.5%	Sodium perborate
Nonpolluting	6-12%	Nonionic surfactant	8-20%	Sodium silicate
	0.5-1.5%	CMC	20-70%	Sodium carbonate
	0.3-0.6%	Fluorescent whitener	0-45%	Sodium chloride
	0.5-1%	Anticake	0-25%	Sodium borate
			0-15%	Sodium perborate
			0-45%	Sodium sulfate

Source: From Ref. 8.

Abbreviations: LAS, linear alkyl sulfonate; CMC, carboxy methyl cellulose; STP, sodium tripolyphosphate.

According to Table 4, the principal elements that would be released to the environment as a result of the discharge of synthetic detergents are: carbon, nitrogen, phosphorus, and sulfur. Each of these elements can exist in numerous forms, and in each case the potential transformations which can occur exist within a natural cycle for that element. The cycles for these elements are based on the chemically and biologically induced transformations which are favored for each of these elements and, in many cases, will include portions of the general biological cycles previously described in Sec. II. C.

Our environment is composed of numerous biological and chemical cycles of this nature and the quality of our environment is in fact determined by these cycles. None of these cycles are singularly independent, but rather are mutually interdependent—the products of one cycle serving as inputs to other cycles. In a closed ecological system such as this planet or a segment of the planet (e.g., a lake or river), there must be a balance between the inputs and outputs of each of the cycles as there is neither an infinite sink for, nor source of, any component. The quality of the environment thus is established by controlling the magnitude of the respective cycles and maintaining the proper balance between each of the cycles.

If one considers each of the cycles as individual rings, the interrelationship between the various cycles can be visualized as a system of interconnected rings or wheels wherein each ring can contact or interrelate with a number of other rings (see Fig. 18). Although such a model is greatly simplified, it demonstrates the interdependence of the numerous factors which serve to establish a given environmental quality. Furthermore, such a model clearly illustrates how a given pollutant can affect numerous aspects of the environment. In order to determine accurately the impact of a given pollutant on the environment, it is desirable to understand as many of the fundamental cycles as possible. The magnitude that an analysis of this nature can assume can become appreciable and, unfortunately, there are currently many gaps in the available data. Thus, as previously discussed, the impact of pollution frequently is still measured solely in terms of simplifying yardsticks such as heterotrophic activity (i.e., oxygen depletion) and autotrophic activity (i.e., productivity).

One currently considered method of attempting to quantify numerous environmental parameters as they relate to water quality is the development of a Water Quality Index [9]. Such an index, as proposed, might range from

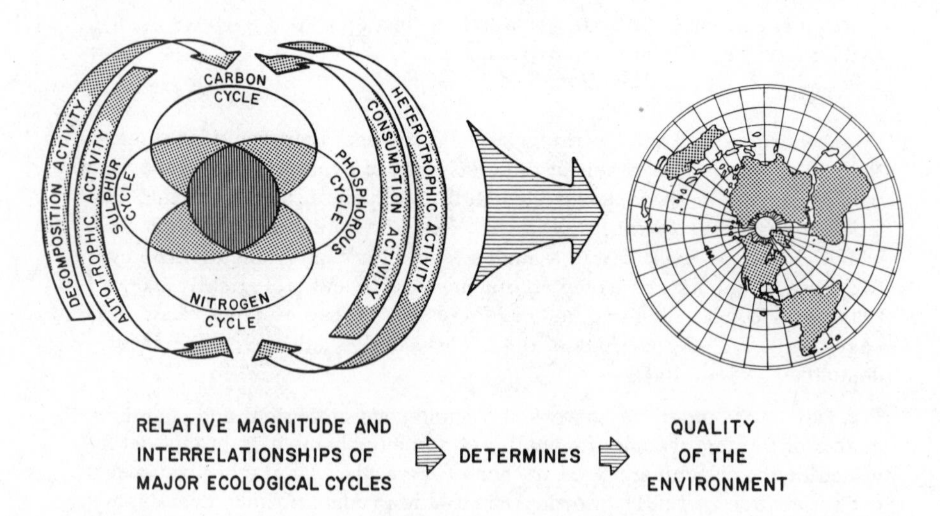

FIG. 18. Schematic representation of the interrelationships among the various key cycles of nature and their effect in determining environmental quality.

0 (poor) to 100 (excellent) and be based on a proportionate weighted evaluation of a number of water quality parameters. Such parameters as dissolved oxygen, BOD, fecal coliform density, pH, nitrate, phosphate, temperature, turbidity, and total dissolved solids have been suggested for inclusion. Development of such a Water Quality Index would allow for immediate classification of the quality of a water, similar, for example, to the immediate recognition a number on the Richter scale has with respect to the magnitude of earthquakes.

However, the most precise assessment of the impact of pollution on the environment can be made only by determining the effect of a given discharge on the inherent ecological cycles and balances. Although certain threshold levels of various pollutants can be tolerated, it is generally the excessive discharge of a given pollutant which one wishes to curtail because of the magnitude of the changes that the given discharge induces in the natural cycles and balances.

Summary information relative to the carbon, nitrogen, phosphorus, and sulfur cycles follows, based on information which is currently available. The environmental impact of each of these elements and other possible minor synthetic detergent constituents are also discussed.

1. Carbon

The potential carbonaceous input from synthetic detergents to the environment can be significant. The element carbon occupies an intermediate position between the electropositive and electronegative elements of the periodic table. Carbon can, therefore, exist in several pure elemental forms as well as in highly reduced or oxidized forms. It can exist in solid, liquid, or gaseous phases in pure or combined states and it can circulate through the organic, inorganic, living, and nonliving cycles. Carbon is the most ubiquitous element within living protoplasm since it forms the principal skeletal matrix of carbohydrates, proteins, and lipids. In addition, the carbon skeleton of intermediate metabolites provides the primary fuel for the combustion processes in respiring organisms.

The earth's crust contains many forms of inorganic carbon as well as the great volume of degraded forms of previously living carbon. The 2% of the earth's atmosphere, which is composed of inorganic carbon dioxide, serves as a source of carbon substrate for autotrophic organisms and a sink for this metabolite from respiring organisms. It is this delicate concentration balance of atmospheric carbon dioxide which is being threatened increasingly by excessive energy consumption in the developed countries. At the present time, the increased carbon dioxide production from more numerous combustion systems or from increased activity of respiring organisms can be absorbed by the world's oceans. If the seas' absorption capacity is exceeded, however, the concentration of carbon dioxide in the atmosphere will increase, which will result in an insulating effect or "greenhouse effect" which will serve to raise the prevailing ambient temperature. On the other hand, if the world's carbon dioxide production can be sufficiently curtailed so that the numerous carbon forms can continue to remain in balance with one another, then the various inputs and outputs can flow continuously from one biospheric position to another. The interchanges characteristic of the carbon cycle are illustrated in Fig. 19.

The principal inputs from detergent components to this cycle would originate from the organic fraction of the surfactant molecule. This wastewater pollutant, if left untreated, feeds into the organic carbon position and from there can be transported throughout the cycle. The other minor carbon-containing components such as soil solubilizers, alcohol solvents, anticaking agents, and dyes could also feed into the cycle at the same position.

The principal inorganic carbon component would originate from the carbonate builder present in the formulation. The inorganic carbonate is utilized only slightly by autotrophic microorganisms in a biological waste treatment plant and passes through virtually intact. The untreated carbonate, released in the treatment plant effluent, enters the carbon cycle at the

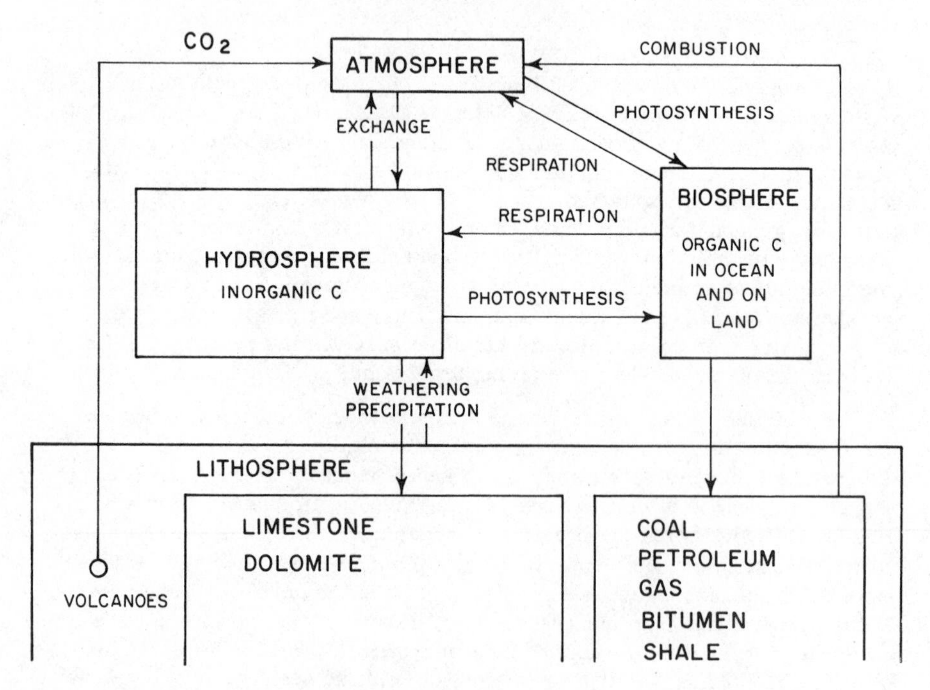

FIG. 19. The circulation of carbon in nature. From Ref. 10.

inorganic carbon position, stimulates the growth of autotrophic microorganisms, and correspondingly is transported throughout the cycle.

Historically, the organic surfactant molecules were not degraded by bacteria in biological wastewater treatment plants or in natural receiving water bodies. These compounds, which caused the detergents to be known as "hard" detergents, in general contained a highly branched hydrocarbon chain at one end of the organic fraction and a benzene ring at the other connected by a quarternary carbon. Swisher [11] demonstrated that the combination of a highly branched chain at one end and an aromatic ring at the other, made these molecules perhaps 5 to 10 times less degradable than the straight-chain hydrocarbon counterparts. He used river water samples and their indigenous bacteria to demonstrate the comparative biodegradation rates in a natural system. These "hard" compounds accumulated in the environment because the rate of input was greater than the rate of biodegradation. The nonbiodegradable branched-chain alkyl benzene sulfonate molecules (ABS) were detectable in some ground waters used for municipal drinking water supplies. Therefore, the U. S. Public Health Service set a maximum limit of ABS to be present in municipal drinking water supplies equal to 0.5 mg/liter [12]. Extreme cases of high accumulated ABS concentrations became prevalent, however; e.g., in 1959, the Illinois River contained concentrations as high as 1.3 mg/liter, or almost three times the minimum concentration at which foaming can be detected [13].

By June, 1965, the detergent manufacturers had completely eliminated the foaming problem by converting to formulations which contained linear hydrocarbon chains attached to the benzene ring. Since concentrations of linear alkyl sulfonates (LAS) are detectable by the same analytical test as are ABS concentrations, continuous measurements of the environment prior to 1965 and after the conversion to "soft" detergents can be compared. Studies conducted by Sullivan and Evans [13] utilizing the same analytical technique for both ABS measurements prior to 1965 and LAS measurements after 1965 compare these concentrations directly. The authors point out that these residual surfactant concentrations are lower after the detergent industry conversion to LAS even though the consumption of synthetic detergents increased each year. A later communication by Sullivan and Swisher [14] utilized the same analytical technique as before to measure the concentration of both ABS and LAS in the Illinois River. They simultaneously used a chromatographic analytical technique which was specific for LAS only. Comparison of the result of the two analyses indicated that perhaps only 10 to 20% of the measurable surfactant concentrations were caused by LAS, the "soft" detergent component. The remaining surfactant material, presumably, was primarily undegraded surfactant material remaining in the environment since before 1965. Indeed, an analysis of surfactant concentration in wastewater in Suffolk County, L. I. [15], showed that a sample from 1966 which contained biodegradable LAS was degraded to below the 0.5 mg/liter concentration limit set by the U. S. Public Health Service within a 2-month period; however, a comparable sample from 1963 which contained the nonbiodegradable ABS had not degraded to the acceptable limit of 0.5 mg/liter within a 5.5-year period.

Although the linear surfactants are more easily degraded within a treatment plant, the relative toxicity of any undegraded fraction can be appreciable. Thatcher and Santner [16] compared the median tolerance level (TL_m), or that concentration of test material which causes death to half a population, for ABS and LAS. They found that for five species of common fish, LAS was two to three times more toxic and the early degradation products were the most toxic. Pickering and Thatcher [17] studied the toxicity of a typical biodegradable surfactant on fat-head minnows. They found that toxicity effects on the minnows were minimal at concentrations up to approximately 0.5 mg/liter of LAS. At increasing concentrations above this level, the minnows' survival, growth, maturity, production of eggs, hatching of eggs, and survival of the newly hatched fry were increasingly jeopardized. Such results were obtained in fully aerated test chambers in order to eliminate any effects from diminished levels of dissolved oxygen which would result from any biodegradation occurring during the test period. The level of dissolved oxygen cannot be controlled in receiving water bodies, however, so that one might consider an LAS pollution input to possess the following doubly lethal effect on higher aquatic forms: (1) Before it is biodegraded within the receiving water body, it is moderately toxic to some fish and their offspring at concentrations higher than the U. S. Public Health Service tolerable limit of 0.5 mg/liter. (2) After it has been degraded within the receiving water body, the resulting diminished level of dissolved oxygen may be so low that higher forms of aquatic life cannot survive. This combination of factors often occurs in polluted water bodies and has been described in Fig. 8 and 9.

2. Nitrogen

The element nitrogen also exists in many forms in all phases of the biosphere. It exists in elemental form as a gas in the atmosphere. It occurs in various inorganic forms with oxidation states from the most reduced compound ammonia (NH_3) to the most oxidized ionic form nitrate (NO_3^-).

All of these oxidation states fill important roles in the interchange between nonliving and living forms. Domestic wastewaters contain large fractions of nitrogen-containing urea. Nitrogen-using organisms can break this and similar organic nitrogen compounds down to inorganic ammonia. Ammonia subsequently imposes an oxygen demand on a system when it is converted to amino acids which serve as protein precursors, and to purines and pyrimidines which form the functional groups of nucleic acids. These transformations from inorganic to organic forms and reversion back to inorganic forms occur primarily by biochemical means. Energy is required to convert each form stepwise throughout the biological pathways. Energy in the form of lightning also may convert atmospheric nitrogen to ionic forms which are washed from the air by rainwater and introduced to the ecosystem. The interrelations and conversions from different nitrogen forms are illustrated in Fig. 20.

The principal nitrogen-containing components in detergent formulations are the organic nitrogen-containing groups in cationic surfactants. This type of bactericidal cationic surfactant is utilized by a minor fraction of the detergent market, namely hospitals and food processing establishments; and, although nitrogen-containing cationic surfactants are used widely as fabric softeners, their contribution to the nitrogen cycle is usually minimal

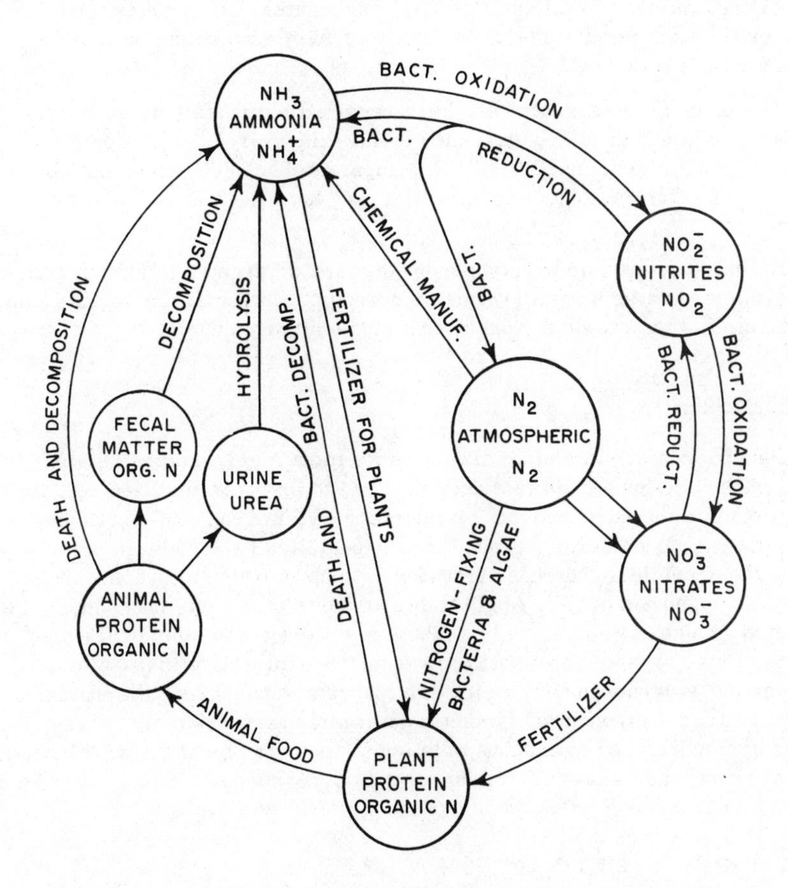

FIG. 20. The circulation of nitrogen in nature. From Ref. 6.

compared to other man-made nitrogen inputs such as fertilizer runoff. Since the cationic surfactants are not only nonbiodegradable, but lethal to bacteria, this type of surfactant in large volumes would be detrimental to municipal biological waste treatment plants. In addition, if large amounts of these wastes were to be released uncontrolled into the environment, not only would the receiving water body be rendered antiseptic but also the higher forms of aquatic life would be killed.

The ultimate fate of these compounds is obscure; their effects are considered to be inhibited by anions such as anionic surfactants or other large organic anions [18] and by the dissolved cations in the water which have been described previously as being associated with hardness characteristics [19]. However, sequestering agents such as phosphates, polyphosphates, amine carboxylic acids—ethylenediamine tetraacetic acid (EDTA) and nitrilotriacetic acid (NTA), carbonates, and borates tend to enhance the bactericidal effects of the quaternary ammonium cationic surfactants.

The nonionic detergents have been reported alternately as nonbiodegradable [20] and as easily degraded by mixed cultures [15]. Conflicting results such as these are difficult to compare because differing test conditions and different nonionic compounds may have been used in the experiments.

A common, but minor, component in many detergents is a foam stabilizer which contains an amide functional group. The amide nitrogen would contribute to the organic nitrogen input of the nitrogen cycle.

3. Phosphorus

Phosphorus exists almost always in its most highly oxidized state, i.e., phosphate ion. This fact does not lead to an oversimplified role for phosphorus in the environment; on the contrary, instead, this fact complicates the case considerably since the phosphate ion need undergo primarily only minor chemical and phase transformations in order to become transported throughout a cycle. Appreciable amounts of organic phosphates are excreted in human wastes and contribute significantly to municipal wastewaters; these organic phosphates are easily hydrolyzed to inorganic phosphate. Polyphosphates, which are inorganic condensed phosphates, can be a large component of synthetic detergents (see Table 4). These forms are hydrolyzed in aqueous solution to the ortho form by the following typical reaction as shown for tetrahydrogen pyrophosphate:

$$H_4P_2O_7 \xrightarrow{(+H_2O)} 2H_3PO_4 \qquad (7)$$

Soluble inorganic phosphate is taken up by living systems, primarily bacteria, phytoplankton and aquatic plants, and converted to high-energy phosphate esters which later can serve as energy sources for thermodynamically unfavorable biochemical reactions required for metabolism. Other inorganic phosphate ions may, in addition, become complexed with metal ions such as calcium, iron or manganese and remain in solution or may be adsorbed on the solid surfaces of clays or muds present in the bottom of surface water bodies. Inorganic phosphate under certain pH conditions may also precipitate with cations such as calcium, iron, manganese, or aluminum to form solid sources of phosphate. The phosphate cycle illustrated in Fig. 21 shows these transformations.

As shown in Table 4, the phosphate builders often comprise an appreciable fraction of the detergent formulation and can represent a significant input to the environment unless advanced wastewater treatment is administered to remove these substances. Evaluation of the exact impact of this particular input on the environment has been the subject of extensive controversy during recent years, particularly with regard to phosphate's potential nutrient-limiting role in the environment. Analysis of the elements

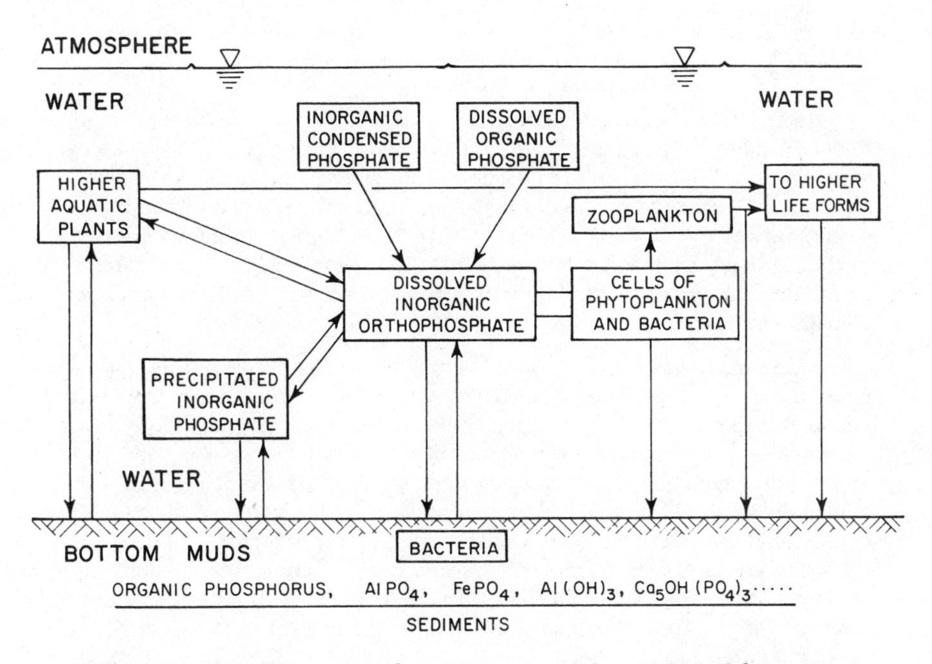

FIG. 21. Phosphorus transformations in a lake. Adapted from Hayes and Phillips [21]; from Ref. 10.

necessary for algal growth, as shown in Fig. 4, indicates the following: (1) carbon dioxide is available either from waste products of aquatic respiring organisms or from atmospheric absorption, (2) nitrogen is available from either storm water runoff of agricultural and urban lands or from waste products of respiring organisms, as well as to certain algal species (i.e., predominantely blue green) which are capable of fixing atmospheric nitrogen, and (3) hydrogen and oxygen are available as ions or in the dissolved molecular form in aqueous systems. In this context, phosphate and requisite trace-metal concentrations are generally considered as the more probable limiting elements with respect to algal growth.

Thus, the argument has frequently been given that the phosphate content of synthetic detergents has served to increase the frequency of algal blooms in the nation's surface waters. A critical analysis of each particular case poses the following questions:

1. Is phosphate the limiting nutrient which, if absent, would preclude excessive algal blooms?

2. How much of the phosphate present in the environment resulted from detergent input?

3. Are the prospective phosphate substitutes more harmful to the environment than are the phosphate builders?

As the subsequent discussion indicates, investigations have yielded different answers to each of these questions. This is simply because each water body exhibits its own characteristics and no two are alike. Thus no *a priori* statement can be made that phosphate builders are either harmful or not harmful to the environment. This decision can be made only after performing a technical evaluation of all the inputs to and outputs from a particular water body and elucidating its inherent biological and chemical characteristics.

In the past, however, various environmental researchers have investigated the complex problem of nutrient balance and biological productivity, and have reported the prevailing conditions in either natural or laboratory situations. Subsequently, proponents of either side of the phosphate controversy quote such statistics, as if applied universally, to prove a case in point. In addition, often the data are taken completely out of their original context and extrapolated from a specific case to a generalization. Such manipulation is not uncommon when simplistic solutions to environmental problems are presented. The following discussion is an attempt to describe the complexity of the phosphate situation in the context of the preceding three questions.

a. Phosphate as the limiting nutrient

Since the introduction of high-phosphate detergents and high-phosphate fertilizers, the concentrations of phosphates in the nation's surface waters has risen dramatically above pre-World War II levels. Ferguson [20] estimates that of approximately 1 billion pounds of human-generated phosphates released into the environment annually, approximately 50% originates from domestic wastewater and 50% originates from storm runoffs. Such large phosphate inputs are not evenly distributed in the surface waters and increasingly conflicting data have been reported concerning limiting phosphate concentrations and resulting algal blooms. Continuous monitoring of a lake in southern Michigan (Stone Lake, Cassopolis, Mich.), for example, indicates that this lake contains an average phosphate concentration of 5 mg/liter [22]. Such a high concentration of phosphate would hardly be limiting since the prevailing pH and alkalinity factors of the lake would limit the inorganic carbon concentration to a lower ratio than the previously described 106:1 ratio needed for protoplasmic synthesis. Such an excessive phosphate concentration normally would support a continuous algal bloom; but an algal bloom occurs in Stone Lake only during the summer months when temperature, light, major nutrient, and/or minor trace element concentrations are ideal. At the opposite extreme, MacKenthun et al. [23] reported that intense algal blooms occur in Lake Sebasticook, Maine, even though the soluble phosphorus concentration never exceeded 0.01 mg/liter. Such disparate extremes in phosphorus concentration data serve to veil the main point of the objection to the release of phosphates into the environment, i.e., that a release of unnatural phosphorus material amounting to 1 billion pounds annually creates, at the least, however unequally distributed, a potential for enhanced algal reproduction. Shapiro [24] explains that such minimal phosphorus concentration levels which support excessive algal growth, as those discussed by Kuentzel [25] are meaningless since such measurements are made <u>after</u> the phosphorus has been removed from the water and incorporated into the cells. He emphasized that algal "growth is proportional to supply, not to the phosphorus left in solution after growth." This very important point is often lost in the heat of the argument against the need to limit phosphorus inputs to the environment. In addition, it has been shown recently that data intended to relate nutrient concentrations to biological activity must include analyses of phosphorus concentrations in the liquid phase and in the indigenous biomass at times before, during, and after the monitored algal bloom in order to obtain accurate phosphorus balances for such phenomena [26]. The need for obtaining complete and accurate data in these investigations is further indicated by the additional complicating fact reported by Borchardt and Azad [27] that algae can take up and store phosphate concentrations in excess of that necessary for cell synthesis.

A report issued to the U. S. -Canada International Joint Commission on the Pollution of Lakes Erie and Ontario and the St. Lawrence River, in November 1969 [28] concludes from the work of several independent scientists from both countries that a low phosphate input is necessary to control the rapidly accelerating eutrophication of this system. This report recommends a double-edged effort at removing phosphates from detergents and treating municipal wastewaters to remove phosphates which originate from nondetergent sources. The report estimated that if such controls are implemented, the degree of eutrophication of Lake Ontario, because of its great depth and lack of biological activity in the bottom muds, can be reduced so much that by 1986 it could be considered an oligotrophic lake, or almost pure; Lake Erie, because its Western Basin is presently in an advanced state of eutrophication, because it contains a great deal of biological activity in its bottom muds, and because it is so shallow, cannot be so dramatically reclaimed in such a short period of time. However, as the report claims, the renovation of this system is primarily a matter of economics and, again, if the controls were implemented Lake Erie could be changed from a highly eutrophic lake to a slightly eutrophic lake by 1986. Presumably, greater recovery would occur in time.

b. Detergent phosphate contribution to the environment

Ferguson [20] estimates that human-generated sources of phosphorus account for 57 to 74% of phosphorus input to the environment. Of these sources, 44 to 56% originate from domestic wastewater. The report to the U. S. -Canada International Joint Commission on Pollution of Lakes Erie and Ontario and the St. Lawrence River [28] claims that detergent builders add 70% of the phosphorus to domestic wastewater. This commission urged that the doubly effective measures of eliminating phosphate builders from detergents and treating domestic wastewaters to remove phosphate be accepted. The argument follows: If a typical wastewater contains 10 mg/liter of phosphate and the 70% due to detergents is eliminated and an 80% phosphate removal treatment is imposed, the phosphate concentration of the effluent being released to the environment would be 0.6 mg/liter. Such a reduction in phosphate release into the environment would diminish the amount of eutrophying biological activity in lakes and streams appreciably.

c. Prospective phosphate substitutes

As early as 1967, the possible necessity to replace phosphate builders was noted [29]. The relative performance compared to phosphate builders and biodegradability were the principal factors considered for prospective substitutes. At first, the prime emphasis was given to organic carboxylic acids and amino polycarboxylic acids. Such compounds as citric acid, NTA, and EDTA were strongly considered. However, consider the roles of phosphate builders in detergents to be as follows [30]:

1. They combine with hardness cations such as calcium and magnesium to reduce the amount of necessary detergents to perform cleansing.
2. They complex with dissolved iron and manganese so that rust deposits and graying of fabrics will be prevented.
3. The negatively charged tripolyphosphate-dissolved ions surround grease and dirt particles and help to suspend them.
4. The highly charged negative ions attract the hydrophilic ends of the surfactant molecule and thereby help to induce micelle formation.
5. They help buffer the water at an ideal pH so that there is not excessive acidity or alkalinity in the water which would prevent good cleansing or cause dermatological responses.

The above three compounds would possess the ability to perform the first four roles—although perhaps they could not perform all four roles simultaneously and as effectively as tripolyphosphate—but significant is the fact that they would not serve to buffer the water at an ideal pH.

These are all organic compounds and as such would be considered as organic input were they to be fully accepted and utilized by the detergent industry as phosphate substitutes; however, at the present time none has any particular favor with the industry and one has been prohibited by the U. S. Food and Drug Administration [31] because its degradation products are considered to be potentially harmful.

The "nonpolluting" formulas contain chiefly sodium carbonate and other inorganic builders in place of any organic phosphate substitute. The basic caustic chemical character of the carbonates can raise the pH to such a high level that concentrated washing solutions have been known to cause damage to laundry facilities and clothing [32, 32a]. In addition, any significant pH change which might be introduced on wash days as a slug to municipal biological-wastewater treatment plants could cause an adverse effect; conventional physical-chemical treatment would not remove carbonate routinely from wastewaters; only the very specific, expensive type of advanced treatment processes such as ion exchange or reverse osmosis would remove carbonates, if this removal were desired. The impact of carbonate detergents on the environment has been reported recently by Jungermann and Silberman [33].

4. Sulfur

The element sulfur can exist in pure elemental form or in intermediate oxidation states from the most reduced form hydrogen sulfide (H_2S), which can cause significant odor problems, to its most highly oxidized state sulfate (SO_4^{-2}). The latter form causes a cathartic effect in the human body and, therefore, its recommended concentration in waters meant for human consumption is limited to 250 mg/liter [12]. The former reduced oxidation state is utilized by living systems to form organic sulfides which are important function-determining and structure-determining components of proteins. The interrelated transformations are illustrated in Fig. 22.

The major detergent input to this cycle originates from the sulfonate or sulfate functional group of the anionic surfactant molecule and from soluble inorganic sulfate builders which are included in the formulation. Only a small fraction of this concentration is removed by biological wastewater treatment plants. The major fraction is released to the environment and the significance of this input has not been investigated thoroughly. Sulfate ion has appeared to have a variable effect on different fish species and lower fauna species whereas sulfide ion is highly toxic to fish.

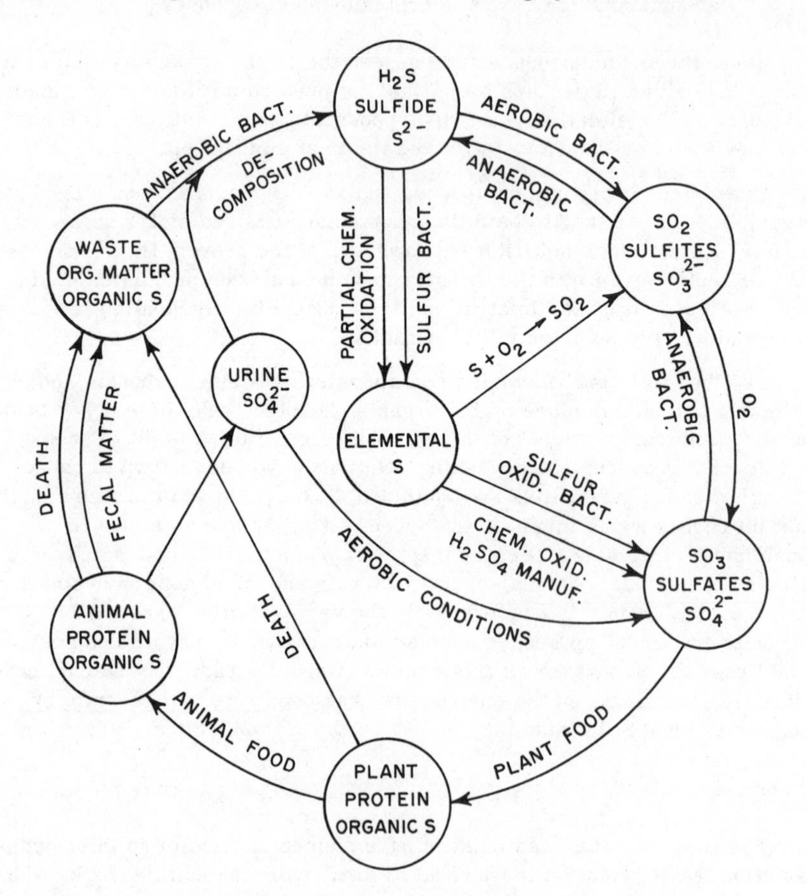

FIG. 22. The circulation of sulfur in nature. From Ref. 6.

5. Miscellaneous Inorganic Components

There are other inorganic compounds that have been used or proposed as builders in synthetic detergents. These compounds may have different individual effects, but, in combination with each other and the other previously mentioned compounds, will have deleterious effects. Since no appreciable amount of these components are removed by conventional waste treatment processes, they tend to accumulate in the environment and increase the existing level of total dissolved solids in waters. The existing water supplies do not have an infinite capacity to dissolve solids and, if advanced wastewater treatment controls are not implemented to remove these components, the drinking water tolerances for various elements will be exceeded frequently.

The U. S. Public Health Service has set a limit for total dissolved solids in drinking water supplies not to exceed 500 mg/liter [12]. Currently, this limit is exceeded only in certain areas of the United States. In the future, though, as multiple use and reuse of fresh waters becomes more prevalent, concentrations of these components can increase to excessive levels.

Consequently, the public must decide according to models such as the one shown in Fig. 1 whether to require advanced treatment for removal of these inorganic ions and thus achieve a purer environment and, ultimately, purer drinking water at increased cost. Conversely, the inorganic ions can be released into the environment with a corresponding sacrifice in environmental quality and drinking water quality.

The inorganic ions in question consist of carbonates, sulfates, chlorides, silicates, borates, perborates, sodium, and heavy metals. The carbonates and sulfates which have been discussed previously and the remaining inorganics deserve individual mention. Chlorides impart a brackish taste to water and, since they normally are untreated in wastewater treatment plants, their presence is utilized as an indication of pollution inputs in freshwater systems. Their concentration is limited to 250 mg/liter in domestic drinking water supplies [12]. Chlorides are not involved in any biological cycle as such, but their presence serves to provide ionic strength in suspending medium for living systems which need to overcome internal osmotic pressures.

Silicates serve as a macronutrient for diatomaceous algae. This type of algae is very prevalent in the biotic system and serves to convert soluble silicate forms into solid skeletal structure in their cells. The cells may be free floating throughout the lifespan and then settle out to the bottom muds of the water body. The dead silicaceous algae may then become part of the permanent bottom muds or may be degraded by decomposer bacteria which resolubilize the silicate. Thus the biological silica cycle transports any silicate through the ecosystem.

The borates and perborates are utilized in trace amounts by most green algae and plants. Any significant boron input to natural waters would remain in excess of the indigenous biological requirement since it is needed in such trace amounts. The excess borates and perborates are toxic to certain plants in concentrations greater than 2 mg/liter and are thought to be innocuous to plants and animals at concentrations less than 0.1 mg/liter.

Sodium ion concentrations can become significant, from a public health aspect, due to their potential deleterious effects on people with circulatory ailments. For this reason, the Advisory Committee appointed in 1971 by the Environmental Protection Agency to review the 1962 Public Health Service Drinking Water Standards has recommended that sodium concentration in drinking water should be reported to the physicians in the utility service area by the reporting agency for the benefit of patients requiring low-sodium diets. Sodium ion is also thought to enhance the ability of algae to assimilate luxury amounts of phosphate in natural systems [34]. This enhancement, in specific cases, may lead to increasing rates of eutrophication.

IV. WASTEWATER TREATMENT

A. General

Wastewater treatment plants depend upon adequate collection systems to collect and deliver wastewater to the plant. Frequently, the collection system is inadequate since it was designed and built to transport far smaller volumes than those encountered in today's urbanized areas. Often storm water and domestic wastewaters are transported in combined lines which overflow to nearby natural water systems if they become overloaded during periods of heavy precipitation.

In many urban areas, the cost to install adequate lines to transport wastewater may appreciably exceed costs to build efficient wastewater treatment facilities. However, the wastewater treatment plant can be, at a minimum, only as good as the delivery system which serves it. This aspect of pollution abatement procedures cannot be neglected.

Modern wastewater treatment plants are designed to utilize biological, chemical, and physical principles in order to remove pollutants. Wastewater treatment plants are generally designed to remove pollutants in the following order: (1) suspended impurities, (2) dissolved organic material and (3) phosphatic constituents. These three stages are referred to generally as primary, secondary, and tertiary treatment respectively, particularly when considering the treatment of domestic wastewaters. The capability of wastewater treatment plants currently must provide secondary treatment. Most states currently require that at least secondary treatment be provided for domestic wastewaters and tertiary treatment in those cases where waste discharges can result in eutrophication of our freshwater bodies.

The impact of nitrogenous species on receiving waters has recently received much attention and many states have or are currently considering establishing effluent standards which would require additional treatment for this constituent in most wastewaters. Wastewater treatment at and beyond this degree is generally referred to as advanced wastewater treatment although (depending on the method of phosphate removal) tertiary treatment of wastewater is also frequently considered as an advanced wastewater treatment process. In this regard, "conventional" wastewater treatment would thus consist of primary and secondary treatment processes. These distinctions between "conventional" and "advanced" wastewater treatment are receiving widespread usage and have been established essentially from the treatment of domestic wastewater. Compatible industrial wastewaters may be treated in a similar manner; due to their great variability, however, the character and composition of each individual industrial waste must generally be considered individually.

Experience has shown that it is most beneficial to remove the settleable solids initially in the primary stage in order to minimize their interference with subsequent treatment processes and equipment. Removal of soluble organic carbon has conventionally taken place in the secondary stage. Historically, this removal was accomplished biologically by feeding the polluted water to microbes which would utilize organic carbon as substrate.

More recently, physical-chemical processes have found application for removal of these pollutants. Many variations in this type of treatment are utilized today and new techniques are being developed constantly. In addition, the remaining processes following secondary treatment are varied and no particular order of application exists.

The purpose of this discussion is to present a state-of-the-art explanation of wastewater treatment in order to indicate current capabilities of preventing pollutants from entering the environment. An attempt will be made to describe each unit operation but all the numerous combinations, modifications, and variations that are possible in a total treatment scheme cannot be delineated here.

It should be noted that with each stage of additional treatment of the wastewater, the cost of treatment is increased (see Table 5). The cost of sewer lines required to transport municipal wastewaters to the treatment

TABLE 5

Approximate Cost Data for Wastewater Treatment and Sludge Treatment and Disposal Processes

A. Wastewater treatment

Unit process	Plant size (1 MGD)	(10 MGD)	(100 MGD)
	Total treatment cost (cents per 1000 gallons)[a, b]		
Primary treatment	20–25	10–15	5–10
Secondary treatment			
Trickling filter	28–33	14–19	8–13
Activated sludge	13–18	7–12	4–9
Physicochemical treatment			
Coagulation	12–17	9–14	7–12
Carbon adsorption	30–35	13–18	5–10
Filtration	12–17	4–9	1–6
Tertiary treatment			
Phosphorus removal	6–9	4–7	2–5
Nitrification	8–11	4–7	1–4
Denitrification	7–10	4–7	1–4
Ion exchange	25–30	14–19	Not reported
Electrodialysis	38–43	23–28	14–19
Reverse osmosis	67–72	48–53	36–41
Chlorination (for disinfection)	1–3	0.5–2.5	0.5–1.5

B. Sludge treatment and disposal

Unit process		Total cost (dollars per ton dry solids)[a]
Thickening:	Gravity	3–9
	Centrifugation[d]	6–37
	Air flotation	10–26
Dewatering:	Vacuum filtration[d]	15–93
	Centrifugation	9–64
	Sand-drying beds	6–40
Digestion:	Anaerobic	10–36
Incineration:	Combustion	14–69
	Wet oxidation	61–75
Disposal:	Lagooning	2–10
	Landfill (dewatered)	3–4
	Land disposal (wet)	8–60

Abbreviation: MGD, million gallons per day.

[a]All costs adjusted to June 1973

[b]Capital costs amortized at 8% for 30 years.

[c]Coagulation includes biological nitrification and clarification.

[d]Includes chemical conditioning costs.

plant location and the cost for disposal of sludges from each state of treatment must also be considered. Such considerations have direct influence on the relationships presented in Fig. 1. In establishing steady-state conditions within a metropolitan area, decisions must be made about how extensive the wastewater treatment should be, what level of quality of environment should be maintained and what costs must be borne by the community.

B. Primary Treatment

Primary treatment generally consists of a number of physical type treatment processes combined to remove settleable and suspended material from wastewater. A typical primary treatment system is shown in Fig. 23. In the flow scheme described in this figure, the wastewater is first screened to remove the large objects carried in the wastewater. The screens are generally bar type with clear spaces of 1/2 to 1-1/2 in. Because of its abrasive nature, grit (sand, coffee grounds, etc.) is generally removed next in a grit chamber. Grit chambers generally are designed with flowing-through velocities in the neighborhood of 1 ft/s and detention times of 1 min in order to allow the grit to settle out on the bottom of the chamber. The wastewater then flows to a communitor which is a shredding device generally consisting of a slotted cylinder through which the wastewater flows. Solids too large to pass the slot, usually 1/4 in., are reduced in size by cutting members which act like shears as the screen cylinder rotates. The solids are subsequently removed by sedimentation. Sedimentation tanks are designed to provide relatively quiescent conditions (e.g., 1- to 2- hr detention times) for the suspended impurities to settle out of the wastewater.

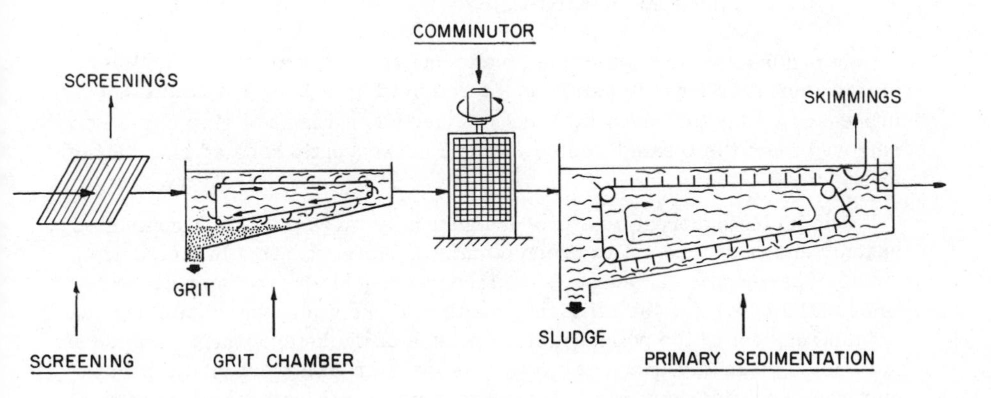

FIG. 23. Flow diagram of a typical primary wastewater treatment system.

The settled material is pushed by flights to a sludge hopper where it is removed by pumps. The flights are frequently used as surface skimmers in their return cycle to push the floating grease and oil to a trough for their removal. The efficiency of primary sedimentation can be enhanced with chemical coagulants. Such treatment is successful in removing 50 to 75% of suspended matter which accounts for 40 to 50% of the biodegradable pollution.

C. Secondary Treatment

1. Biological Treatment

Biological waste treatment processes provide environmental conditions conducive to the controlled growth of selected microbial species which remove undesired pollutant(s) from a given wastewater before it is released into the environment. The pollutant serves as a food or nutrient source (substrate) to the microorganism. Equaltion 8 indicates the overall summary reaction for a microbial process.

$$\text{Substrate} + \text{microbes} \xrightarrow{\substack{\text{satisfactory} \\ \text{environmental} \\ \text{conditions}}} \text{more microbes} + \text{end products} \qquad (8)$$

As shown by this equation, the pollutant (i.e., substrate) is utilized as follows:

1. conversion to microbially produced end products, and/or
2. incorporation into (or onto) the cellular protoplasm of newly synthesized microbial species

In order to achieve effective treatment, the products of step 1 either must be nonpollutional in nature or be removed by subsequent treatment processes. The increased biomass produced as a result of step 2 must be removed from the treated wastewater and subsequently handled as a sludge material.

The biological treatment process generally takes place in a controlled reactor and is an attempt to preempt natural biodegradation in the environment. The reactor, as such, is designed to provide environmental conditions satisfactory for the efficient growth of those microorganisms desired for the removal of the pollutant(s). Because of the heterogeneous nature of the wastes generally encountered at a wastewater treatment plant, pure cultures of microorganisms, as such, are never encountered in these processes. Rather, the growth of predominate classes or species of microorganisms develops in which the respective concentrations are determined by wastewater characteristics and environmental conditions of the biological reactor. The metabolic activities of the microorganisms follow such relationships as those described in Eqs. 1 to 3.

A concentrating or separating device usually follows the reactor (see Fig. 24) in order to discharge an effluent as free of microorganisms as possible. A concentrated suspension of microorganisms is then available either for recycling and/or subsequent sludge handling techniques.

The secondary treatment of wastewater by biological means has classically been designed for the removal of organic carbon. The process is one in which heterotrophic microorganisms are utilized in an aerobic environment in order to oxidize the organic material to carbon dioxide while synthesizing new cellular protoplasm (Eq. 1). In general terms, this process is commonly referred to as satisfying the biochemical oxygen demand of the wastewater (see Table 3). The design of a secondary biological treatment system is generally based on the volume of the wastewater and the concentration of the biodegradable organic material in the wastewater, so that a given degree of BOD removal is achieved. Three major types of biological treatment processes are currently utilized for secondary treatment operations:

1. lagoons (vis, oxidation ponds)
2. trickling filters
3. activated sludge

Numerous types of process modifications exist for each of these secondary treatment operations. Generally, however, activated sludge processes are capable of achieving the greatest degree of treatment but require the greatest skill in operation and are, therefore, the most expensive to operate. In contrast, lagoons are the least efficient biological treatment process but

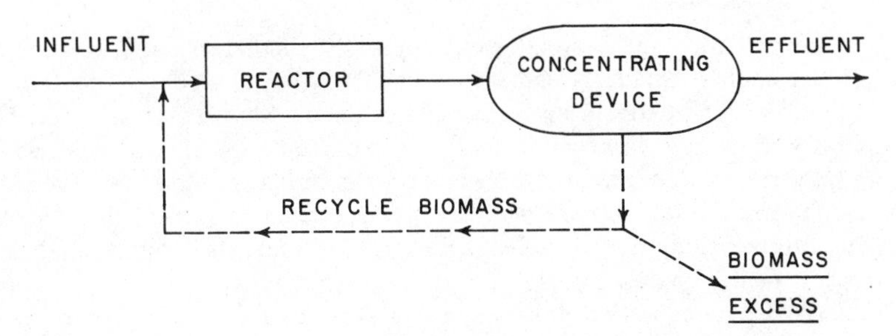

FIG. 24. Flow diagram of a typical biological secondary wastewater treatment system.

are the easiest and least expensive biological process to operate, if land costs are not too high.

a. Lagoons

Lagoons are generally constructed as relatively shallow (2 to 5 ft) bodies of water in excavated basins. The term "lagoon" is commonly used interchangeably with "stabilization pond" and/or "oxidation pond". The process basically utilizes heterotrophic and autotrophic microorganisms working together as shown in Fig. 25. The heterotrophic microorganisms are employed to stabilize the organic material in accordance with Eq. 1. The autotrophic microorganisms are developed to supply oxygen to the heterotrophs in addition to that provided by atmospheric reaeration at the lagoon surface and to provide some nutrient removal in accordance with Eq. 2.

Most waste stabilization ponds develop into some type of facultative system as shown in Fig. 25, with aerobic conditions near the surface and throughout most of the depth of the pond. Because of the presence of settleable organic material, however, anaerobic conditions will generally persist near the bottom and metabolic activity in this zone will follow the relationships outlined in Eq. 3. Depending on the design, however, one can have a shallow pond which, for all practical purposes, is completely aerobic or a pond which is so heavily loaded with respect to organic material that except for a small zone at the very top, the entire pond will be anaerobic. Oxidation ponds have been successfully utilized singly or in various combinations for the treatment of both domestic and industrial wastes. Appreciably higher organic loadings have been tolerated in lagoons equipped with aeration devices; both diffused air and mechanical aeration equipment have been employed successfully in such designs. The efficiency of oxidation pond performance is extremely temperature-dependent, and adequate consideration is seldom given to separation and final handling of the microbial species from the effluent of oxidation ponds.

b. Trickling filters

Trickling filters are generally cylindrically shaped tanks, 3 to 10 ft in depth, filled with spherically shaped rocks 1 to 4 in. in size. A central rotating axle which supports extended arms fitted with nozzles is rotated horizontally while the wastewater is sprayed onto the top of the rock bed. The wastewater falls down onto the rock media which provide a surface upon which aerobic heterotrophic microbial growth occurs. The growth is called bioslime and serves to extract the organic material from the liquid wastewater until it grows so thick it can no longer adhere to the rock surface. It is then sloughed off due to hydraulic shear and admixed with the treated liquid phase until it is separated and removed from the liquid phase in a final settling tank.

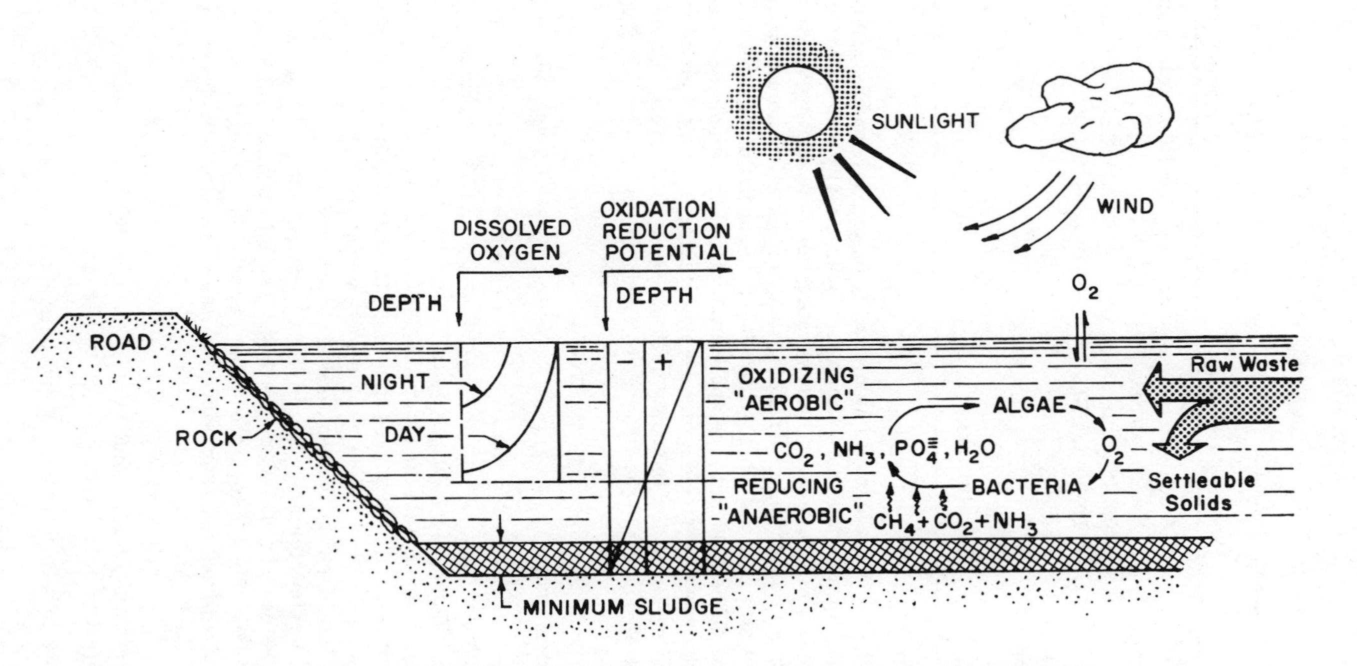

FIG. 25. Cross section of a typical oxidation pond. From Ref. 35.

TABLE 6

Comparisons of Low-Rate and High-Rate Trickling Filters

Parameter	Low-rate trickling filter	High-rate trickling filter
Hydraulic load	2-4 million gallons per acre per day	10-40 million gallons per acre per day
Organic load	1000 ft^3 stone per 10-20 lb BOD_5	1000 ft^3 stone per 20-90 lb BOD_5
Depth	6-10 ft	3-8 ft
Removal efficiency	85%	67-75%
Recirculation	None	1:1 to 4:1
Dosing interval	Intermittently	Continuous

Trickling filters are classified by hydraulic or organic loading as high-rate or low-rate filters. The more important operational parameters for each case are shown in Table 6. The low-rate filter is a relatively simple device and is highly dependable, producing a consistent effluent quality with varying influent strength. In a high-rate filter, recirculation of filter effluent or final effluent permits the application of higher organic loadings. Numerous variations in recirculation flow schemes have been proposed for high-rate trickling-filter design and operation. Generally, recirculation aids in preventing ponding in the filter and in reducing the nuisance due to odors and flies.

c. Activated sludge

The conventional activated sludge process consists of a long narrow rectangularly shaped tank, the length of which is fitted with air diffusers to introduce finely dispersed compressed air. The compressed air provides the oxygen needed to maintain the heterotrophic microorganisms in an aerobic environment and also serves to mix continually the microbial contents with the wastewater. The untreated wastewater and recycled microorganisms are introduced at the head end of the aeration tank. The organic material in the wastewater is stabilized in accordance with Eq. 1 as the wastewater flows down the length of the tank. The microorganisms are generally most conveniently separated from the treated wastewater by sedimentation. Conventional activated sludge systems are normally designed in terms of:

1. volumetric organic loading of 30 to 40 lb of BOD_5 per thousand cubic feet of aeration tank capacity per day

2. a hydraulic retention time of approximately 6 hr and

3. a food/microorganism ratio (lb BOD_5 to lb MLSS) of 0.25:0.50.

The activated sludge process is a relatively complex one to operate and requires a reasonable amount of skill. The process, however, is a very sound one and has appreciable flexibility by virtue of its numerous modifications (e.g., step aeration, contact stabilization, extended aeration, high rate, complete mixing, etc.). The major limitation to the activated sludge process, as is the case with any biological treatment process, is its sensitivity to toxic substances.

d. Effects of detergents on biological treatment processes

The most pertinent inputs to biological systems from synthetic detergents are due to the organic fraction of the surfactant, as well as other minor organic components such as foam stabilizers and solvents. The effects of previous "hard" surfactants have already been discussed. The conversion from "hard" to "soft" detergent surfactants has been accomplished by the use of an alkyl sulfonate which must be degraded within the retention time of any given biological treatment system. Otherwise, the incompletely degraded portion remaining at the end of a treatment period will stimulate increased biological activity and impose an oxygen demand upon the receiving water body.

Davis and Gloyna [7] determined in laboratory experiments that the type of bacteria generally associated with human intestinal wastes and, therefore, indicative of domestic sewage contamination, can easily degrade certain types of surfactant molecules. The implication was that if these bacteria were present in the environment due to some pollution input, they could thrive on incompletely degraded surfactant pollution and perhaps become a health hazard.

The same authors also established that variations in these surfactant molecules present biodegradation unpredictabilities. Variations in formulation as used by different companies, those variations in different detergents from the same company, combinations of various compounds used in one detergent, and the large variation between the cationic, nonionic, and anionic surfactant molecules all present different substrates to be utilized by bacteria. The following degradation conditions may exist:

1. The bacteria may contain the enzymes necessary to degrade the particular molecule.
2. They may not contain the specific enzyme necessary, but may be able to manufacture the necessary enzyme if given enough time.
3. The bacteria may not have the necessary enzymes or be able to induce their production at all so that the substrate cannot be utilized.

Considering the number of common species of bacteria which exist in the environment, each with the above variations in capability for degrading the various surfactant molecules as well as the number of variations in marketed surfactant molecules described above, one could determine an almost infinite number of combinations thereof. Davis and Gloyna [7] confirmed that mixed cultures were able to degrade the various surfactants tested faster than any particular species of algae or bacteria.

Mancy and Okun [36] studied the effects of surface active agents on the activated sludge process and found that smaller, more rigid bubbles were formed during aerations, and the rigidity of the air-water interface seriously impeded the rate and amount of gas transfer from the air bubble to the liquid. Since the biological treatment process is dependent upon the oxygen supply to the microorganisms utilizing the wastewater pollution as substrate, the efficiency of the biodegradation would be affected. Lynch and Sawyer [37] studied the effects of various types of surfactant on the rate of oxygen transfer in an aeration system. They found that the decrease in rate of oxygen transfer to the liquid system was approximately proportional to the chain length of the hydrophobic end of the surfactant molecule. In addition, they found that the oxygen transfer rate was depressed to a greater extent when there was a high concentration of ions due to phosphate builders. Mancy and Okun [38] postulated that the presence of surfactant molecules in an aerated system not only decreased the effective area through which oxygen molecules could penetrate at the air-water interface surrounding a bubble but, in addition, the surfactant molecules offered increased resistance to oxygen penetration because of the surfactant molecule orientation at the aqueous layer immediately adjacent to the interface. These investigators found that the presence of surfactant molecules was insignificant only at very low or very high aeration rates (rates at which oxygen transfer is relatively ineffective). McClelland and Mancy [39] found that concentrations of ABS-type surfactants greater than 5 mg/liter and concentrations of LAS-type surfactants greater than 10 mg/liter seriously hampered biodegradation of other carbonaceous substrates. In addition, ABS-type surfactants were reported to prevent the sludge from settling out of suspension in the subsequent settling tank.

2. Physical-Chemical Treatment

The concept of utilizing physical and chemical processes for the primary and secondary treatment of wastewaters in lieu of the more conventional combination of physical and biological processes is currently being practiced to some degree. This type of treatment is particularly advantageous for systems which experience appreciable fluctuations in pollutant concentrations or which contain pollutants that would be toxic to a biological wastewater treatment plant; also, the process is capable of removing appreciable concentrations of other pollutants in addition to the dissolved organic material normally removed in secondary treatment (e.g., phosphorus, suspended impurities, heavy metals, color, etc.). The treatment is accomplished in a series of unit operations which may be varied in any number of order. Like a biological treatment plant, the physical-chemical treatment plant will produce solid sludges which must ultimately be handled; in addition, the physical-chemical treatment plant must either recover and/or regenerate the chemicals utilized to perform the treatment or, alternatively, new chemicals continually must be purchased.

A flow diagram of a typical physical-chemical treatment plant is shown in Fig. 26. Preliminary treatment usually consists of screening in order to remove the larger objects and can include a complete primary treatment system as shown in Fig. 23. One of the major advantages of physical-chemical treatment schemes is the fact that only a minimal amount of preliminary treatment is absolutely necessary. As shown in Fig. 26, the first unit process after the preliminary treatment involves chemical coagulation in a mixing tank. Inorganic metal ions have been found to be best suited for these applications because of their inherent dual role of acting as both a chemical precipitant and a flocculant. Due to their relatively low cost and general availability, the three most frequently used metals are calcium, iron, and aluminum. After a period of rapid mixing for approximately 1 min and flocculation of 15 to 30 min, the flocculated and precipitated impurities generally are removed quite effectively by ordinary sedimentation or upflow clarification.

The next step, adsorption, utilizes tanks filled with carbon particles over which the liquid wastewater is passed so that the dissolved carbonaceous pollution can be adsorbed onto the surface of the solid carbon particles in the filter bed. This adsorption process may continue until all the adsorption sites on the carbon filtration bed are occupied by the dissolved carbonaceous pollutant molecules from the wastewater. When the filtration bed is saturated, the filter bed must be removed from the treatment stream and the saturated carbon particles treated to remove the adsorbed organic

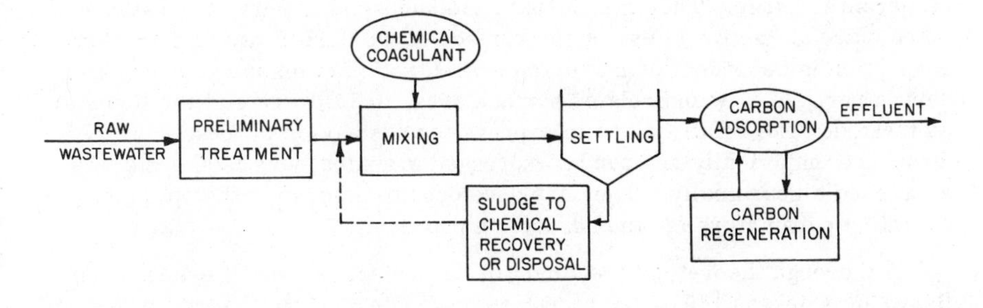

FIG. 26. Flow diagram of a typical physical/chemical treatment plant.

carbon molecules, thereby clearing the adsorption sites on the solid carbon particles. The carbon particles are said to be regenerated and thus can be utilized repeatedly.

a. Chemical coagulation

The inorganic salts of calcium, aluminum, and iron are the most suitable chemicals for chemical coagulation processes. Depending on the conditions, one of the following chemicals is generally selected: ferric chloride, ferric sulfate, sodium aluminate, aluminum sulfate (alum), or calcium oxide (lime). The chemical dose necessary to achieve a preset removal can best be determined experimentally on a small scale previous to the treatment; theoretical doses can be determined from a consideration of the stoichiometry of the precipitation and flocculation processes. The requisite concentration of metal-ion coagulant is influenced by such parameters as pH, alkalinity, temperature, mixing, and concentration of dissolved and suspended species, and normally is in the range of 150 to 400 mg/liter of the above salts.

The clarification following chemical addition depends upon the separation of the solid suspended particles from the suspending liquid phase. The separation is generally achieved by sedimentation or upflow clarification techniques and thus the solid particles must be in an agglomerated state to facilitate the separation. The agglomeration of the individual particles is based on colloidial chemistry relationships and is accomplished in the coagulation tank prior to the clarification tank. Synthetic organic polyelectrolytes have been used successfully in conjunction with the metal-ion coagulant in order to enhance the separation process.

b. Dissolved organic adsorption

Activated carbon has long been utilized to remove color and odors from water. Adsorption of color- and odor-causing substances on the carbon surface takes place because the crystalline carbon possesses unsatisfied lattice charges and the dissolved substances contain a polar or ionic group of opposite charge. The attachment of the undesirable organic substance takes place at "active sites" on the surface of the carbon particles. This adsorption is dependent primarily on concentration of dissolved substances, carbon particle size and related surface area, and flow rate of the liquid to be treated. The relationships to express the adsorption process are developed empirically and can be expressed graphically as isotherms or expressed algebraically. They are developed to determine the optimum operating parameters for the adsorption process.

A thorough theoretical treatment of carbon adsorption filters was published by Cookson [40]. A recent variation in carbon adsorption incorporates a countercurrent flow system, in which the wastewater flows in a forward direction and powdered carbon is pumped in a reverse direction.

Such a flow configuration is called "expanded-bed" design and, although it requires longer columns, it provides greater surface area and, correspondingly, increased adsorption capacity and greater operating efficiency. Weber [41] recently studied operational characteristics of the concurrent and countercurrent designs and found that the conventional concurrent flow design required considerable maintenance in order to produce comparable removal efficiency. Fig. 27 illustrates conventional and expanded designs.

D. Tertiary and Advanced Wastewater Treatment

1. Phosphorus Removal

Quite commonly, wastewaters are overconcentrated with respect to phosphatic pollutants; i.e., there is a greater concentration of this nutrient than the stoichiometric metabolic requirement of a carbon/phosphorus ratio of 106:1. Consequently, only a fraction of the phosphorus in domestic wastewater generally will be biologically removed by conventional secondary wastewater treatment systems. The two most commonly utilized secondary biological treatment processes, trickling-filter installations and activated sludge aeration facilities, have been reported to be capable of removing a maximum of 20% and 40% of the phosphorus concentrations, respectively, under ordinary operating conditions [42]. Therefore, additional treatment

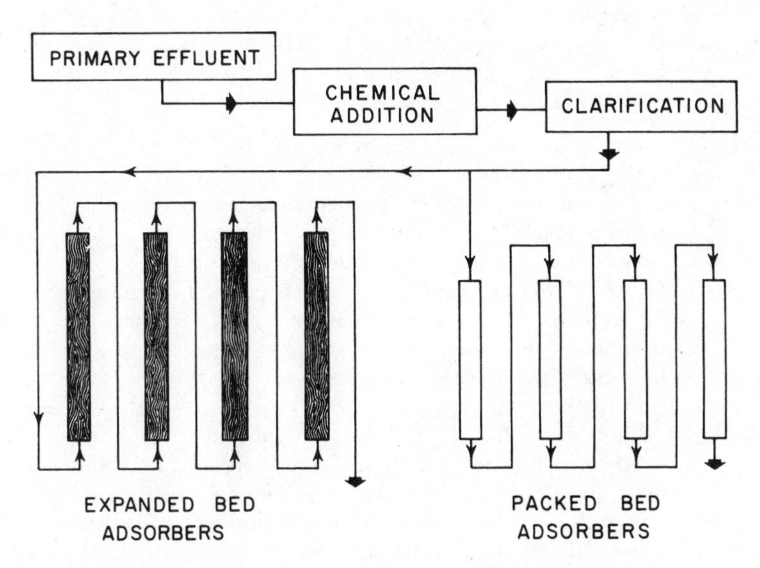

FIG. 27. Flow diagram of expanded-bed and packed-bed carbon adsorption systems. From Ref. 41.

must be utilized if efficient removal of phosphorus is desired following secondary biological treatment. Efficient phosphate removal may be achieved by either microbiological means or chemical precipitation.

a. Microbiological removal of phosphorus

In accordance with Eqs. 1 and 2, microbiological removal of phosphorus can be accomplished by either heterotrophic or autotrophic microorganisms. Because of the illumination requirements, which include provision of artificial illumination during periods of darkness and the use of relatively shallow tank or pond depths, as well as the generally difficult separation characteristics of algae, the use of autotrophic systems for the removal of phosphorus has been rather limited. The use of combined heterotrophic-autotrophic systems (i.e., lagoons or oxidation ponds) as described above are widely used and, depending on their operating conditions, are capable of achieving appreciable percentages of phosphorus removal. Great attention, however, has focused on manipulating the operation of the activated sludge process in an attempt to achieve greater phosphorus removal percentages than stoichiometric values. Ordinarily, the activated sludge microorganisms would contain approximately 2 to 3% of their volatile suspended solids concentration as phosphorus in accordance with the stiochiometry of their elemental composition of microbial protoplasm ($C_{106}H_{180}O_{45}N_{16}P_1$). By proper adjustment of such parameters as aeration rates, method of sludge recycle, and cation concentration of the carriage water, a number of activated sludge wastewater treatment plants have reported phosphate removals appreciably greater than 40% [43-46]. These removals far exceed any 106:1 stoichiometric carbon/phosphorus ratio and, at times, these treatment plants reportedly achieve a 90 to 95% phosphate removal, even in phosphate-rich wastewaters.

Such excessive phosphate removal by an activated sludge process has been termed "luxurious uptake of phosphorus," and much controversy has arisen over the mechanism by which this luxurious removal takes place. In 1965, Levin and Shapiro [45] reported that this uptake was a biological one which was dependent on oxygen transfer into the cell and was relatively independent of carbon concentration. In 1970, Menar and Jenkins [48] reported that in their opinion such luxurious removals were due to a chemical precipitation phenomenon and, although the extent of removal was a direct function of the aeration rate, the removal was oxygen-transfer independent. In their scheme, the aeration process served merely to strip the metabolically produced CO_2 from the sludge suspension. The resultant increase in pH would induce precipitation of some form of calcium phosphate, which must then be removed with the biomass during sludge separation. Later studies [34] appear to support the theory that luxury uptake of phosphorus is a biological mechanism operated by an active transport system, apparently controlled by the concentration of cationic species such as sodium and potassium in the suspending medium.

b. Chemical removal of phosphorus

Many metal ions are particularly effective in precipitating phosphate from solution. Due to their relatively low cost and general availability, the three most frequently used metals for phosphate precipitation are calcium, iron, and aluminum. An examination of Table 7 reveals that all three of these metals form relatively insoluble precipitates with phosphate.

The tendency of aluminum and iron to hydrolyze in aqueous solution creates a competition between the hydroxide and phosphate ions for the coordination sphere of the metal. Thus the efficiency of phosphate removal is dependent upon the relative concentrations of these two anions in solution and is consequently pH dependent. A decrease in pH or, more precisely, OH^-, favors precipitation of the metal phosphate. However, as the solubility of the metal phosphates increases with decreasing pH, an optimum pH exists for the removal of phosphate with metal-ion precipitants. When calcium is used as a precipitant, the competition for calcium is predominately between the phosphate and carbonate anions and, again, phosphate removal is dependent upon the relative concentrations of the anions present, and upon pH. Hydroxylapatite, $Ca_{10}(OH)_2(PO_4)_6$, is the most stable calcium phosphate solid phase as shown in Table 7. In practice, therefore, the pH and alkalinity of a wastewater determine to a large degree the relative efficiency of phosphate removal by precipitation with metal ions.

TABLE 7

Relatively Insoluble Precipitates with Phosphate

Number	Reaction	Log equilibrium constant (25°C)
1	$Fe^{3+} + PO_4^{3-} = FePO_4(s)$	23
2	$3Fe^{2+} + PO_4^{3-} = Fe_3(PO_4)_2(s)$	30
3	$Al^{3+} + PO_4^{3-} = AlPO_4(s)$	21
4	$Ca^{2+} + 2H_2PO_4^- = Ca(H_2PO_4)_2(s)$	1
5	$Ca^{2+} + HPO_4^{2-} = CaHPO_2(s)$	6
6	$10\ Ca^{3+} + 6PO_4^{3-} + 2OH^- = Ca_{10}(OH)_2(PO_4)_6(s)$	90

Source: From Ref. 49, p. 1005.

The chemical precipitation of phosphorus is a two-stage process: Firstly, the phosphorus must be made insoluble through the use of metal ions in accordance with reactions such as those shown in Table 7; and, secondly, the precipitate must be removed from suspension by sedimentation. The second step thus requires additional chemical for conditioning or flocculating the precipitate for its subsequent separation. The addition of various synthetic organic polyelectrolytes, or the addition of excess metal ions serves identical purposes in bringing about agglomeration of the insoluble metal phosphate particles into a settleable mass. Chemical removal of phosphorus generally has been designed as a separate third-stage treatment system, particularly when using lime. In the case of alum and ferric salts, successful removals have been achieved by chemical addition to existing secondary biological treatment units as well as to existing primary systems.

2. Nitrogen Removal

Appreciable consideration is being given currently to the removal of nitrogen from wastewaters because of its potential eutrophying characteristics and because of the oxygen-demanding characteristics of the more reduced forms of nitrogen (i.e., ammonia and organic nitrogen species). In addition, the presence of ammonia nitrogen in effluents has been shown to be toxic to aquatic life and an interference to the chlorination process. For these reasons, many states have already promulgated effluent standards limiting the total concentration of ammonia-nitrogen and/or total nitrogen in wastewaters. Nitrogenous species can be moved from wastewater either by biological means or by physical-chemical methods.

a. Biological removal of nitrogen

The biological transformations inherent to the nitrogen cycle (see Fig. 20) form the basis for nitrogen removal by biological treatment, although some removal results from nitrogen incorporation into cellular protoplasm. A major advantage of achieving biological nitrogen removal is that all forms of nitrogen ultimately are removed from solution in the nonpolluting form of nitrogen gas. The following two successive steps are involved in the biological removal of nitrogen:

1. nitrification—the oxidation of ammonia forms to nitrate
2. denitrification—the subsequent reduction of nitrate to nitrogen gas

Nitrification. The oxidation of ammonia to nitrate is a two-step process if carried to completion. Initially, ammonia is oxidized to nitrite by a generus of strict aerobic autotrophic bacteria (*Nitrosomonas*) that utilizes ammonia as its sole source of energy. The stoichiometry of this reaction is shown in Eq. 8. The second step, the conversion of nitrite to nitrate, is accomplished by the Nitrobacter generus which is a specific group of autotrophic bacteria that utilizes nitrite as its sole energy source. The stoichiometry of this reaction is shown in Eq. 9. The overall nitrification reaction is represented by the summary Eq. 10. Equation 10 clearly illustrates the oxygen demand of ammonia nitrogen; in accordance with this equation, 1.0 mg/liter of ammonia nitrogen (an N) requires 4.6 mg/liter of dissolved oxygen for complete nitrification.

$$2NH_4^+ + 3O_2 = 2NO_2^- + 2H_2O + 4H^+ \qquad (8)$$

$$2NO_2^- + O_2 = 2NO_3^- \qquad (9)$$

$$NH_4^+ + 2O_2 = NO_3^- + 2H^+ + H_2O \qquad (10)$$

Denitrification. Biological denitrification is achieved under anaerobic conditions by heterotrophic microorganisms which utilize nitrate as a hydrogen acceptor and require an organic energy source. Denitrification must be considered as a two-step process. For example, if methanol is selected as the organic carbon source to provide energy for bacterial synthesis, the following equations represent the microbially induced nitrogen transformations characteristic of denitrification.

$$NO_3^- + \frac{1}{3} CH_3OH = NO_2^- + \frac{1}{3} CO_2 + \frac{2}{3} H_2O \qquad (11)$$

$$NO_2^- + \frac{1}{2} CH_3OH = \frac{1}{2} N_2 + \frac{1}{2} CO_2 + \frac{1}{2} H_2O + OH^- \qquad (12)$$

$$NO_3^- + \frac{5}{6} CH_3OH = \frac{1}{2} N_2 + \frac{5}{6} CO_2 + \frac{7}{6} H_2O + OH^- \qquad (13)$$

Biological denitrification has been accomplished successfully both in stirred reactors in which microorganisms were maintained in suspension with the wastewater under anaerobic conditions and in column or packed bed filtration-type systems in which microbial growth takes place on the available surface area.

b. Chemical removal of nitrogen

Physical-chemical methods to remove nitrogen forms from wastewater are in various stages of development. The best-developed processes are ammonia stripping and breakpoint chlorination. In addition, ion-exchange methods can be utilized.

Ammonia stripping [50]. Ammonium ions exist in equilibrium with ammonia and hydrogen ions in accordance with the following equation:

$$NH_4^+ \rightarrow NH_3 + H^+ \tag{14}$$

With increasing pH above 7.0, the equilibrium is displaced to the right until virtually all the nitrogen will exist in the form of molecular ammonia. Figure 28 illustrates the relationship between pH and percentage of ammonia at various temperatures. Since ammonia is volatile, it can be removed by stripping the water with air. General operating parameters would be to adjust the pH to the range of 10.0 to 10.5 and utilize air-liquid loadings in the range of 600 to 900 ft^3/gal and aeration times in the neighborhood of 1/2 to 1 min. Although the concept of ammonia removal by air stripping is technically and economically sound, practical applications involve some restrictions. Cold-weather operation requires special consideration, blower-type operation in below freezing temperatures, for example, is not possible, and the practice of transferring ammonia from a wastewater to the air is questionable—particularly near large bodies of water.

Chlorine oxidation of ammonia [50-51]. It has long been recognized that chlorination during water purification treatment has resulted in oxidation of ammonia. In water, at ammonia nitrogen concentrations usually below 1 mg/liter, hypochlorous acid reacts with ammonia to form various chloramines via the following equations:

$$Cl_2 + H_2O \rightarrow HOCl + H^+ + Cl^- \text{ (hypochlorous acid)} \tag{15}$$

$$NH_4^- + HOCl \rightarrow NH_2Cl + H_2O + H^+ \text{ (monochloramine)} \tag{16}$$

$$NH_2Cl + HOCl \rightarrow NHCl_2 + H_2O \text{ (dichloramine)} \tag{17}$$

$$NHCl_2 + HOCl \rightarrow NCl_3 + H_2O \text{ (trichloramine)} \tag{18}$$

In breakpoint chlorination, chlorine is added to process waters until a point is reached where the total dissolved residual chlorine has reached a minimum (the breakpoint) and the ammonia-nitrogen content has disappeared. Palin [52] has suggested that two reactions describe the destruction of ammonia-nitrogen:

$$NH_2Cl + NHCl_2 \rightarrow N_2 + 3H^+ + 3Cl^- \tag{19}$$

$$2NH_2Cl + HOCl \rightarrow N_2 + 3H^+ + 3Cl^- + H_2O \tag{20}$$

In practice, breakpoint chlorination of wastewater does not quite follow the simple paths represented by Eqs. 19 and 20. The reactions become very much more complicated. In wastewaters, not only may the ammonia-nitrogen concentration be more than an order of magnitude higher than those

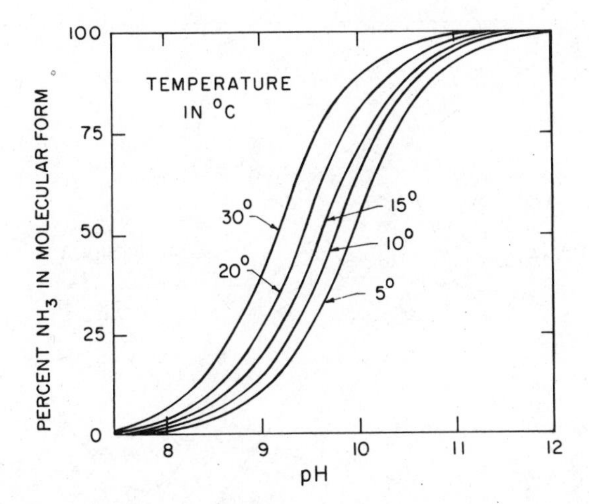

FIG. 28. Relation between pH and percentage of ammonia in aqueous solution. From Ref. 50.

normally encountered in natural waters, but also numerous other dissolved and suspended species may be present in concentrations to affect significantly the basic chemical reactions between chlorine and ammonia nitrogen. A typical breakpoint chlorination curve for a secondary effluent is shown in Fig. 29. The breakpoint, in this case, can be observed to occur between the 8:1 and 9:1 weight ratio of chlorine/ammonia nitrogen. Palin [52] has offered an explanation of the typical breakpoint curve. In the ascending branch of the curve, chlorine is present in combination with ammonia as mono- and dichloramine. The peak of the curve represents the point where all the ammonia initially present is combined to form the chloramines. In the presence of excess chlorine, the chloramines are unstable and react with free chlorine; the net result is that the amount of available chlorine is decreased. This accounts for the descending leg of the curve. When all the chloramines have been converted to a form that will no longer react with free chlorine, the further addition causes a proportionate rise in residual chlorine. This accounts for the second ascending branch of the curve. The simple overall reaction of chlorine with ammonia can be expressed by the following equation:

$$2NH_3 + 3Cl_2 \rightarrow N_2 + 6H^+ + 6Cl^- \qquad (21)$$

From Eq. 21 it can be seen that the final products of the reaction at the breakpoint are nitrogen gas and hydrochloric acid. Also, this is the equation that is used normally to calculate the stoichiometric weight of chlorine needed to oxidize ammonia nitrogen.

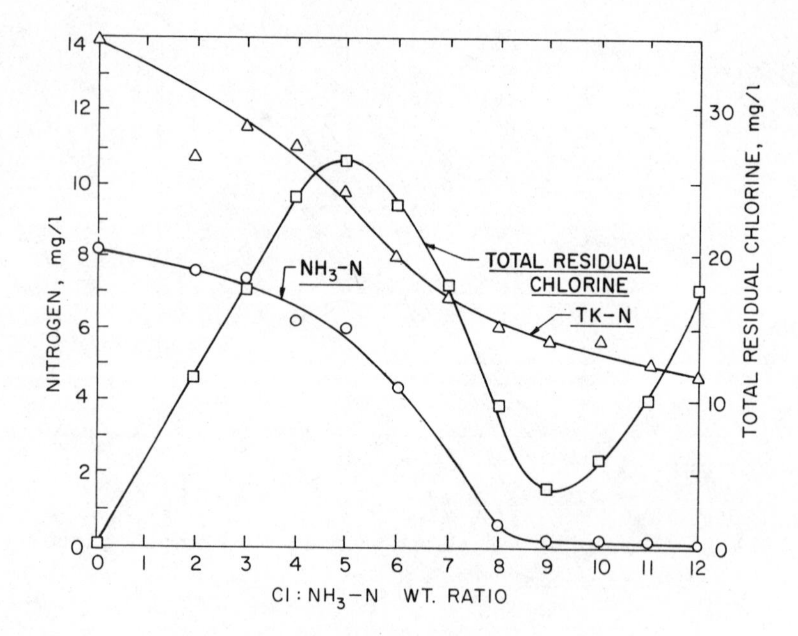

FIG. 29. Chlorination of secondary effluent: temperature—22°C; NH_3 - N - 8.1 mg/liter; suspended solids—40 mg/liter; total alkalinity—90 mg/liter $CaCO_3$; pH range during chlorination—6.5 to 7.5; breakpoint between the 8:1 and 9:1 weight ratio of Cl:NH_3 - N. Reprinted with permission from Environmental Science and Technology, 6, 627 (1972). Copyright by the American Chemical Society.

Figure 29 also indicates the reduction of both total organic nitrogen concentration and ammonia-nitrogen concentrations with increasing chlorine concentration. Nearly all ammonia nitrogen is removed at the breakpoint, but only a slight reduction of the organic nitrogen concentration occurs. Measurements by Pressley et al. [51] indicate that ammonia nitrogen removal occurs by oxidation of the ammonia chiefly to nitrogen gas with only small amounts of nitrate-nitrogen and nitrogen trichloride being formed.

The process basically is selective for ammonia nitrogen and, for the most part, will not remove the other nitrogen forms significant in water (i.e., organic, nitrite, and nitrate). Other wastewater treatment processes that would be used prior to breakpoint chlorination (e.g., activated sludge, carbon adsorption, etc.), however, successfully remove much of the organic nitrogen; nitrite- and nitrate-nitrogen concentrations are present only under highly oxidized conditions. Potential products of N_2O, O_2 (from the decomposition of N_2O), NO, and NO_2 do not occur. The pH of the solution greatly influences the amount of chlorine required to reach the breakpoint for a given initial ammonia-nitrogen concentration. Pressley et al [51] reported that the minimum chlorine dosage required to reach the breakpoint occurred in the range of pH 6 to 7. The pH of the solution will also greatly influence the relative concentration of each of the chloramine species and the formation of the undesirable side reactions of nitrate and nitrogen trichloride.

Ammonia removal by selective ion exchange [50, 53]. The fact that ammonia exists in solution predominately as a positively charged cation (NH_4^+) at pH values less than about 9.5 permits it to be removed by conventional ion-exchange methods. Synthetic ion-exchange resins are generally considered to be uneconomical for the removal of ammonium ion, however, because of the selectivity of these ion-exchange resins for other cations, principally divalent ions, which predominate in wastewaters and are removed preferentially. The selectivity of synthetic ion-exchange resins follows the Hofmeister or lyotropic series ($Ca^{2+} > Mg^{2+} > K^+ > NH_4^+ > Na^+$) which makes these exchanges inefficient in removing ammonia in the presence of divalent ions. Recent reports [53, 54] indicate that these problems may be minimized by using an ammonia-selective zeolite such as clinoptilolite. Clinoptilolite is a naturally occurring zeolite with its selectivity derived from the size of the pores and from the exchangeable cation sites which are selective for $K^+ > NH_4^+ > NA^+ > Ca^{2+} > Mg^{2+}$. The combination of selectivity and an exchange capacity of about 2.0 meq/gram appears to make clinoptilolite a useful exchanger for ammonium ion in wastewater. The loaded clinoptilolite can be regenerated by any basic solution at a pH value exceeding 11.0. Ammonia can be removed effectively from the regenerant by air stripping.

3. Removal of Other Constituents

By the time wastewater has been subjected to primary, secondary, and tertiary (for nutrient removal) treatment, it has become reasonably high in terms of water quality. In some cases, however, further consideration must be given to removing residual suspended solids, refractory organic material, and/or additional dissolved inorganic constituents. Numerous processes have been suggested for each of these purposes and many have been employed successfully in various installations. Microscreening and filtration (diatomaceous earth or granular media) processes, for example, have found successful application for the removal of suspended solids. In addition to the use of activated carbon (granular or powdered) for the removal of organic material, oxidative processes (e.g., ozone, molecular oxygen, catalytic oxidation with molecular oxygen, chlorine, potassium permaganate, electro-chemical treatment, etc.), and foam separation techniques have also all been used successfully.

Currently, and no doubt to an increasing degree in future applications, the removal of appreciable quantities of dissolved inorganic constituents from wastewaters will be necessary. A number of different processes have been investigated for these purposes. Among those processes presently considered to have definite application in this regard are: ion exchange, electrodialysis, reverse osmosis, and distillation. The application of each of these four processes for the treatment of wastewaters has been studied in some detail during the past years and numerous publications are available on each. Two relatively early government publications (1965 and 1968) on advanced waste treatment [55, 56] serve as good summary reports on these and other related advanced wastewater treatment processes and much of the material which follows on the above four treatment processes is taken from these two publications.

It is possible to produce a final water of virtually any desired quality (e.g., drinking water, deionized water, etc.) from any of the processes. The construction and operation of each of these processes, however, will remain a relatively expensive proposition. Based on the technological and economic information currently available, it appears that imposition of any requirement which will necessitate the removal of dissolved inorganic constituents from wastewater by these processes can result in total treatment costs in the neighborhood of $0.75 to $1.25 per thousand gallons of wastewater treated (see Table 5). It should be noted that each of these advanced wastewater treatment processes generally requires some degree of pretreatment (depending on the composition of the raw wastewater). Such process deterrents as fouling of resins, membranes, and/or heating elements from the presence of such impurities as suspended solids and/or dissolved organic material are common. Hence, at least primary- and secondary-treatment-type processes are required prior to the unit and frequently carbon adsorption and filtration must also be included. The cost of this pretreatment must also be included in the total cost figures accounting to some degree for the high cost range. In addition, it should also be noted that a concentrated solution or brine results from each process and consideration must be given to its subsequent handling—the cost of which is often <u>not</u> included in cost figures such as those reported above.

a. Ion exchange

Ion exchange is a well-known method for softening or for demineralizing water. Although softening can be useful in some instances and ion exchange can be useful for ammonia removal, the most likely application for ion exchange in wastewater treatment is for demineralization. Demineralized water produced by this process is of higher quality than generally is necessary, however; therefore, only part of the total stream generally need be treated.

Many natural materials and, more importantly, certain synthetic materials have the ability to exchange ions from an aqueous solution for loosely bound ions in the material itself. Cation-exchange resins, for example, can replace cations in solution with hydrogen ion from the resin. Similarly, anion-exchange resins can either replace anions in solution with hydroxyl ion or absorb the acids produced from the cation-exchange treatment. A combination of these cation- and anion-exchange treatments results in a high degree of demineralization.

Since the exchange capacity of ion-exchange materials is limited, they eventually become exhausted and must be regenerated. The cation resin is regenerated with an acid; the anion resin is regenerated with a base. Important considerations in the economics of ion exchange are the type and amounts of chemicals needed for regeneration. Regeneration of the resins is an expense which cannot be avoided, and the cost of chemicals required for this purpose will represent a significant portion of the operating costs. Unless these regenerating chemicals, which themselves could create secondary pollution, are recovered and reused, this cost can be prohibitively high.

b. Electrodialysis

The principle of operation of an electrodialysis system is to impose an electric potential across a cell containing the wastewater. This will induce the dissolved ions to migrate to the oppositely charged electrode. Alternate placement of selective membranes (i. e., cation permeable, anion permeable, respectively) between the electrodes as shown in Fig. 30 will result in alternate compartments becoming more concentrated with ionic species, while the intervening compartments become better (purer) in water quality due to smaller concentrations of dissolved constituents being present. A water quality approaching that of demineralized water can be achieved, but only partial demineralization is practical, however, because electrical power requirements become excessive if the ion concentration is reduced too much.

c. Reverse osmosis

A natural phenomenon known as osmosis occurs when solutions of two different concentrations are separated by a semipermeable membrane such as cellophane. With such an arrangement, water tends to pass through the membrane from the more dilute side to the more concentrated side and produce concentration equilibrium on both sides of the membrane. The ideal osmotic membrane permits passage of water molecules but prevents passage of dissolved materials. The driving force that impels this flow through the membrane is related to the difference in concentration between the two solutions and an osmotic pressure difference between the two compartments results. The principle is illustrated in Fig. 31A. If the liquid on the more concentrated side is allowed to rise in a standpipe as water passes through the membrane, as shown in Fig. 31B, the hydrostatic pressure on the right side of the membrane gradually increases until finally it

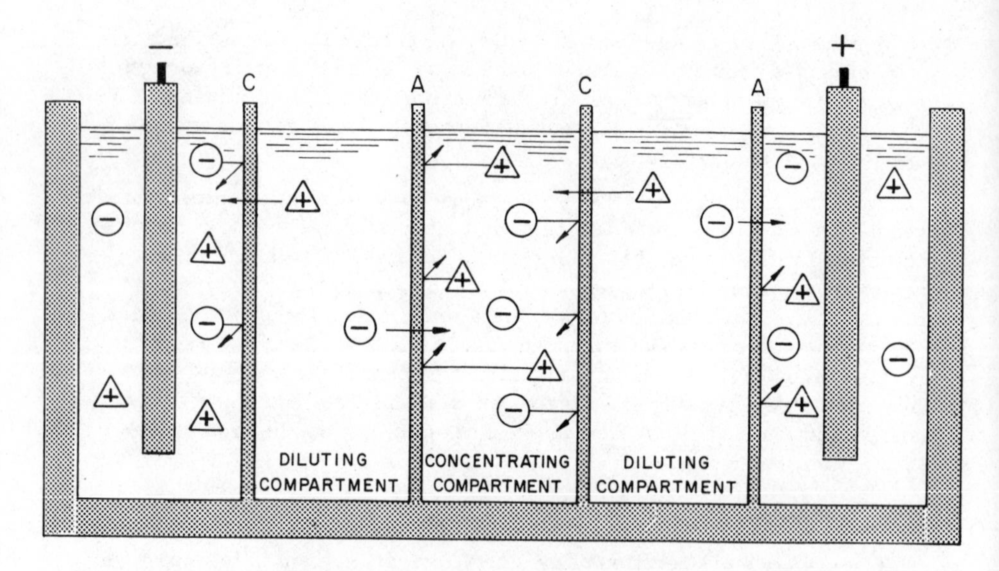

FIG. 30. Schematic representation of the principles involved in electrodialysis: A-anion-permeable membrane; C-cation-permeable membrane; ⊖-anion; △+-cation. From Ref. 55.

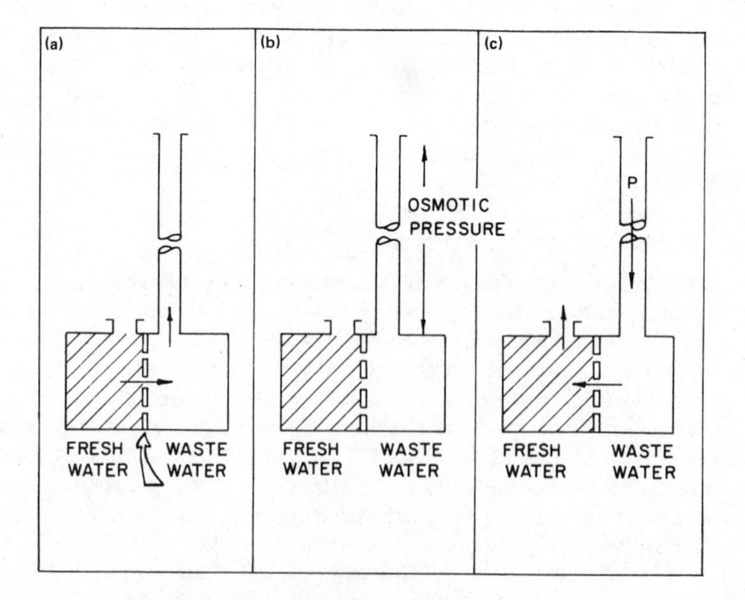

FIG. 31. Schematic representation of the principles involved in (a) normal osmosis, (b) osmotic equilibrium, and (c) reverse osmosis. From Ref. 55.

equals the osmotic pressure. At this point, flow through the membrane ceases. If, then, the pressure on the more concentrated (right) side of the membrane is increased, as shown in Fig. 31C, the flow of water through the membrane reverses; that is, water moves from the more concentrated compartment to the less concentrated compartment. This method of wastewater treatment is known as reverse osmosis.

Sheets of several types of material have the property of preventing the passage of minerals while allowing the passage of water in which the minerals were dissolved.

d. Distillation

Distillation has long been known as a process for producing "pure" water. It is capable of removing both organic and inorganic dissolved contaminants with relatively high efficiency; suspended solids are eliminated and even microorganisms are destroyed by the boiling temperatures employed. Volatile impurities in the feed pass over into the distillate, but several "polishing" techniques are available to eliminate this problem.

Compared with the "organic only" or "inorganic only" removal processes, distillation remains a relatively expensive process even though much recent work in the desalinization field has been aimed at decreasing costs. The prospects of lower cost nuclear energy or combination water renovation/power plants might improve the economics significantly in the future.

For distillation to be economically practical, the latent heat of evaporation must be reused many times. One type of system that is commonly used to achieve this type of heat economy is the multiple-effect evaporator. A schematic arrangement of a multiple-effect evaporator is shown in Fig. 32. In this system, each effect is maintained at a slightly lower pressure and, hence, a lower temperature than the preceding one. This allows the steam produced in one effect to be used as the heating medium for the next. The result is that for each pound of steam supplied to the first effect, there is produced a number of pounds of product water approximately equal to the number of effects. Other types of distillation equipment available include multistage-flash evaporators and vapor-compression distillation.

D. Sludge and Brine Handling

Either a concentrated suspension of solid material and/or a concentrated solution of dissolved ionic or molecular constituents will result from the operation of a wastewater treatment process. This material consists of the impurities removed from the wastewater and/or the excess microorganisms synthesized during a biological treatment operation. Concentrated suspensions of solid materials, which can consist of any combinations

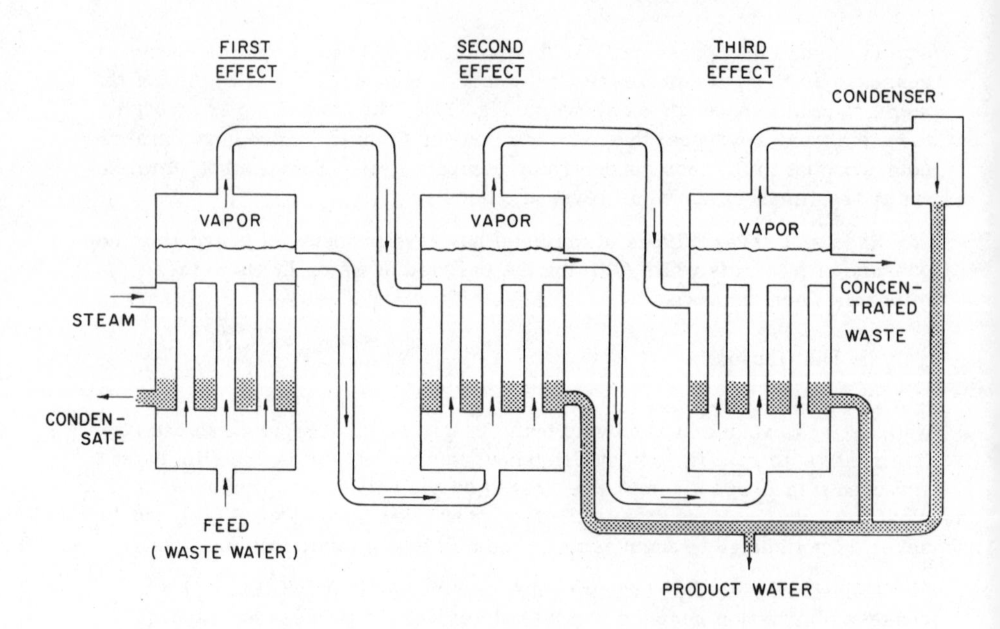

FIG. 32. Schematic arrangement of multiple effect evaporator. From Ref. 55.

of physical, chemical, and biological masses, are called sludges and concentrated solutions of dissolved constituents are called brines. Since the evaporation of either sludge or brine yields a solid residue, the management of these materials is referred to as "solids handling." The proper handling or management of these materials represents the biggest problem currently associated with wastewater treatment. For all practical purposes, the technology exists to remove virtually any type of pollutant from wastewater and reclaim a wastewater to any desired quality; the sludges and brines resulting from this treatment, however, can no longer be disposed of indiscriminately and forgotten. Consideration, rather, must be given to the proper management of these excess sludges and brines in order to preclude the potentially undesirable impacts of discharging these materials into the environment.

No method can be developed for the <u>complete</u> treatment or elimination of these solid materials, since such a process would violate the fundamental laws of thermodynamics. Development and utilization of acceptable procedures of sludge and brine handling, thus, becomes a most significant factor in the proper design and operation of every wastewater treatment process. Sludge and brine handling considerations currently attempt to satisfy one or more of the following types of objective:

1. reduction in sludge or brine volume
2. destruction or stabilization of sludge solids or brine constituents
3. recovery of desired elements
4. utilization of the sludge or brine material
5. destruction or control of hazardous components (pathogenic microorganisms, heavy metals, poisons, etc.)

Sludge- and brine-handling practices can be grouped into three major categories: (1) treatment, (2) disposal, and (3) utilization. Table 8 summarizes the more commonly used processes in each of these categories. Selected processes from each of these categories are generally combined into a total scheme in order to achieve the most effective degree of solids handling for a given situation. Of increasing significance is the consideration of sludge as a resource having beneficial uses for recycling, rather than as a waste product to be discarded. For economic and legal reasons, future sludge handling considerations will no doubt depend upon more utilization schemes with a corresponding reduction in disposal activities.

1. Sludge Treatment

The more common sludge treatment processes can be grouped into the following major categories: (1) concentration, (2) digestion, (3) dewatering, and (4) heat drying. Sludge concentration and dewatering result in sludge-volume reduction, while sludge digestion has the major function of solids destruction and stabilization. Heat drying and combustion can accomplish both volume reduction and solids destruction. Numerous combinations of two or more of the treatment processes shown in Table 8, Sec. A have been designed and operated successfully in order to achieve a given degree of sludge treatment.

a. Sludge concentration

Sludge concentration or thickening is one of the most widely used and significant wastewater treatment processes available. Its purpose is to separate the sludge from the suspending liquid in a reasonably concentrated form because the cost effectiveness of every means of sludge treatment, utilization, and disposal depends on the suspended solids concentration of the sludge. Gravity settling basins are most often employed for this purpose.

Thickening can be accomplished in separate sludge thickening tanks or in clarifiers; a more concentrated sludge is generally obtained, however, with separate thickeners. Mechanical and dissolved air floatation thickeners are also commonly used to thicken sludges from various sources in wastewater treatment plants.

TABLE 8

Summary of Common Sludge Treatment, Disposal and Utilization Processes

---

A. Common sludge treatment processes
   1. Concentration
      a. Clarifier thickening
      b. Separate concentration
         i. Gravity
         ii. Flotation
   2. Digestion
      a. Aerobic
      b. Anaerobic
   3. Dewatering
      a. Drying beds
      b. Vacuum filtration
      c. Pressure filtration
      d. Centrifugation
   4. Heat drying and combustion
      a. Heat drying
      b. Incineration
         i. Multiple hearth
         ii. Fluidized beds
      c. Wet oxidation

B. Common final sludge-disposal methods
   1. Landfilling
   2. Lagooning
   3. Deep well disposal (subsurface injection)
   4. Discharge to sea

C. Some sludge utilization processes
   1. For biological sludges
      a. Food stuff
      b. Auxiliary fuel
      c. Fertilizer
   2. For chemical sludges
      a. Substance recovery
      b. Complementary chemical usage
      c. Refinement for marketing
   3. For physically inert sludges
      a. Filler for construction materials
      b. Soil conditioning

---

b. Sludge digestion

Such sludges as those from biological waste treatment processes that have a high organic content can be subjected to digestion processes which serve to break down the sludge particles and decrease the sludge mass. This treatment can be accomplished aerobically, thereby requiring oxygen and large tank sizes; the process, however, is generally carried out anaerobically, a slower procedure requiring a longer retention time. Anaerobic treatment eliminates the expense of the oxygen requirement and, furthermore, it produces methane—a useable end product which can serve as an energy source at various points in the wastewater treatment plant.

Aerobic sludge digestion. Aerobic digestion is used to stabilize waste sludge solids by long-term aeration. The most common application of aerobic digestion is in reducing the bulk of waste activated sludge. The excess biological sludge is aerated in the absence of any external organic carbon source. Under these conditions, a fraction of the microorganisms will lyse and release their cellular constituents back into solution. The released materials serve as substrate for the remaining microorganisms which, under the aerobic conditions, will convert the released organic material to new cell mass and carbon dioxide and water in accordance with Eq. 1. The overall reaction under these conditions is a net loss in cell mass, equivalent to the fraction of the cellular mass which has been oxidized. The microbiological term for the metabolic process or growth phase is known as endogenous respiration. The longer the cell residence time, the greater will be the amount of cell destruction achieved, provided that sufficient oxygen is supplied.

Anaerobic sludge digestion. Anaerobic digestion is a widely used process for the treatment of primary sludges and excess biological sludges—particularly from activated sludge and trickling filter operations. The process normally is conducted in a covered tank which is heated to maintain the sludge in the range of 35°C, the optimum temperature for mesophilic sludge digestion. The biochemistry of the metabolic reaction taking place is in accordance with Eq. 3, a two-step reaction wherein the sludge is converted first to organic acids by acid-forming bacteria and is converted subsequently to methane and carbon dioxide by the methane-forming bacteria.

The units designed for this process normally are cylindrically shaped with a cone-shaped bottom ranging in size 100 ft or more in diameter and up to 50 ft in depth. The process can be conducted either under what is known as standard rate conditions (conventional digestion) in which stratification of the digester contents occurs, as shown in Fig. 33, or it can be conducted under high rate conditions in which the digester contents are mixed (see Fig. 34). The purpose of the mixing is to utilize the entire contents of the

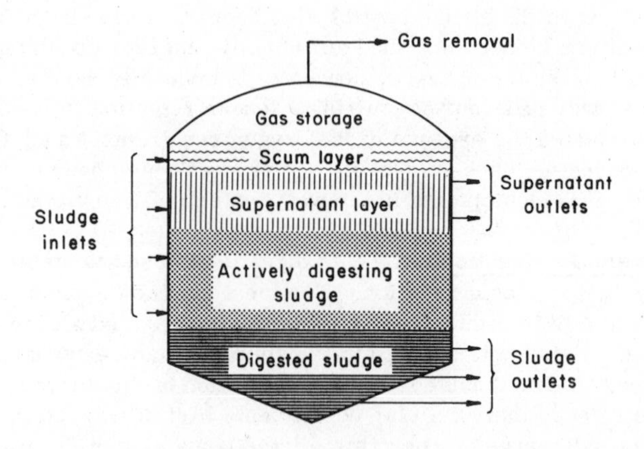

FIG. 33. Schematic diagram of conventional digester used in the single-stage process. From Ref. 57.

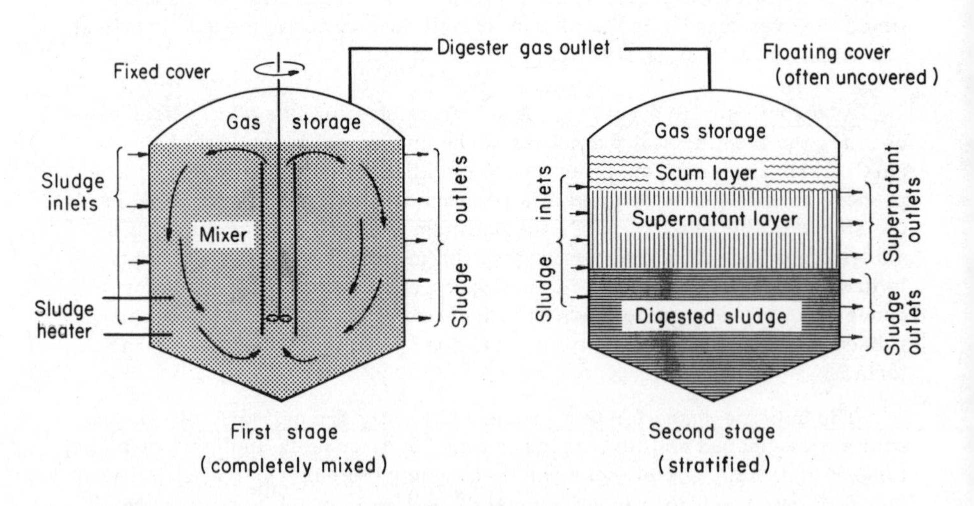

FIG. 34. Schematic diagram of two-stage digestion process. From Ref. 57.

digester for active digestion; a second stage digester is generally provided under these conditions primarily for separation of the digested sludge and digester supernatant liquid. Additional digestion can be achieved also in the secondary digester, depending on the retention time of the sludge. Mixing is generally achieved in the first stage either by gas recirculation, pumping, or draft tube mixers. The high-rate digestion process differs from the conventional single-stage process in that the solids-loading rate is much greater (e.g., 0.05 to 0.15 lb of volatile solids per cubic foot per day for high-rate as compared to 0.02 to 0.05 lb of volatile solids per cubic foot per day for conventional single-stage) and shorter solids retention times are practicable (e.g., 10 to 15 days for high-rate compared to 30 to 90 days for single-stage conventional).

c. Sludge dewatering

The methods available for sludge dewatering can be classified as either filtration-type methods (i.e., gravity, pressure, vacuum) or centrifugation methods. The selection of a sludge-dewatering method depends on many factors. Generally, sand-drying beds have been used for smaller, domestic wastewater treatment operations where sufficient land usually is available at a reasonable cost. Dewatering of sludge organisms by this method requires pretreatment by anaerobic digestion in order to prevent odors and render the sludge dewaterable. Climatic conditions will also greatly affect this method. The method consists principally of pumping digested sludge uniformly onto sand strata, allowing the sludge to dry over a period of time, removing the dried cake (usually by manual means), and cleaning and scarificating the bed, after which the process can be repeated. Digested sludge normally is applied to a sand-drying bed at a depth of 8 to 12 in.; sludge drying times of 1 to 3 months normally are allowed—depending on the sludge and climatic conditions.

Vacuum filtration has the inherent advantage of working well on a wide variety of sludges and producing a relatively dry filter cake which can be incinerated. The solids capture is very good by this method, although sludge preconditioning with chemical coagulants is normally required for effective operation. Vacuum filtration normally is conducted on a slowly rotating drum filter. The filtration drum consists of a series of cells which run the length of the drum and can be placed under vacuum as required. The drum is covered with a suitable filter medium (e.g., wool, cotton, synthetic filter cloth, steel mesh, etc.). The slurry to be dewatered is contained in a tank through which the filter drum rotates. The filter cake initially is formed on the filter medium as the drum rotates through the slurry tank and then is dried by liquid transfer to air drawn through the cake. At the end of the cycle, the dried cake is removed from the drum by a knife-edged scraper or equivalent device. A schematic drawing of a typical vacuum filtration system is shown in Fig. 35.

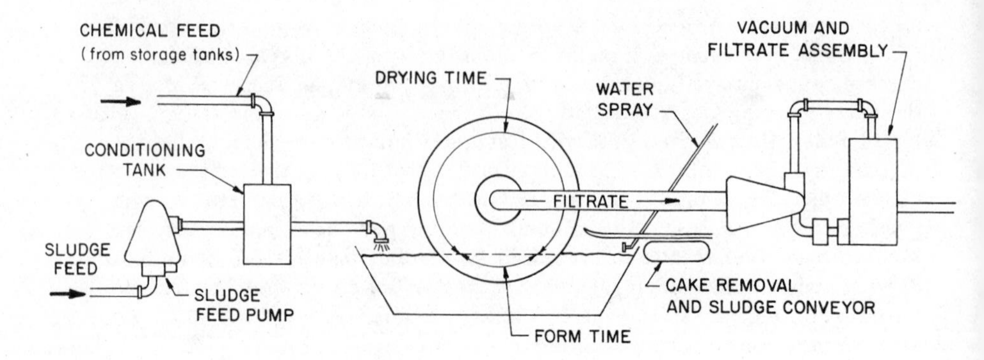

FIG. 35. Schematic drawing of vacuum filter system.

The operation of pressure filtration systems is similar to that of vacuum filtration in that a porous medium is used to separate the solids from the suspending liquid, i. e., as the solids are captured on the medium, they build up to form a cake through which the liquid is filtered. Sludge pumps provide the pressure to force the water through the cake—rather than pulling the water through by vacuum. Filter-press operation classically has been on a batch basis and limited mostly to European installations. Recently, however, a number of filter presses have been developed which are capable of continuous operation and produce reasonably dry filter cakes. One system (see Fig. 36) consists of two vertical moving belts running together at an acute angle, thus forming a narrow, tapering pressure space into which the sludge to be filtered is pumped. Dewatering takes place in the following two-stage sequence as the material moves continuously downwards: (1) liquid stage which utilizes normal hydrostatic pressure, and (2) solid stage which requires application of low pressure.

Continuous-flow centrifuges are now utilized in many industrial and municipal sludge dewatering operations. Although the resulting centrate is generally of poorer quality than the corresponding filtrate would be from sludge filtration, the process has the advantage that the capital cost is low compared to other mechanical dewatering equipment, operating and maintenance costs are moderate, the unit is totally enclosed, odors are minimized, and operation is relatively simple. The most effective centrifuges for sludge dewatering are continuous-flow horizontal solid-bowl machines (see Fig. 37). These machines are cylindrically shaped and fitted with a truncated cone section on one end. Sludge is continuously fed into a rotating bowl through a stationary pipe along the axis of rotation. As the solids enter the centrifuge, they are acted upon by the centrifugal force which

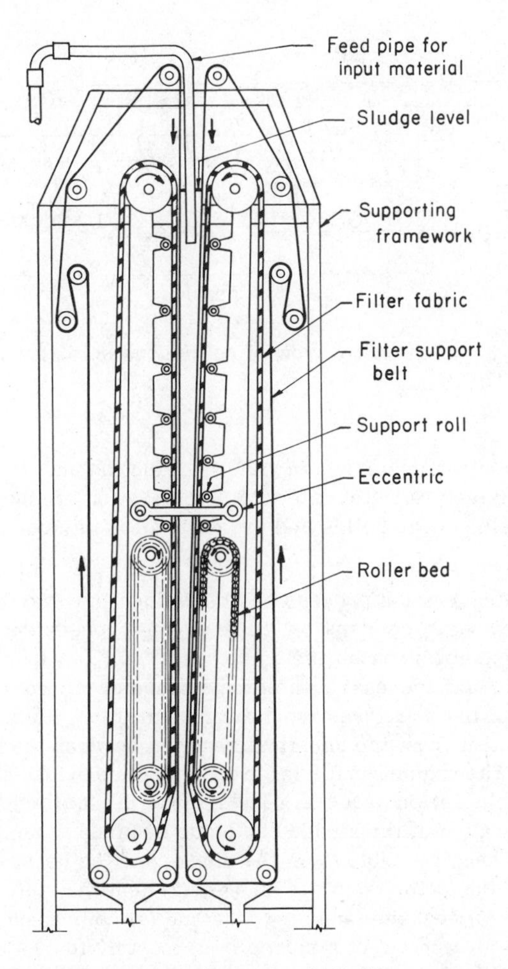

FIG. 36. Schematic drawing of moving belt filter press.

deposits them against the wall of the bowl where they are scraped to the conical section by a screw conveyor and ultimately discharged. The liquid effluent (centrate) is discharged from effluent ports after passing the length of the centrifuge under centrifugal force.

d. Heat drying and combustion

The incineration of sludge is becoming more and more widely practiced as sufficient land for other treatment and disposal methods is more difficult to obtain. Although the process is relatively expensive and can create an air pollution problem if it is not properly operated and controlled, sludge

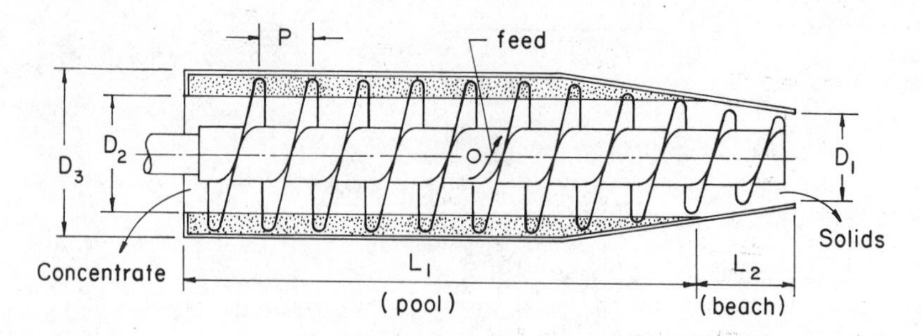

FIG. 37. Schematic view of continuous solid/bowl centrifuge. From Ref. 58.

incineration can achieve both volume reduction and solids stabilization. The major factors which affect the incineration of sludge are the calorific or heat value of the solids and the residual moisture content in the sludge cake.

The incineration process can be divided into two distinct operations: (1) drying and (2) combustion. The drying process entails the raising of sludge temperature from the ambient to 212°F, evaporating the water from the sludge, and increasing the temperature of the water vapor to the temperature of the exit gases from the incinerator. The combustion process liberates heat from the chemical reaction between the sludge fuel and oxygen. The organic material in the sludge provides the source of fuel for the incineration process. The amount of auxiliary fuel required for the incineration of the sludge can be determined from a heat balance between the heat available from the sludge and the heat required to: (1) evaporate the water from the sludge, (2) heat the solids to ignition temperature, (3) heat the incoming air, and (4) compensate for heat losses. If the organic content of a sludge is at least 40 to 50%, the dewatering process will generally provide material for a self-sustaining combustion reaction, once the burning of auxiliary fuel raises incineration temperatures to the ignition point.

The products of sludge incineration from municipal wastewater-treatment plants primarily are ashes and stack gases. If the operating conditions are satisfactory, the ash is primarily sterile inorganic material, while the composition of the stack gases will depend to a great extent on the type and operation of installed air cleaning devices. Results of mass spectrographic analyses of stack gases from municipal sludges produced under good operating conditions indicate that the gas is composed predominately of nitrogen (82%) with smaller percentages of carbon dioxide (9%) and oxygen (8%); hydrocarbons, oxides of nitrogen, hydrogen sulfide, and oxides of sulfur can be present under poorer operating conditions.

The incineration processes available for the burning of sludges include multiple hearth furnaces, rotary kiln furnaces, and fluidized bed reactors. Wet oxidation processes in which the organic constituents in the sludge are oxidized at a temperature between 350 and 500°F, and at a pressure between 1000 $lb/in.^2$ and 2000 $lb/in.^2$ are also available.

2. Sludge Disposal

In the final analysis, however, a residue will always result from any sludge treatment process. Thus, in order to achieve complete sludge treatment, this residual material ultimately must either be disposed of effectively or utilized in some fashion. As indicated in Table 5B, sludge disposal techniques currently practiced consist principally of land or sea disposal techniques, depending on geographical location.

a. Landfilling

Sanitary landfills have been utilized as final disposal sites for virtually every kind of raw and treated sludge. If sludge is to be delivered to the landfill by truck rather than by pipeline, the economics will usually indicate that dewatering for volume reduction will result in justifiable savings. Operation of a landfill receiving sludge should be similar to one receiving solid wastes, i.e., the wastes should be deposited in a designated area, compacted in place with a tractor or roller, and covered with a 12-in. layer of soil. The sanitary landfill method is most suitable if it is used also for disposal of refuse and other solid wastes of the community. With daily coverage of deposited sludge, nuisance conditions such as odors and flies are minimized.

Sanitary landfills for sludge disposal have the inherent disadvantages of requiring large land areas, possibly polluting the groundwater, and having their operation affected by the weather. The use of sanitary landfills for sludge disposal from medium- to large-size urban cities is uncommon because land is too expensive or simply not available at these locations. In those cases where landfilling is practiced currently, more expensive but more compact sludge treatment-utilization techniques are being substituted at a rapid rate.

b. Lagooning

Lagooning is currently a widely practiced sludge disposal technique—particularly at industrial sites. Because lagooning is cheap and simple, it will continue to be used widely, provided that sufficient land is available at a reasonable cost. Various types of sludge lagooning operations exist. Lagoons have been used as mechanical thickeners, storage tanks, digestion tanks, drying lagoons (from which sludge is periodically removed), and permanent lagoons in which sludge remains for an indefinite period. Lagoons are constructed generally in areas where the soils are porous unless groundwater contamination is a threat. Sludge from domestic wastewater treatment plants can generally be dewatered to about 55 to 60% moisture content in a two-to-three-year period in a lagoon; a residual moisture content in the range of 60 to 70% is more common in most operations, however. Lagoons are subject to the same general disadvantages as landfills, i.e., large land areas are required, possible groundwater contamination can result and, in general, they are not feasible for medium-to-large-size urban areas because land is either too expensive or simply not available. In addition, unpleasant odors can result frequently from lagooning operations.

c. Deep-well disposal

Subsurface injection is suggested frequently for the disposal of sludges, but is currently seldom approved of in most states. Although the method is in general appreciably less expensive than are conventional methods of sludge treatment or utilization, the potential hazards associated with its use, particularly with respect to groundwater pollution, generally preclude its acceptance. Both natural cavities, such as caverns, depleted mines, wells, etc., and man-made ones have been used for sludge disposal. Numerous states have legislation which prohibits the filling of natural cavities, however. The production of man-made cavities requires expensive excavation and/or blasting (including nuclear blasting) and presents problems and costs associated with disposal of excavated material and with radiation hazards when working with nuclear materials. Subsurface injection is governed by strict controls. State regulatory agencies, in general, view deep-well disposal only as a temporary means of disposal. Approval for new systems and/or approval to continue discharging through existing systems generally necessitates a clear demonstration that such techniques will not interfere with the present or potential use of subsurface water supplies, or otherwise damage the environment.

d. Discharge to the sea

Sludge disposal by discharge into the sea has been practiced in past years by numerous cities located on coastlines. The practice has proven to be less costly than alternative sludge disposal methods. Both pipeline discharge and barging have been employed successfully—with some barging distances totaling 400 miles round trip. The usual practice has been to digest anaerobically the sludge before ocean discharge in order to preclude the problems associated with the disposal of raw sludge.

This method has the advantage, like deep well disposal, of providing a method for completely removing the sludge from the wastewater treatment plant and, as such, represents a complete disposal method. Major considerations have been employed to ensure the removal of floatable materials before discharge and to select discharge sites to preclude any sludge deposits on beaches or other shoreline areas. The suitability of allowing this sludge disposal practice (as in the case of deep well disposal) has been questioned on numerous grounds and its future use also may very well be limited or curtailed completely. In this regard, such questions have been raised concerning: (1) the impact of the sludge discharges on the aquatic environment, (2) the accumulation of bottom deposits, (3) the potential health hazard, and (4) the wasting of a land resource into the sea. Congress has mandated that the ocean dumping of sewage sludge, dredge spoil, and certain industrial wastes be curtailed by December 31, 1981.

3. Sludge Utilization

The utilization of sludges from manufacturing and waste treatment processes has become more of a necessity in recent years. Such a trend will no doubt continue as disposal of increasing amounts of sludge becomes less feasible. "Recycle" is now a catchword which has been introduced and widely adopted by industry and the public. The brine wastes from analogous processes are more difficult both to dispose of and to utilize because the water content is so high; the large volume added by the water to the chemical content causes greater transportation costs for disposal or utilization. The number of potential uses for sludges and brines is limited only by the imagination and recycling costs. In the future, costs of recycling waste products may be added to the initial costs of manufactured goods. Table 8, for example, lists a number of utilization methods for various sludge types.

The South Lake Tahoe Wastewater Reclamation Project has pioneered the concepts of chemical recovery from sludges produced during the wastewater treatment process. In the total treatment scheme operated at that facility, lime is added to the stream in order to raise the pH so that calcium phosphate precipitation occurs and ammonia stripping can be carried out [61]. The lime is recovered from the calcium phosphate sludge by an elaborate but efficient recalcination process, so that the recovered lime may be added to the chemical feed of the next treatment cycle (see Fig. 38). Depending on the initial calcium and alkalinity concentrations of the water and the recovery method selected, the quantity of "recovered" lime actually may be greater than that added to the water and thus may completely eliminate the need for additional chemical. Regardless, utilization of such techniques, then, not only eliminates an otherwise necessary sludge disposal problem but also provides necessary chemical for the wastewater treatment process.

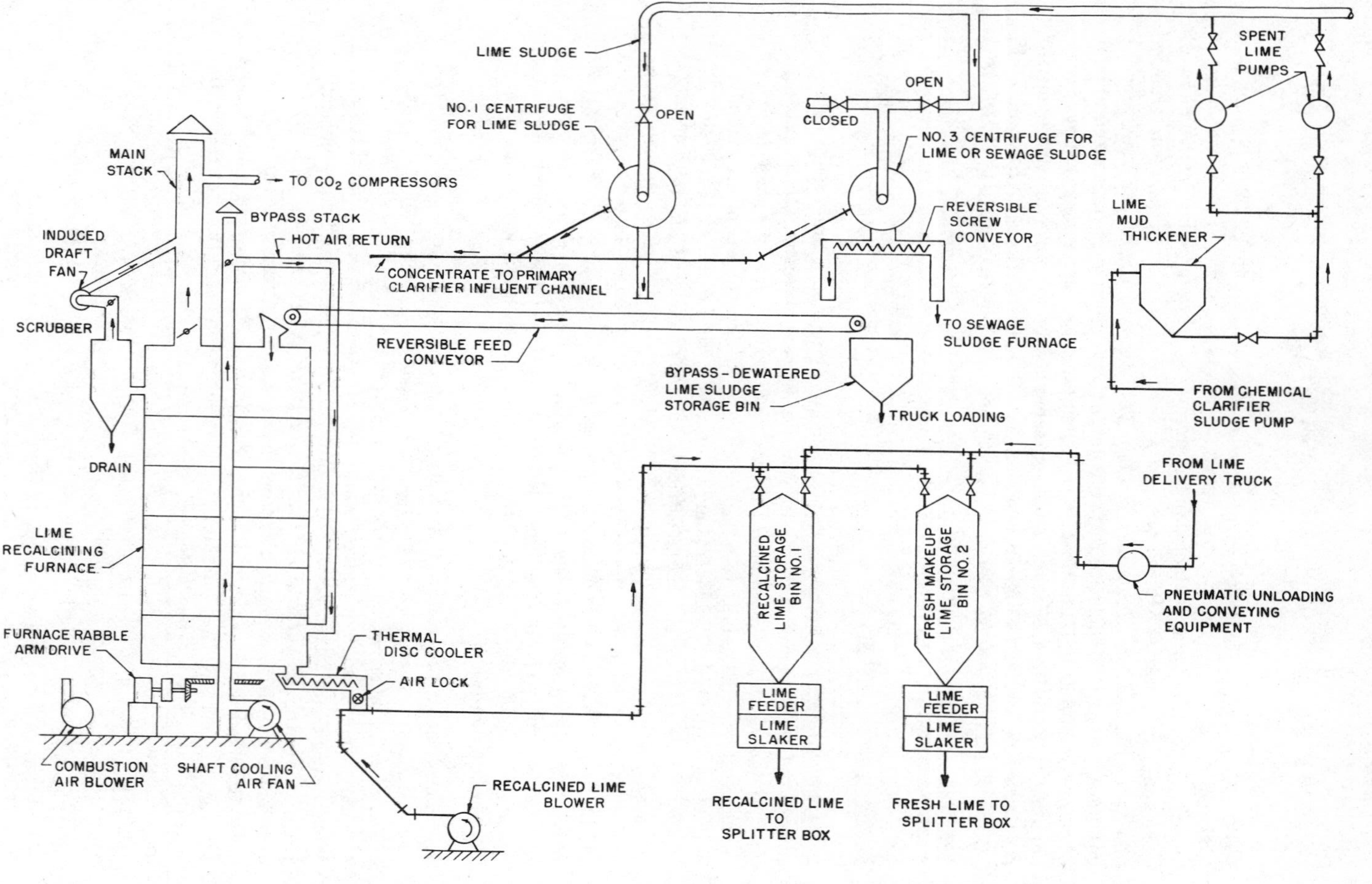

FIG. 38. Flow diagram of lime sludge recalcination system. From Ref. 61.

Recovery attempts to obtain heavier metals from sludge have been unsuccessful or economically unfeasible at the present time. A method utilizing high-temperature plasmas for the elemental recovery of all sludge constituents has been described by Eastland, but such exotic processes seem futuristic [62].

Studies have been conducted on a laboratory scale to show that closed-cycle food-wastes systems can be operated on a self-perpetuating basis [63, 64]. These systems involve the synthesis of algae from the sludges of wastes; the algae, in turn, serve as food for the primary consumer.

It has been suggested that dried sludge could be used as an auxiliary fuel, since disposal techniques by incineration can be self-supporting if the organic content of the sludge is sufficient. In plants which dispose of sludge by incineration, instead of being wasted, the heat produced could be utilized as an energy source within the plant, e.g., to heat sludge digestors.

Sludge "farming" has become a widely touted procedure for the utilization of waste sludge. The Chicago Metropolitan Sanitary District has operated such a facility very successfully [65, 66]. The cost of transporting the sludge to an outlying district has incurred increasing costs, however, and pipelines have been suggested for the transport of large volumes of sludge [67]. The advantageous aspects of this type of utilization are that the inorganic fraction of the sludge serves to condition the soil and the organic and nutrient fractions fertilize the soil. With such waste recycling, marginal or depleted fields can be made more productive. Certain aspects of this technique require additional investigation at this time, however, before widespread use of such methods can be endorsed; these considerations include both the fate of heavy metals and pathogenic microorganisms in the farming operation. Dried sludges from activated sludge operations in a number of cities (e.g., Milwaukee, Chicago, Winston-Salem, N.C.) have been bagged and sold as a form of fertilizer with varying degrees of success [68].

Other uses for waste sludge have been suggested or demonstrated. Gasification of sludge to produce low-molecular-weight hydrocarbons to serve as fuels is one example [69]. The slag produced from incinerated sludge has been studied as a road building and concrete manufacturing aggregate [70].

Many of these waste sludge utilizations are not economically or technologically feasible at the present time. As costs of ultimate disposal increase or when certain questionable disposal techniques are no longer permitted, however, these procedures will become more attractive. In addition, the ultimate need to recycle wastes which currently are disposed of in some manner but could be utilized as a resource will become more prevalent as resources become more scarce and expensive.

V. SUMMARY

This chapter has provided information relative to the impact of synthetic detergents on the environment. There is no clear-cut, universally applicable, treatment of this topic. A valid assessment of this situation can only be made after determining, as accurately as possible, the current status of each of the many related factors for each geographical location under consideration. Because of the number, complexity, interrelationship, and extreme variations possible in each of the pertinent factors, it is suggested that an analysis of this nature can be conducted most conveniently by utilizing computer simulation and modeling techniques.

The following major considerations were submitted as the most pertinent to the evaluation:

1. Identification of the potential pollutionary constituents present in synthetic detergent formulations
2. Elucidation of the fundamental biological, chemical, and biogeochemical interrelationships inherent to the environment
3. Determination of the effect of the potential pollutionary constituents on the biological, chemical, and biogeochemical cycles of the environment
4. Delineation of the quality of the environment desired
5. Consideration of the capabilities of the wastewater treatment systems available and the economic and environmental aspects associated with the use of each type of process

Since synthetic detergents are utilized primarily in water, their main pollutionary impact will be to the nation's waters. Their impact on the land or the air cannot be entirely neglected either. Disposal of synthetic detergents on the land can occur, for example, by direct or indirect wastewater discharges; utilization of wastewater treatment processes results in the production of concentrated sludge suspensions or brine solutions which require subsequent treatment and/or disposal generally to the air and/or land.

Two of the basic thermodynamic laws state briefly that: (1) matter can neither be created nor destroyed, and (2) every change in form involves an energy transformation. Clearly, then, the elemental constituents of synthetic detergents cannot be eliminated, only changed in form (with a corresponding energy transformation), and are limited to an analysis which considers what form the various elements of synthetic detergents are in when they are released into the environment. The impact of these constituents on the environment can be assessed with respect to where these elements enter their respective natural cycles, how these "inputs" affect the functioning of each of the individual cycles, and how their effect on the balance or interrelationships between the various cycles ultimately dictates environmental quality (Fig. 18).

One or more of the following immediate reactions may occur as a result of releasing synthetic detergents to the aquatic environment:

1. The components may react chemically with other dissolved or suspended ions or compounds.
2. The components may be taken up biologically and, therefore, stimulate metabolic activity of any indigenous organisms.
3. The components may remain in solution, thus increasing the concentration of total dissolved species.

In any case, given sufficient time and regardless of their immediate fate, the various components ultimately will enter into the appropriate chemical and biological cycles inherent to the biosphere. It would be impossible to enumerate (i.e., model) all of the possibilities that could occur for each of the constituents present in synthetic detergents. Nevertheless, by employing a sensitivity analysis, it is possible to review the various alternatives and model only those which are the most significant. Attempts to be overly rigorous in the analysis frequently make the system so complicated that no effective or viable model can be developed; hence the value of a computer-aided systems analysis is completely lost and a value engineering judgment cannot be made. By properly selecting and modeling the key elements of a system, a model of sufficient accuracy can be synthesized fairly readily, which will yield adequate information for a most valid assessment.

In this regard, characteristics of typical anionic, nonionic, and cationic detergent formulations were reviewed and grouped in terms of their organic and inorganic constituents. The elemental constituents considered to be the most significant were carbon, nitrogen, phosphorus, and sulfur. The chemical cycles for each of these elements were presented and the potential input locations for discharges from synthetic detergent formulations to each of these cycles were noted.

When considering biological effects, pollution has its most direct effect at the unicellular levels of the plant and animal kingdom, which are represented by algae and bacteria, respectively. Hence, accurate depiction of the metabolic activities of these species is one of the most important considerations requisite to this analysis. The metabolic activities of algae are described in terms of autotrophic photosynthetic relations, and the metabolic activities of bacteria are described in terms of aerobic and anaerobic heterotrophic relations. In the integrated ecological system of the biosphere there is a tendency to approach a steady state between production and destruction of organic material. Pollutional discharges at localized points of the biosphere, however, will cause an imbalance between the heterotrophic-autotrophic activities.

The organic fraction of a synthetic detergent formulation will have its most significant effect on the environment by stimulating heterotrophic activity—provided that the organic material is biodegradable. Previously manufactured synthetic detergents whose organic formulations were biologically undegradable tended to persist in the environment, caused foaming problems, and accumulated in water with each use. Alterations by the manufacturers to biodegradable detergents has eliminated these problems and the organic formulations can be removed from wastewaters by conventional biological wastewater treatment processes.

Direct discharge of any of the current synthetic detergent formulations to the environment will result in the organic fraction exerting an oxygen demand on the receiving water. The oxygen demand originates from the biological activity of organisms as they oxidize the organic material as a substrate to carbon dioxide and water in order to gain energy to synthesize new cellular protoplasm. In the absence of any dissolved oxygen in the receiving water, the heterotrophic activity will change to one of anaerobic metabolism, and the major biochemical reaction will be one of decomposition in which the organic material will be reduced to methane and carbon dioxide.

From an environmental impact consideration, the major inorganic constituents of synthetic detergents of concern are nitrogenous and phosphatic components. Biologically, these species are commonly referred to as fertilizing elements and their most significant effect on the environment is in stimulating autotrophic biological activity. The most direct observable effect of this activity is the increased productivity or growth which occurs in the receiving water. The increase in productivity of a water body as a result of increasing nutrient levels is referred to frequently as eutrophication. Hence, the direct discharge of nitrogenous and phosphatic elements to our lakes and streams has been thought to be the cause of eutrophication, or "killing" of these waters. These accusations have been directed particularly at phosphorus in synthetic detergents, as phosphorus was so widely used in detergent formulations. As a result, the sale of phosphate detergents was banned in some areas and appreciable research and developmental efforts have gone into attempting to find acceptable phosphate substitutes. As indicated throughout this chapter, the accurate assessment of the impact of phosphorus (or any other inorganic constituent) on the environment is a complex problem and no universally applicable treatment is possible. An accurate assessment can be made only after evaluating all the relevant factors and their interrelationships for a given geographical area.

Some of the most relevant facts regarding the phosphate controversy were summarized by Steinfeld, the former Surgeon General of the United States from December 1969 to January 1973 [60]. The following facts were submitted:

1. The health and environmental effects of the alternatives to phosphates are either partially unknown (as with NTA) or known to be largely undesirable and even hazardous.
2. Eutrophication is an immensely complex process that cannot be linked to a single element such as phosphorus; other elements such as manganese, iron, zinc, carbon, silica, and molybdenum are also required for growth.
3. Sewage treatment, not detergent reformulation, is the answer to phosphorus in wastewater.

The most significant impact of synthetic detergent discharges to the environment is observed in their stimulatory effect on heterotrophic and autotrophic microbial activities. Any changes from the steady-state activities of these microorganisms caused by synthetic detergent discharges will be transmitted to the higher trophic levels which are dependent on the lower levels for growth, and ultimately the food chain and health in humans can both be affected. Since an appreciable loss of energy occurs with each transformation up the ecological ladder and there are a number of transformations which must be made, there must be an appreciable effect on the lower levels before human health and well-being will be influenced significantly. This should not be considered as a startling statement, as the human is known to be a relatively strong species capable of tolerating appreciable changes in environmental quality. On the other hand, such a statement should not be interpreted to mean that one can deteriorate the environment indiscriminately with uncontrolled discharges. Unfortunately, there is close to a direct response at the unicellular level to each additional pollutional discharge, and a degradation in environmental quality results which is closely proportionate to the amount of pollution added.

Although human health may not be directly influenced with each degree of additional degradation, nevertheless humans generally lose a specific use for the environment with each reduction in quality. The world environment is being degraded and uses are being lost; yet the degradation has not reached the level where human life is being threatened. Humans have survived and could continue to survive with reduced environmental quality, but the question is asked whether or not environmental degradation should be allowed to continue. If environmental degradation should be checked, what quality of the environment should be maintained?

The answer to this question must be given in the future as a value judgment made in a rational manner. The decision-making body must be knowledgeable on environmental matters, be provided with the pertinent information on which to base its decision, and have descriptive and viable predictive models to aid it in its decision-making process. Equitable judgments with respect to establishing and maintaining a given environmental quality can only be achieved by implementing proper environmental planning and management techniques. As has been the case so frequently in the past, either the complete neglect of or arbitrary establishment of a few standards can no longer be tolerated if we are truly to maintain and manage our environment.

It must be determined if there really is a need for or a justified benefit from the use of a given product or substance; and if so, what is the best way to use it, and what environmental control measures should be employed. Provided that the use of a given material or substance is believed to be necessary, maintenance of the desired environmental quality must then be achieved by proper environmental controls.

In the case of maintaining or meeting specified water quality standards, adequate wastewater treatment facilities will have to be provided in each case. For the most part, the technology exists to treat a wastewater of virtually any waste characteristics sufficiently to meet established or anticipated water quality standards. The capabilities and associated costs for primary and secondary wastewater treatment processes, for example, have been clearly established and demonstrated over past years for both municipal and industrial wastewaters. Although the more recent tertiary and other advanced systems do not have as much field operational data, sufficient background information exists for such systems as physical-chemical treatment systems, most tertiary systems, and such advanced wastewater treatment systems as ion exchange, electrodialysis, and reverse osmosis, in order to select and design adequately these systems to ensure a reasonable degree of reliability in their operation.

It is possible, by the proper design and operation of units such as those mentioned above, to produce reliably a final water of any desired quality, e.g., literally to produce such high quality effluents as to meet drinking water standards. As a general rule, the higher the final water quality desired, the greater will be the number of processes or sophistication of the treatment system required and the greater the cost of the system. The prospect of releasing a water of extremely (or even reasonably) high quality is subject to much more deliberation than has been the case in the discharge of lower quality waters, and in many cases the high quality waters are not released to the environment but, rather, are recycled. Considering the increase in water quality standards anticipated in future years and the fact that zero pollutional discharge is currently the goal of the future, the employment of water-wastewater recycling processes, by necessity, will become widespread.

Regardless of the type or sophistication of wastewater treatment process(es) employed, a sludge suspension and/or brine solution will result. This material consists of the impurities which have been removed from the wastewater and/or the biological or chemical species that were produced as a result of treating the wastewater. The greater the amount of impurities to be removed, the greater will be the volume of sludge or brine which results. The ultimate fate of this material is the biggest problem in wastewater treatment.

This sludge or brine material is produced as a result of the treatment and represents the pollutants in a concentrated form that have been removed from the wastewater. This material must now be handled in some fashion. No miraculous solution has been nor may ever be developed to eliminate this problem. As mentioned earlier, like any matter, this material is governed by the laws of thermodynamics and as such it can be changed in form, but it cannot be either created or destroyed.

The methods available for sludge or brine handling can be grouped into three categories: (1) disposal, (2) treatment, and (3) utilization. Sludge or brine disposal has been widely practiced in previous years, utilizing such techniques as deep-well injection, burial, and ocean dumping. These techniques are currently meeting with less and less favor, however, and in the future such practices, at best, will be very limited. Sludge treatment methods fall into four categories: (1) concentration, (2) digestion, (3) dewatering, and (4) heat drying and combustion. Frequently, processes from a number of these categories are combined, all with the specific purpose of reducing the volume of sludge which ultimately must be handled. Yet, in any case, some material will still remain from each process which ultimately will require further handling.

Considering the fact that the sludges and brines from wastewater treatment plants cannot be handled completely by treatment methods, and that attempts to employ sludge or brine disposal methods will no doubt meet with increasing resistance in the future, it becomes evident that more attention must be given to utilizing these materials in the future. The possibility of having to employ this concept to a greater degree in the future should not be met with a great deal of resistance, although it has been and no doubt will continue to be—particularly in the United States. The concept is fundamentally sound and has as its basis the return of sludge or brine constituents to their natural cycles in the biosphere. It is frequently stated, "pollutants are simply resources out of place." Proper utilization procedures would dictate, therefore, that any utilization method employed ensures that the elements are reintroduced into the environment in a manner compatible with their natural cycles.

The actual uses which ultimately can be made of a sludge or brine depend to a large degree on such factors as their composition, their potential health hazards, and the economics involved. Analysis of the ranges in characteristics of sludge composition typically encountered indicates that there are four major properties of sludges which can be utilized: (1) utilization of their calorific values, (2) utilization of their nutritional values, (3) recovery or reuse of materials contained in the sludge, and (4) utilization of the material as a builder or filler material. Some of the utilization processes currently being used or under development include: use of the calorific value by sludge incineration and as an animal foodstuff; use of the nutritional values as agricultural fertilizers; reuse of chemical sludges in process streams; and use of inert sludges in concrete and concrete-related products.

In summary, assessing and regulating the impact of synthetic detergents on the environment is no different in a broad sense than the considerations involved in this regard for any other product. Product production and marketing should be based on what would best meet the needs of society with a consideration for the product's environmental impact. Where waste streams must result, wastewater treatment processes must be provided which properly treat the wastewater to such a level that it can be discharged to the environment in consonance with existing water quality standards and/or be recycled. The sludges and/or brines resulting from the treatment processes must be properly treated and ultimately utilized in such a fashion as to be reintroduced into the environment in a manner compatible with the natural cycles existing for their elemental constituents. Achievement of these ends will necessitate that sufficient monetary funds be provided and it must be based on proper planning, management, and regulation.

GLOSSARY

Definitions of some of the commonly used terms in the measurement of pollution (see Table 3).

Biochemical Oxygen Demand. The quantity of oxygen utilized by bacteria in the biochemical oxygen of organic material under aerobic conditions in a specified time and at a specified temperature (usually 5 days at 20°C).

Chemical Oxygen Demand. The quantity of oxygen required for checmial oxidation of organic material to carbon dioxide and water.

Coliforms. A group of microorganisms that are used as indicators of pollution as their source, either from (a) intestinal tracts of man and animals, or (b) soil or plants. The group includes a number of organisms of which the following are generally considered the most important: *Aerobacter aerogenes*, *Aerobacter cloacae*, and *Escherichia* coli. The coliform group is defined as all aerobic and facultative anaerobic, gram stain-negative, non-spore-forming bacilli which ferment lactose with gas formation within 48 hr at 35°C.

Color. The color imparted to water from the presence of natural metallic ions (iron and manganese), humus and pest materials, plankton, weeds, and industrial wastes. Color is determined from a standard platinum-cobalt solution in which one unit of color is that produced by 1 mg/liter platinum in the form of chloroplatinate ion.

Fecal Coliform. A test to differentiate the organisms of the coliform group into those of fecal origin and those derived from nonfecal sources. The test differentiates between those organisms of the intestinal tract of warm-blooded animals rather than from other sources.

Heavy Metals. Metals that can be precipitated by hydrogen sulfide in acid solution, e.g., cadmium, chromium, copper, iron, lead, manganese, nickel, mercury, and zinc. The presence of metals in domestic sewage, industrial effluents, and receiving streams is a matter of serious concern because the possible toxic properties of these materials may affect adversely sewage treatment systems or the biological systems of the receiving streams.

Mixed Liquor Suspended Solids (MLSS). The dry weight (103° C) of material taken from a biological reactor (such as an activated sludge tank) after filtration through a 0.45 micron membrane filter, or equivalent, expressed as a concentration (i.e., mg/liter).

Nitrogenous Species. Four forms of nitrogen are of significance from a water quality standpoint. From the most reduced to oxidized states, these forms are: organic nitrogen, ammonia nitrogen, nitrite nitrogen, and nitrate nitrogen. Each form can be determined independently and the sum represents the total nitrogen concentration.

Phosphate Species. Three forms of phosphorous compounds are of significance from a water quality standpoint: orthophosphates (e.g., $H_2PO_4^-$, HPO_4^{2-}, PO_4^{3-}), polyphosphates (molecularly dehydrated orthophosphates (e.g., $Na_3(PO_3)_6$, $Na_5P_3O_{10}$, $Na_4P_2O_7$), and organically bound phosphorus.

Salt. A term loosely used to denote the dissolved species in a water. The total dissolved solids concentrations can be measured as dry weight after evaporation or approximated by conductivity measurements. More precisely, salt concentrations would refer to the concentration of chloride ion, and sometimes to the sulfate ion concentration as well.

Standard Plate Count. A method for determining the bacterial density of waters. The method is a highly empirical one utilizing tryptone-glucose-extract agar or plate-count agar with incubation at 35°C.

Suspended Materials. The dry weight (103°C) of material retained on a 0.45-μm membrane filter after filtration.

Total Oxygen Consumption. A rapid method (approximately 2 min) of determining the oxygen equivalent of organic material. The procedure is based on injecting the aqueous sample into a catalytic combustion tube, wherein at 900°C the organic material is converted to CO_2 and H_2O. The CO_2 can then be measured by a Golay-type thermal detector or the CO_2 produced can be measured by infrared techniques.

Total Oxygen Demand. The total quantity of oxygen demanded by a sample. This value would include, in addition to the oxygen demand of the organic material, the demand of reduced chemical species such as the nitrogenous oxygen demand of the reduced nitrogen species.

Turbidity. An expression of the optical property of a water which causes light to be scattered or absorbed rather than to pass through the sample in straight lines. Silica dioxide was chosen as the standard where 1 mg/liter SiO_2 is equivalent to one unit of turbidity.

REFERENCES

1. D. A. Okun, Environ. Aff., 2(1), 64 (1972).
2. G. M. Fair, J. C. Geyer, and D. A. Okun, Water and Wastewater Engineering, Vol. 1 Wiley, New York, 1966, p. 6-3.
3. D. K. Todd, The Water Encyclopedia, Water Information Center, Port Washington, N.Y., 1970.
4. H. W. Streeter and E. B. Phelps, Public Health Bull., no. 146 (1925).
5. C. N. Sawyer, J. Water Pollut. Control Fed., 38, 737 (1966).
6. C. N. Sawyer and P. L. McCarty, Chemistry for Sanitary Engineers, McGraw-Hill, New York (1967).
7. E. M. Davis and E. F. Gloyna, J. Water Pollut. Control Fed., 41, 1494 (1969)
8. R. C. Davis, perconal communication, 1971.
9. R. M. Brown, N. I. McClelland, R. A. Deininger, and R. G. Tozer, Water Sewage Works, 117(10), 339 (1970).

10. W. Stumm and J. J. Morgan, Aquatic Chemistry, Wiley Interscience, New York, 1970.

11. R. D. Swisher, J. Water Pollut. Control Fed., 35, 877 (1963).

12. Public Health Service Drinking Water Standards (PHS Publication no. 956), 1962.

13. W. T. Sullivan and R. L. Evans, Environ. Sci. Technol., 2, 194 (1968).

14. W. T. Sullivan and R. D. Swisher, Environ. Sci. Technol., 3, 481 (1969).

15. M. Lieber, Water Sewage Works, 116, R66 (1969).

16. T. O. Thatcher and J. F. Santner, in Proceedings of 21st Annual Industrial Waste Conference (J. M. Bell, ed.), Purdue University, West Lafayette, Ind., vol. II, 1966, p. 996.

17. Q. H. Pickering and T. O. Thatcher, J. Water Pollut. Control Fed., 42, 243 (1970).

18. C. A. Lawrence, Surface Active Quaternary Ammonium Germicides, Academic Press, New York, 1950.

19. T. W. Humphreys and C. K. Johns, Rev. Can. Biol., 11(5), 517 (1953).

20. F. A. Ferguson, Environ. Sci. Technol., 2, 188 (1968).

21. F. R. Hayes and J. E. Phillips, Limnology and Oceanogr., 3, 459 (1968).

22. M. W. Tenney and W. F. Echelberger, Jr., Effects of Domestic Pollution Abatement on an Eutrophic Lake—Summary of Five Year Water Quality Data. (Partial report for FWQA Demonstration Grant No. WPD-126), September 1970.

23. K. M. MacKenthun, L. E. Keup, and R. K. Stewart, J. Water Pollut. Control Fed., 40, R72 (1968).

24. J. Shapiro, J. Water Pollut. Control Fed., 42, 772 (1970).

25. L. W. Kuentzel, J. Water Pollut. Control Fed., 41, 1737 (1969).

26. M. W. Tenney, W. F. Echelberger, Jr., K. H. Guter, and J. B. Carbery, in Nutrients in Natural Waters (H. E. Allen and J. R. Kramer, eds.(, Wiley, New York, 1972, p. 391.

27. J. A. Borchardt and H. S. Azad, J. Water Pollut. Control Fed., 40, 1739 (1968).

28. Anonymous, Environ. Sci. Technol., 3, 1243 (1969).

29. Anonymous, Chem. Eng. News, 45, 18 (1967).

30. P. J. Weaver, J. Water Pollut. Control Fed., 41, 1647 (1969).

31. J. L. Steinfield and W. D. Ruckelshaus, J. Am. Water Works Assoc., 63, 743 (1971).

32. L. Loeb, in Technical Conference on Household Equipment, Association of Home Appliance Manufacturers, Chicago, 1971, p. 26.

32a. A. P. Consdorf, Appliance Manuf., p. 58 (March, 1972).

33. E. Jungermann and H. C. Silberman, J. Am. Oil Chem. Soc., 49, 481 (1972).

34. J. B. Carberry and M. W. Tenney, J. Water Pollut. Control Fed., 45, 2444 (1973).

35. E. F. Gloyna, in Process Design in Water Quality Engineering—New Concepts and Developments (E. L. Thackston and W. W. Eckenfelder, eds.), Jenkins, 1972, p. 45.

36. K. H. Mancy and D. A. Okun, J. Water Pollut. Control Fed., 32, 351 (1960).

37. W. O. Lynch and C. N. Sawyer, J. Water Pollut. Control Fed., 32, 25 (1960).

38. K. H. Mancy and D. A. Okun, J. Water Pollut. Control Fed., 37, 212 (1965).

39. N. I. McClelland and K. H. Mancy, in Proceedings of 24th Annual Industrial Waste Conference (J. M. Bell, ed.), Purdue University, West Lafayette, Inc., 1969, p. 1361.

40. J. T. Cookson, Jr., J. Water Pollut. Control Fed., 42, 2124 (1970).

41. W. J. Weber, Jr., C. B. Hopkins, and R. Bloom, Jr., J. Water Pollut. Control Fed., 42, 83 (1970).

42. E. F. Barth, Paper presented at Advanced Waste Treatment and Water Reuse Symposium, Chicago, Ill., 1971.

43. D. Vacker, C. H. Connoell, and W. N. Wells, J. Water Pollut. Control Fed., 39, 750 (1967).

44. J. L. Witheron, in Proceedings of 24th Annual Industrial Waste Conference (J. M. Bell, ed.), Purdue University, Lafayette, Inc., 1969, p. 1169.

45. W. F. Milbury, D. McCauley, and C. H. Hawthorne, J. Water Pollut. Control Fed., 43, 1890 (1971).

46. R. D. Borgnan, J. M. Betz, and W. Garber, Paper presented at 5th International Water Pollution Research Conference, San Francisco, Calif., 1971.

47. G. V. Levin and J. Shapiro, J. Water Pollut. Control Fed., 37, 800 (1965).

48. A. B. Menar and D. Jenkins, Environ. Sci. Technol., 4, 1115 (1970).

49. T. L. Theis, P. C. Singer, W. F. Echelberger, Jr., and M. W. Tenney, J. Sanit. Eng. Div., Am. Soc. Civ. Eng., 96, SA4, 1004 (1970).

50. J. M. Cohen, in Nutrients in Natural Waters (H. E. Allen and J. R. Kramer, eds.), Wiley Interscience, New York, 1972, p. 353.

51. T. A. Pressley, D. F. Bixhop, and S. G. Roan, Environ. Sci. Technol., 6, 622 (1972).

52. A. T. Palin, Water Wastes Eng., 54, 151, 189, and 248 (1950).

53. J. R. McLaren and G. J. Farquhar, J. Environ. Eng. Div., Am. Soc. Civ. Eng., 99 (EE4), 429 (1973).

54. B. W. Mercer, L. L. Ames, C. J. Touhill, W. J. Van Slyke, and R. B. Dean, J. Water Pollut. Control Fed., 42, R95 (1970).

55. Summary Report - The Advanced Waste Treatment Program, January 1962 to June 1963 (PHS Publication no. 999-WP-24) April 1965.

56. Summary Report - Advanced Waste Treatment, July 1964 to July 1967 (Federal Water Pollution Control Administration Publication no. WP-20-AWTR-19), 1968.

57. Metcalf and Eddy, Inc., Wastewater Engineering: Collection, Treatment, Disposal, McGraw-Hill, New York, 1972.

58. C. G. Moyers, Jr., Chem. Eng., p. 182 (June 20, 1966).

59. Standard Methods for the Examination of Water and Wastewater, 13th ed., American Public Health Association, New York, 1971.

60. J. L. Steinfeld, Reader's Dig., p. 170 (November 1973).

61. R. L. Culp and G. L. Culp, Advanced Wastewater Treatment, Van Nostrand Reinhold, New York, 1971.

62. B. J. Eastlund, Water Pollut. Control, 109, 18 (1971).

63. H. B. Gotaas, W. J. Oswald, and H. E. Ludwig, Sci. Mon., 79, 368 (1954).

64. G. L. Dugan, C. G. Golucke, and W. J. Oswald, J. Water Pollut. Control Fed., 44, 432 (1972).

65. S. M. King, Fert. Solut., 15, 24 (May, 1971).

66. R. B. Ogilvie, Environment, 1, 1 (1971).

67. T. L. Thompson, Chem. Eng. Prog. Symp. Ser., 67, 413 (1971).

68. F. Styers, Am. City, 86, 48 (1971).

69. H. F. Feldman, Chem. Eng. Prog., 67, 51 (1971).

70. Anonymous, Environ. Sci. Technol., 5, 197 (1971).

Chapter 23

RECENT CHANGES IN LAUNDRY DETERGENTS

Beverly J. Rutkowski

Whirlpool Corporation
Benton Harbor, Michigan

I. INTRODUCTION 994

A. Basic Detergent Components and their Functions 994
B. Factors Initiating Detergent Formulation Change 995

II. NONPHOSPHATE DETERGENT BUILDERS 998

A. Organic Replacements for Phosphate Builders 998
B. Inorganic Replacements for Phosphate Builders 1002

III. SURFACTANT CHANGES IN LAUNDRY DETERGENTS 1006

A. Synthetic Surfactants 1006
B. The Use of Soap with Dispersing Agents 1007

IV. PERFORMANCE CHARACTERISTICS OF PHOSPHATE-RESTRICTED DETERGENTS 1009

A. Detergent Types of the 1970s 1009
B. Performance Evaluation Test Method 1011
C. Detergency Characteristics 1012
D. Problem Areas of Nonphosphate Detergents 1015

V. DETERGENTS OF THE FUTURE 1017

REFERENCES 1018

I. INTRODUCTION

The detergent era in the United States began in the mid-1930s with the introduction of relatively simple light-duty detergents. Consisting almost entirely of synthetic surfactants, these first products, although unsuitable for heavy-duty laundering, were moderately successful [1]. However, it was not until 1947 when phosphate builders were combined with surfactants that the considerable potential of synthetic detergents for home laundering began to be realized. Consumers were quick to see the superiority over soap of these phosphate-built products and by 1953 had made them the dominant product for home laundering.

Following the introduction of phosphate builders, and until 1970, further changes in laundry detergent formulations were generally of an evolutionary type, motivated by small performance improvement or cost reduction factors. With few exceptions, heavy-duty laundry detergents of this period were well performing phosphate-built products differing only in surfactant type (high sudsing anionics and low sudsing nonionics) and to a small extent, in phosphate content (40 to 60%). In 1970, however, laundry detergent formulation changes became more dramatic and, although not inventive in a sense similar to the introduction of phosphate builders, some were equally revolutionary.

Motivation for these more recent formulation changes stemmed not from performance improvement considerations, but almost exclusively from certain external pressures on the detergent industry. As a result, laundry detergents became comparatively poorly performing products with widely differing compositions and performance characteristics.

The purpose of this chapter is to describe recent changes in laundry detergent formulations in terms of their initiating forces, their effects on detergent performance characteristics and their impact on detergent formulations of the future. Since detergent manufacturers jealously guard their formulations and frequently announce formulation changes simply as "improvements," the source of information contained here is quite varied, coming from research publications, observations of the marketplace, detergent supplier technical bulletins, news releases, independent laboratory analyses and evaluations of detergent products and other sources.

A. Basic Detergent Components and their Functions

It is well known that heavy-duty laundry detergents typically contain two major ingredients, surfactant and builder, and a number of other ingredients essential to an acceptable product, although of lesser importance to detergency. These miscellaneous ingredients include anticorrosion agents, antisoil redeposition agents, fluorescent whitening agents, and perfume, etc. The two major components most essential to detergency—surfactant and builder—have undergone recent and significant change. Previous chapters in this volume [2-4] describe in considerable detail the roles of surfactant and builder in fabric detergency, and so these ingredients are discussed only briefly here for the purpose of emphasizing their importance to an adequately performing product.

1. Surfactant Functions

The term surfactant is derived from the words "surface active" which aptly describe the dominant physical action of this detergent component. Frequently, surfactants are called simply the "active" ingredient in a detergent. Detersive action of surfactants, the most important portion of a detergent, is based on their fundamental characteristic to adsorb or concentrate at soil-fiber-water interfaces. This phenomenon of adsorption leads to two basic actions, reduction of surface tensions and stabilization of surfaces, which in turn are responsible for the various surfactant functions. Lowering of surface tensions permits wetting out of soil and fabric, penetration of wash solution into soil-fabric interstices, and dispersion of soil aggregates into small particles. Stabilization of surfaces results in suspension of particulate soils, emulsification of oily soils, and solubilization of oily soils.

2. Detergent Builder Functions

Detergent builders are of nearly equal importance to surfactants since their major function is to prevent divalent calcium and magnesium water-hardness ions from interfering with surfactant action. Additional important builder functions are to provide and maintain optimum alkalinity of the washbath for neutralization of acidic soils, to provide electrolyte which enhances surfactant activity, to deflocculate, disperse, and suspend soil particles, and to emulsify oily soils.

B. Factors Initiating Detergent Formulation Change

External pressure on the detergent industry causing change in detergent formulations was a combination of human emotionalism, government interference, consumerism, energy shortages, an inflationary economy, and, to a small degree, technological advancement.

1. The Environmentalist Movement and Phosphate Legislation

The initial impetus to the most significant detergent formulation change began in 1965 and stemmed from environmentalists' charges that detergent phosphates were ecologically harmful. In 1967, Stewart R. Udall, then U.S.

Secretary of the Interior, asked the detergent industry to research and develop a suitable substitute for "reasonable, gradual replacement" of phosphate builders [5]. Detergent manufacturers, although not in agreement that their products were harming the environment, but already heavily investing in builder research, promised their cooperation. However, the issue was prematurely forced in late 1969 when the environmentalist movement, greatly influenced by emotional and political factors, achieved sufficient momentum in many areas of the United States to rally successfully against detergent phosphates. Resultant restrictive legistlation and a typically quick response of detergent manufacturers to matters of public concern precipitated the most significant and far-reaching change in laundry detergents: the reduction and, in some areas, elimination of phosphate builders [6].

By 1971, six states in the United States had enacted legislation limiting phosphate content to an 8.7% phosphorus (P) level (35% sodium tripolyphosphate). Such reduced phosphate legislation, however, became unnecessary. In 1971, two major detergent manufacturers voluntarily reduced the phosphorus content of their laundry products nationwide to the 8.7% P level, and the third followed suit shortly after. By 1973, two states—New York and Indiana—and three local areas—Dade County (Florida), Chicago (Illinois), and Akron (Ohio)—had laws in effect completely banning phosphates. In 1977, three additional states—Minnesota, Michigan, and Vermont—also banned detergent phosphates.

Although phosphate restrictive legislation on the federal level has not yet been enacted, there is some probability that it will be, at least for those states bordering the Great Lakes [7]. Should this occur, more than 40% of households in the United States will be forced to use phosphate-free laundry detergents.

2. Consumerism

Regulatory restrictions and consumer demand for effective, safe products and concern for all consumer products entering the environment placed severe restrictions on a detergent manufacturer's choice of phosphate replacement materials. These restrictions demand that new detergent ingredients undergo extensive, rigorous examinations for human and environmental hazards. Because such examinations require years in time and millions of dollars in costs, detergent manufacturers had little choice initially than to replace the phosphate in their products with materials with a known history of safe use, while they continued their investigation for more effective materials. Frequently, a change in builder necessitated a change in a detergent's surfactant system, placing further constraint on a manufacturer's choice of detergent ingredients.

3. Raw Material Shortages and Rising Costs

The uncertain future of phosphate detergents during the peak of the anti-phosphate movement in the early 1970s caused phosphate suppliers to hold off on needed production facility expansion. However, demand for phosphates, especially for fertilizer phosphates, continued to increase. As a result, the period from 1973 to 1975 was marked by critical phosphate shortages with subsequent allocations to detergent manufacturers. This undoubtedly prompted the further reduction in phosphates noted in a number of laundry detergent brands. By 1976, the average phosphorus content of laundry detergents was reduced from 8.7 to 7.6%, and by 1978 the average had dropped to 6%.

Raw material shortages during this period were not limited to phosphate builders. Many detergent chemicals, including surfactants, carboxymethylcellulose, and sodium silicate, were also in short supply. Sparked by these shortages and compounded by an escalating inflationary economy, costs for detergent raw materials increased markedly. To cope with this added restriction, detergent manufacturers reduced certain ingredients in their products to the lowest practical level and substituted less costly ingredients. Generally, these were sodium sulfate, sodium carbonate, or sodium chloride.

4. Energy Considerations

Rising manufacturing costs, particularly for energy, were an additional economic burden on detergent manufacturers and motivated further formulation changes for cost reduction. Of some significance is the fact that during this period a number of new heavy-duty liquid laundry detergents, requiring far less energy to manufacture than spray-dried or even dry-blended powders, were introduced nationally. A trend in consumers towards cold-water washing, definitely related to increasing energy costs, is perhaps a primary reason for the success of these new products. Liquid detergents avoid the reduced-solubility problem of powdered detergents in cold water.

5. Technological Advancement

It is perhaps somewhat ironic that technological advancement instigated so few of the recent formulation changes in laundry detergents. In fact, only the introduction of synthetic fibers motivated a change—the greater use of nonionic surfactants and an increase in surfactant level. All other changes could be summarized as resulting from legislative restrictions, biomedical and environmental factors, or energy and cost considerations.

II. NONPHOSPHATE DETERGENT BUILDERS

Prior to 1970 and the start of the reduced and nonphosphate detergent era, the most commonly used builder ingredients in heavy-duty home laundry detergents were sodium tripolyphosphate (STPP), sodium silicate, and sodium carbonate (soda ash). By far the major builder was STPP, the single largest raw material of the detergent industry. Sodium silicate, used at much lower levels, was present primarily as an anticorrosion agent, although it does possess some builder functions. Soda ash was used in some formulations for added alkalinity or simply to reduce costs. The ability of soda ash to absorb liquids without becoming tacky made it a useful material for dry-blended powdered detergents based on liquid nonionic surfactants. Typical detergent formulations of the late 1960s, shown in Table 1, illustrate the relative amounts of phosphate, silicate, and carbonate commonly used.

A. Organic Replacements for Phosphate Builders

The detergent industry began its quest for suitable phosphate replacements in the mid-1960s at the first suggestion that phosphates might be a factor in environmental problems. Because a major builder function is the inactivation of water-hardness ions which can inhibit surfactant action, first attention was directed to organic materials capable of sequestering or complexing divalent metal ions. Hundreds of compounds were screened with most being found unsuitable because of unfavorable cost, performance,

TABLE 1

Typical Detergent Formulations of the Late 1960s

	High suds	Low suds
Surfactant	18-25%[a]	8-10%[b]
Sodium tripolyphosphate	40-60	40-60
Sodium silicate	5-9	5-9
Sodium carbonate	0-3	0-30
Sodium sulfate	10-25	0-25
Fluorescent whitener	0.7-0.75	0.2-0.3
Carboxymethyl cellulose	0.2-1.0	0.2-1.0

[a]Linear alkylbenzene sulfonate.
[b]Alcohol ethoxylate.

environmental, or toxicological factors. A few, however, showed some promise of success. Table 2 lists those materials.

1. Sodium Nitrilotriacetate

For a time, sodium nitrilotriacetate (NTA) was the prime candidate for at least partial replacement of STPP. Detergency performance of NTA approached that of phosphate, and human and environmental toxicity studies indicated that it was a safe material [8]. In early 1970, a number of products containing NTA appeared on the market and at least one detergent manufacturer was preparing actively for the switch to NTA. However, in December 1970, at the "request" of the U. S. Surgeon General, detergent manufacturers "voluntarily" agreed to desist in the use of NTA in laundry detergents until further health studies were completed. The request was based on a report from the National Institute of Environmental Health Sciences, which indicated that NTA complexes of certain heavy metals such as cadmium and mercury may produce mutagenic or teratogenic responses in pregnant mice [9]. The U. S. government's position against NTA was restated in September 1971, in May 1972, and again in November 1974. This federal investigation of NTA, regardless of its eventual outcome, effectively eliminated the use of NTA as a laundry detergent ingredient in the U. S. In Canada, NTA has been used in laundry detergents since the early 1970s and remains in use at the present time.

2. Sodium Citrate

Sodium citrate, the trisodium salt of citric acid, has a long history of safe use, meets all human and environmental toxicity requirements, and

TABLE 2

Organic Compounds Investigated as Potential Detergent Phosphate Replacements

Sodium nitrilotriacetate
Sodium citrate
Sodium carboxymethyltartronate
Sodium carboxymethyloxysuccinate
Sodium oxydiacetate
Sodium benzopolycarboxylate
Sodium polyalphahydroxyacrylic acid

was one of the first materials at which detergent manufacturers looked when phosphate restrictions took effect. Although it is a much less effective sequestrant for water-hardness ions than are STPP of NTA, it is used as a phosphate replacement in heavy-duty liquid laundry detergents. Although citrate-based powdered laundry products reportedly provide acceptable detergency [10], none has been marketed. Marginal detergency performance and unfavorable economics perhaps limited the use of citrate as a laundry detergent ingredient.

3. Monsanto's Builder M

"Builder M", developed by Monsanto Industrial Chemicals Company as a phosphate replacement, is composed of ether polycarboxylates, principally 2-oxa-1,1,3 propane tricarboxylate, commonly called sodium carboxymethyltartronate. According to Monsanto, an extensive human and environmental safety assessment indicated no problems with this potential phosphate replacement material [11].

The primary function of Builder M, similar to sodium citrate, is sequestration of water-hardness ions. In addition, it possesses some buffering and soil dispersion abilities [12]. In regard to detergency performance, it has been demonstrated that Builder M, although not equal to STPP, is superior to sodium carbonate or sodium silicate as the primary builder in a powdered laundry detergent. A nonphosphate product based on Builder M was found to be comparable to products containing 35% STPP (8.7% P) [13]. Because of high water solubility, Builder M can be used at typical builder levels to formulate liquid detergents with similar performance characteristics. Of significance is the finding that Builder M has some advantage in cleaning polyester fabrics at lower water temperatures.

Monsanto first made Builder M available to detergent manufacturers for laboratory scale testing in 1975. Pilot plant production for larger scale testing followed in 1976 and at that time at least one major detergent manufacturer test marketed a product built with Builder M. In 1977 Monsanto predicted that they would have a plant producing 100 to 150 million pounds per year in operation by 1980 [14]. Long-term confirmatory tests of the human and environmental assessment will have to be completed and found satisfactory before large-scale use of Builder M is possible. Economic factors and/or the development of a more suitable phosphate replacement will also influence the commercialization of Builder M.

4. Lever Brothers' CMOS

Sodium carboxymethyloxysuccinate (CMOS), an ether tricarboxylate like Builder M and empirically isomeric to sodium citrate, was developed by Lever Brothers as their phosphate replacement. The chemical name of CMOS is 2-oxa-1,3,4 butane tricarboxylate. Figure 1 shows the chemical

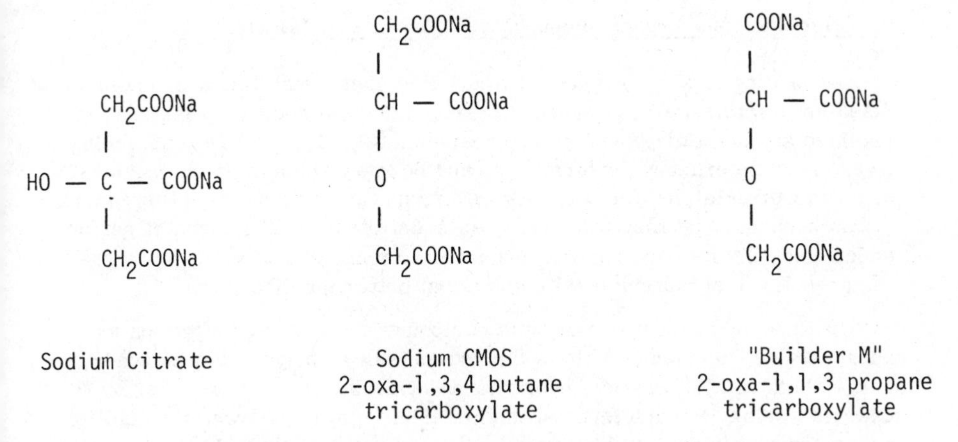

FIG. 1. Chemical similarities of three polycarboxylate-type phosphate substitutes.

similarities of CMOS, Builder M, and sodium citrate. Human and environmental safety assessments of CMOS, as reported by Lever Brothers to the U. S. government (Committee to Coordinate Toxicological and Related Programs, U. S. Department of Health, Education, and Welfare), show CMOS to be a safe and acceptable material [15]. Lever Brothers are so confident in the safety of CMOS that they have obtained a patent covering its use as a food additive [16].

Similar to Builder M, CMOS is a weak calcium sequestrant compared to STPP, although slightly superior to sodium citrate [17]. Published detergency data indicate that CMOS can be formulated into acceptable nonphosphate laundry detergents for use under prevalent U. S. washing conditions [18]. Such formulations are based on higher than normal surfactant levels and/or on the use of other builders such as sodium carbonate or sodium silicate in combination with CMOS. As was noted for Builder M, a significant characteristic of CMOS is its advantage over other phosphate replacement materials in cool-water washing. In addition, the essentially nonhygroscopic nature and good water solubility of CMOS would appear to make it a potential builder for both heavy-duty powdered and liquid detergents.

In 1977, CMOS was available to detergent manufacturers in quantities sufficient only for laboratory scale testing. Lever Brothers, holder of the U. S. patent covering the use of CMOS in detergent formulations [19], gave a manufacturer (Ethyl Corporation) an option to license CMOS and make it available to all detergent manufacturers on a royalty-free basis. Acceptance of CMOS by other detergent manufacturers, thereby providing a profitable market, is perhaps a major key to its utilization as a detergent ingredient.

5. Miscellaneous Organic Phosphate Replacement Materials

According to early reports, two organic compounds listed in Table 2, disodium oxydiacetate and benzopolycarboxylic acid showed considerable promise as potential phosphate replacements [20, 21]. Good sequestering pwer, good detergency performance, and no known human or environmental hazards were claimed for both. However, no further reports followed and it was assumed that they were found to be deficient in some area of performance, cost, safety, or technical feasibility for manufacture, or were rejected in favor of materials with somewhat better qualifications.

In Europe, at least one organic compound has received attention as a phosphate replacement. This is the sodium salt of polyalphahydroxyacrylic acid, which in 1977 was in a consumer test market in Belgium. Solvay & Cie (Brussels), manufacturer of the material, claim good water solubility, sequestering power, and biodegradability for their product [14].

B. Inorganic Replacements for Phosphate Builders

In the early 1970s, when it first became necessary for detergent manufacturers to reduce or eliminate phosphorus from their products, a suitable replacement with efficacy and safety characteristics similar to those of phosphates was not available. With no alternative, detergent manufacturers reformulated their products to contain environmentally safe, but less effective materials. In most cases these replacement materials were sodium carbonate, sodium silicate, or a combination of the two. Both of these inorganic compounds had been used for many years in detergent formulations, but at relatively low levels and never as a primary builder.

1. Sodium Carbonate and Sodium Silicate

Of the three major U. S. detergent manufacturers, only Lever Brothers formulated a nonphosphate powdered detergent with sodium carbonate as the major builder. Lever Brothers used moderately high carbonate levels in these products. In their liquid nonphosphate detergents, Lever Brothers used trisodium citrate. Colgate-Palmolive Company entered phosphate-banned areas with silicate-built products. Alkalinity of these products was comparable to that of low phosphate detergents since high SiO_2/Na_2O ratio silicates were used. Procter & Gamble Company, who had already started their switch to NTA when it was effectively although not legally restricted, for a time withdrew from phosphate-banned areas. They later re-entered these areas with formulations based on a carbonate-silicate builder system combined with a surfactant system less sensitive to water-hardness ions. Typical nonphosphate detergent formulations of the major detergent manufacturers are shown in Table 3 [21a].

TABLE 3

Typical Nonphosphate Detergent Formulations of the Major Detergent Manufacturers in the Mid-1970s

Lever Brothers Co.		Colgate-Palmolive Co.		Procter & Gamble Co.	
Sodium carbonate	55%	Sodium sulfate	35%	Sodium sulfate	29%
Sodium sulfate	16	Sodium silicate	26	Sodium carbonate	20
Alcohol ethoxylate	9	Alkylbenzenesulfonate	20	Sodium silicate	20
Sodium silicate	8	Sodium soap	6	Alkylbenzenesulfonate	15
Sodium alcohol sulfate	1	CMC, water, other	13	Alcohol ethoxylate sulfate	6
Perborate	2			Sodium toluene glycol	1.5
Borax	1			Polyoxyethylene glycol	1.5
CMC, water, other	8			CMC, water, other	7

From Ref. 21a.

In the early days of phosphate restrictions, a number of smaller detergent manufacturers marketed laundry products of very high alkalinity. Builder components of these products were extremely high levels of sodium carbonate and, in a few cases, high carbonate plus sodium meta silicate—the most alkaline of the silicates. The appearance on the market of these more hazardous detergents prompted a statement by the U. S. Surgeon General, in coordination with officials from EPA, FDA, and the Council on Environmental Quality, expressing concern over the alkaline nature of phosphate replacements. Fortunately, the more hazardous of these products were withdrawn, although several with carbonate levels of 60 to 70% are still marketed. Typical minor brand nonphosphate detergent formulations which exhibited highly alkaline characteristics are shown in Table 4 [21b].

In general, the builder characteristics of sodium carbonate and sodium silicate are somewhat similar. Both soften water by precipitation of hardness ions and add alkalinity. In addition, sodium silicate has some soil dispersive ability. Neither material, however, approaches the efficacy and safety of phosphate builders.

2. Trisodium Imidobissulfate

A new material, possibly with some potential as a phosphate replacement, has been developed in Japan by the National Chemical Laboratory for Industry in Tokyo [22]. This material, trisodium imidobissulfate (TSIS), has the chemical formula: $(NaSO_3)_2NNa \cdot H_2O$. Although no data are available, the developers claim that TSIS overcomes the performance, biomedical, and environmental disadvantages of conventional builders. It was reported in late 1977 that Kao Soap Company and Lion Fat Company, both

TABLE 4

Highly Alkaline Nonphosphate Detergent Formulations Marketed in the Early 1970s[a]

	X	Y	Z
Surfactant	11%	14%	7%
Sodium carbonate	70	60	34
Sodium silicate (2:1)	9	-	-
Sodium meta silicate $\cdot 5\ H_2O$	-	25	16
Sodium sulfate	-	-	39
Water and other	10	1	4

[a]See Ref. 21b for additional detail.

of Tokyo, were close to using a "special" ingredient based on undisclosed technology in reduced phosphate formulations [14]. It seems likely that this special ingredient is TSIS.

3. Zeolite

Perhaps the most unusual phosphate replacement to be used in laundry detergents is sodium aluminosilicate, more frequently called zeolite, a water-insoluble inorganic material with ion-exchange properties. Zeolite was developed concurrently by Henkel KGaA in Düsseldorf, West Germany, and Procter & Gamble in the United States. In late 1975, Procter & Gamble introduced a laundry detergent containing zeolite into several test market areas. By late 1976 they announced their plans to reformulate all their laundry detergents to low-phosphate (3% P) and nonphosphate products based on Zeolite. Because of limited zeolite production facilities, a full conversion is not anticipated before the early 1980s.

Zeolites are defined as crystalline aluminosilicates. Although they are naturally occurring, it is more economical to manufacture them synthetically. There are three frequently used type A zeolites: types 3A, 4A, and 5A. All have the empirical formula: $Na_2O \cdot Al_2O_3 \cdot 2SiO_2 \cdot 4 \cdot 5H_2O$. Type A zeolites have a three-dimensional structure of aluminate and silicate tetrahedra linked together by sharing all oxygen atoms. Within this structure is an interconnected network of channels or pores containing ion-exchangeable cations and water molecules. The openings to these pores are of uniform size, characteristic for each of the three type A zeolites. Type 4A, the zeolite most applicable to detergent applications, has a pore size of about 4.2 Å. Types 3A and 5A have pore openings of 3 Å and 5 Å, respectively [23]. When type A zeolites are dehydrated by heating, the intracrystalline water is reversibly removed, leaving the crystalline structure intact. Such dehydrated or "activated" alluminosilicates are used extensively in the petroleum industry as components of cracking catalysts and as adsorbents [23]. The hydrated zeolites, although possessing little adsorptive ability, are relatively good ion-exchange materials. It is this characteristic of zeolites—the ability to soften water by ion exchange—that makes them applicable to laundry detergents [24].

According to published information, one of the difficulties in the development of zeolite as a laundry detergent builder was optimization of its particle size [24]. It had to be small enough for rapid ion-exchange kinetics, a function of surface area, and to be invisible if trapped in clothing during the wash; yet it also had to be large enough for adequate removal in subsequent wastewater treatment. Procter & Gamble's research indicated that a particle size range of about 1 to 10 μm satisfies these three requirements. Extensive human and environmental safety testing, also conducted by Procter & Gamble, has uncovered no undesirable characteristics. Since zeolites are a GRAS (generally regarded as safe) type of material, the possibility that presently unknown but important safety factors may surface in the future appears to be minimal.

III. SURFACTANT CHANGES IN LAUNDRY DETERGENTS

A. Synthetic Surfactants

For many years the dominant detergent type in the United States has been the relatively high-sudsing, spray-dried products based on an anionic surfactant, linear alkyl benzene sulfonate (LAS). (A more slowly biodegradable nonlinear alkyl benzene sulfonate was used prior to 1965.) Certain of these LAS products also contain a smaller amount of another anionic surfactant, usually alcohol sulfate or alcohol ethoxy sulfate. A second detergent type in the United States, far less popular than LAS products, has been low-sudsing detergent based on nonionic alcohol ethoxylates. With an excellent combination of high performance, competitive price, ease of processing, and a long history of safe use, LAS remains a major surfactant in laundry products and is likely to continue as such for the foreseeable future. However, beginning in the early 1970s, the use of LAS became relatively static while that of other surfactants began to increase steadily. The reason for this change was that a number of developments occurred in a relatively narrow time span, significantly affecting the use of LAS as the major or sole surfactant in home laundry detergents. Each of these developments, listed below in chronological order, revealed certain deficiencies in LAS and opened the door to other surfactants possessing those attributes that were lacking, or less than optimum, in LAS.

1. The increased preponderance of synthetic fibers, principally polyester, in the home washload
2. The switch to reduced phosphate and nonphosphate detergents containing nonsequestrant-type phosphate replacements
3. A continuing increase in consumer acceptance of heavy-duty liquid laundry detergents
4. The trend to cool- and cold-water washing as a cost and energy saving measure

Synthetic fibers, especially polyester, characteristically have hydrophobic surfaces, which tenaciously retain oily or greasy soils. Compounding this cleaning problem for synthetics, warm-water rather than hot-water washing is required since higher temperatures cause penetration of oily soil into synthetic fibers from which it is virtually impossible to remove. Cooler water is also recommended for better wash and wear characteristics. Because of these influences, the use of alcohol ethoxylates, which have superior oily-soil removal performance and cool-water performance, has steadily increased.

Perhaps the most serious deficiency of LAS is that it is highly sensitive to calcium and magnesium ions of hard water. Although it does not form the troublesome curds associated with the use of soap in hard water, it is nonetheless inactivated by hard water. For this reason, LAS functions with maximum effectiveness only in soft water or in adequately sequestered systems. Alcohol sulfate, a cosurfactant with LAS in some formulations, is similarly inactivated by water-hardness ions. Since current reduced and nonphosphate powdered detergents use nonsequestering carbonate and/or silicate phosphate replacements, LAS suffers significant performance loss in such formulations. To overcome this deficiency, detergent manufacturers reformulated some of their products to contain a blend of LAS with the more hardness-resistant alcohol ethoxy sulfates or alcohol ethoxylates. Alcohol sulfate has been practically eliminated from laundry formulations. At present, a typical high-sudsing powdered formulation contains 7 to 10% LAS and 10 to 14% alcohol ethoxy sulfate, or 10 to 14% LAS and 1 to 6% alcohol ethoxylate. A citrate-based heavy-duty liquid detergent contains 14 to 17% LAS and 4 to 7% alcohol ethoxylate [25]. Higher levels of these surfactants are also being used to compensate for the less effective carbonate and silicate builders.

Detergency performance of the relatively new unbuilt heavy-duty liquid detergents, which continue to receive increasing consumer acceptance, is based solely on a high active-ingredient level. LAS, with its limited water solubility and intolerance to hard water, is unsuited for this type of product except as a relatively low level coactive in blends with more soluble and more hardness-resistant surfactants. Ethoxylated alcohol has been the surfactant of choice for this application. The surfactant system of a typical unbuilt liquid detergent is a blend of 10 to 18% LAS and 30 to 40% alcohol ethoxylate [25].

A probable reason for the continuing consumer acceptance of liquid detergents is the soaring cost of energy. Because the major cost in home laundering is water heating, up to 95% of the total cost, many consumers have turned to the use of unheated (cold) water for laundering as a cost-saving measure. Liquid detergents, which present no solubility problem in cold water, have this advantage over powdered detergents, especially the poorly soluble nonphosphate products. The trend to the use of unheated water for laundering is likely to continue, paralleling the increasing cost of energy.

B. The Use of Soap with Dispersing Agents

The addition of phosphate builders to synthetic surfactants shortly after World War II was the principal reason why soap lost the position it had held for centuries as the traditional home laundering product. With its intolerance to hard water and limited solubility in cold water, soap could not compete with the improved performance of phosphate-built detergents. It was not until the early 1970s, when detergent phosphates became suspect as contributors to ecological problems and when their reduction or elimination seemed a distinct possibility, that serious interest in soap as a laundering agent was reviewed. At this time, the Eastern Regional Research Center (ERRC) of the U. S. Department of Agriculture initiated research studies on the use of soap as a major ingredient in laundry detergents. The objective of this work was a phosphate-free or reduced-phosphate product based on a soap, i. e., the alkali salt of a fatty acid, modified to eliminate limitations imposed by water hardness and temperature. The ERRC approach, suggested by Linfield [26], was the addition of lime-soap dispersing agents (LSDA) to a laundry soap.

As described by Stirton and co-workers [27], LSDAs are surface active molecules with a bulky hydrophilic group. It was theorized that LSDAs function through formation of a mixed micelle with soap molecules. A simple explanation suggests that bulky LSDA hydrophiles maintain the curvature of the LSDA-soap micelle in hard water so that the hydrophilic groups of the LSDA and soap molecules remain exposed to the hard water. This is in contrast to micelles containing only soap molecules which invert in hard water, turning the hydrophobic chains of the soap outward and thereby forming lime soap curds and scum. The formation of a mixed micelle of LSDA-soap molecules also serves to solubilize soap, a phenomenon which enhances the solubility of soap in cold water.

The research conducted at ERRC in the area of soap and lime soap dispersants as laundering agents was extensive, covering a period of more than 7 years. The basic finding of this research was that compositions containing tallow soap, an LSDA, and an inorganic builder were effective laundering agents comparing favorably with typical phosphate-built powdered detergents [28, 29]. A number of LSDAs were examined including sulfated ethoxylated fatty alcohols, sulfated fatty alkanolamides, alpha sulfo fatty esters, 2-sulfoethyl fatty esters, 2-sulfoethyl fatty amides, and nonionics derived from tallow alcohols. Nonionics proved to be good LSDAs, but detergency of such systems was poor. Amphoteric surfactants were also examined and found to be highly efficient LSDAs offering good detergency when combined with a tallow soap [30]. A disadvantage of amphoterics, however, is their inherently higher cost.

At least 20 major classes of LSDAs were studied at ERRC with several thoroughly investigated for their biological and ecological characteristics and feasibility for production as well as their laundering performance. A summary of this work, provided by Linfield [31], would indicate that the modified soap concept of a laundry detergent offers a reasonable alternative to phosphate-built synthetic detergents. Soap-LSDA combinations appear to equal or exceed present-day synthetic detergents in all areas of performance and safety and to undergo rapid and complete biodegradation. Future limited supplies and the present trend to increasing cost of petrochemicals, the source of synthetic surfactants, also make tallow-derived detergents an attractive alternative. Linfield noted that several Japanese firms have manufactured and successfully marketed soap-LSDA products. However, lack of soap-making facilities in the United States would prevent their utilization in the near future.

IV. PERFORMANCE CHARACTERISTICS OF PHOSPHATE-RESTRICTED DETERGENTS

The prime objective of all researchers working in areas related to fabric cleaning is to assist homemakers in obtaining the best possible laundering with a minimum of labor, cost, and fabric wear. To realize this objective, comparative studies of representative samples of current home laundry detergents have been conducted annually for a number of years. The primary purpose of this work has been to compile information on which to base realistic detergent usage recommendations for washing-machine use manuals and other customer educational material. A further purpose has been to maintain an awareness of reformulated or new detergent products which might present potential problems or perhaps suggest washing-machine cycle or machine modifications for better performance. This continuing annual study has resulted in considerable information which could be called a profile of the chemical, physical, and performance characteristics of U. S. laundry detergents. The following discussions concerning reduced and nonphosphate laundry detergents are drawn from this profile and from the various studies which led to its development.

A. Detergent Types of the 1970s

Virtually all heavy-duty laundry detergent formulations intended for clothes washing marketed in the United States prior to 1970 contained a high level (40 to 60%) of phosphate builder combined with either a high-sudsing anionic or a low-sudsing nonionic surfactant system. Although a few were in liquid form or pellet form, most were powders, and laundry detergents of this period were generally classified only as "high sudsers" or "low sudsers". As a result of changes in detergent formulations discussed earlier in this chapter, laundry detergents can no longer be classified simply according to their sudsing potential. By the mid-1970s there were seven distinct types of laundry detergent on the U. S. market, as follows:

1. High-Phosphate Powders (10 to 16% Phosphorus)

Detergent products of this type were almost extinct by 1975. The few brands (only one major brand) available are sold in those areas of the United States with no laws restricting detergent phosphate content.

2. Low-Phosphate Powders (5.0 to 8.7% Phosphorus)

This is the dominant detergent product type in the U. S. It is sold in all areas where detergents are restricted to 8.7% phosphorus and in nonrestricted areas. There are many major and minor brands.

3. Carbonate Powders (20 to 65% Sodium Carbonate)

Detergents containing sodium carbonate as a phosphate substitute make up the dominant product type sold in total phosphate-banned areas. There are many brands available, some of which are sold nationally as well as in phosphate-banned areas.

4. Silicate Powders (20% Low Alkali Silicate)

Colgate-Palmolive is the only major detergent manufacturer which marketed silicate-built nonphosphate powders. These products were sold mostly in total phosphate-banned areas until 1978 when Colgate-Palmolive reformulated their nonphosphate powders, replacing sodium silicate with sodium carbonate.

5. Phosphate Liquids

Only two phosphate-built liquid products—Lever Brothers' liquid ALL and WISK, containing 4.8 and 3.5% phosphorus, respectively—represent this type of detergent product. They were sold in all areas where phosphate detergents were not banned until 1978 when production of liquid ALL ended. A year later phosphate-built WISK was also discontinued.

6. Citrate Liquids

The nonphosphate counterparts of Lever Brothers' liquid ALL and WISK were both sodium-citrate-built products until 1976. At this time ALL was reformulated into an unbuilt liquid leaving WISK as the only major brand product in this group.

7. Unbuilt Liquids

This new product type, containing no inorganic builders, was introduced in 1972. By 1976 there were at least seven brands sold nationally.

Powdered laundry detergents for many years held the lion's share of total detergent sales. However, from 1970 to 1976, liquid laundry detergents considerably enlarged their share of the market, with an average annual growth rate of approximately 17%. Although this rapid rate of growth later slowed, by the end of 1978 liquids held nearly 20% of total laundry detergent sales.

B. Performance Evaluation Test Method

A complete performance assessment and ranking of individual detergent products was not the intent of this investigation, but rather the objective was a characterization of detergent performance patterns over a broad range of water hardness and wash temperature conditions. A knowledge of such patterns assists in the prediction of direction and extent of detergent performance change with change in wash conditions and is a useful guide for establishing detergent type and dosage recommendations for specified laundering conditions. It is believed that laboratory scale test methods, using sensitive, realistic soils and fabrics are sufficiently reliable for determining detergent performance patterns and, within certain limitations, for estimating their relative performance levels.

The detergency data discussed here were obtained using a test procedure based on a Terg-O-Tometer as the washing device. Home laundering conditions were simulated by washing clean test swatches together with soiled test swatches. The soiled test fabrics, a pure cotton and a pure polyester, were prepared by first exposing them to an aqueous slurry of a 50:50 mixture of sieved vacuum-cleaner dirt and a colored kaolinite-type clay, drying the particulate soiled fabric, and then applying a fatty soil. The fatty soil, triolein, was solvent applied at 5% fabric weight level.

The laundry detergents evaluated and discussed here are typical of products available in 1978. They include eight major-brand phosphate powders, their nonphosphate counterparts and six major-brand heavy-duty liquids. As leading products of the three largest detergent manufacturers—Procter & Gamble, Colgate-Palmolive, and Lever Brothers—they represent approximately 75% of all laundry detergents sold in the United States. The detergency performance of each of these products was evaluated at several water-hardness levels and wash temperatures. Although soil redeposition, fluorescent whitening, and soil removal measurements were made, only the latter are discussed here, being the most informative. Soil removal data were calculated from reflectance measurements of the soiled test fabrics taken before and after washing and by applying the Kubelka-Munk equation. For correlation of data obtained at different times and with different batches of soil cloths, AHAM* Reference Detergent was always included in each test series. Soil removal data of the detergents under test were then calculated and reported as a percentage of the soil removed by the reference detergent. Wash test conditions used in the experimental studies are summarized in Table 5.

---

*Association of Home Appliance Manufacturers, 20 N. Wacker Drive, Chicago, IL 60606.

TABLE 5

Detergent Evaluation Test Conditions

Washing device	Terg-O-Tometer, 100 strokes per minute
Wash water	1 liter at three hardness levels[a]
Wash temperature	38°C and 10°C
Wash time	10 min
Rinse water	7.6 liters, same hardness as wash water
Rinse time	5 min, no agitation
Detergent concentration	Powders - 0.20% Built liquids - 0.25% Unbuilt liquids - 0.125%
Replications	Three

[a]Hardness levels: 0 ppm—Softened tap water, original hardness of 145 ppm; 150 ppm—Tap water with added hardness to 150 ppm; 300 ppm—Tap water with added hardness to 300 ppm.

C. Detergency Characteristics

The relative soil removal values from laboratory studies conducted in 1969, 1974, and 1978 of the eight major brand powdered detergents are shown in Table 6. In 1969, phosphate-builder content of these brands averaged approximately 12% phosphorus. By 1974, phosphorus content of each had been reduced to 8.7% with further reductions following, so that by 1978 their average phosphorus content was 7.6%. A comparison of the phosphate detergent soil removal data of 1969 with those of 1974 and 1978 reveals an overall continuing trend to lower detergency performance. It is of interest to note that, based on average values, this trend approximately parallels the reduction in phosphorus content of the eight detergents.

Also shown in Table 6 are the relative soil removal values obtained in 1978 of the nonphosphate formulations of these eight brands of powder. Seven of the eight nonphosphate powders show somewhat less soil removal from cotton, while all show very significantly less soil removal from polyester than do their phosphate-built counterparts. If it can be assumed that the soil removal data for 1969 represent good detergency, then it is apparent that laundry detergents of the 1970s, especially nonphosphate products, must be judged as relatively poor performers.

TABLE 6

Percentage Soil Removal of Major Brand Detergents[a,b]

Detergent brand	From cotton				From polyester			
	Phosphate			Nonphosphate	Phosphate			Nonphosphate
	Year evaluated							
	1969	1974	1978	1978	1969	1974	1978	1978
A	107	94	88	73	111	93	83	61
B	109	89	97	74	113	93	107	68
C	108	85	74	65	114	78	69	49
D	107	85	82	71	100	83	75	53
E	105	98	75	77	110	102	87	70
F	96	89	84	75	81	85	80	67
G	98	92	80	74	86	88	74	63
H	103	89	83	73	92	83	83	69
Average	104	90	83	73	101	88	82	62

[a]All tests conducted in moderately hard water (150 ppm) at a product concentration of 0.20%.

[b]Soil removal values are relative to performance of a standard reference detergent.

A performance characteristic of nonphosphate detergents—other than poor performance—which distinguishes them from phosphate products is their "nonstandard" response to water hardness. Typically, the soil removal ability of phosphate-built detergents, at constant concentration, decreases in waters of increasing hardness. Nonphosphate detergents, however, follow a different pattern. The data shown in Table 7, which are average relative soil removal values of the several detergent types, illustrate the differences between phosphate and nonphosphate detergents relative to change in water-hardness level. It is evident here that both powder and liquid phosphate-built products, when tested at a constant dosage level, perform best in soft water and with decreasing performance in waters of increasing hardness. The nonphosphate powders, all of which were carbonate built, also perform best in soft water but, unlike their phosphate-built counterparts, are relatively insensitive to the degree of water hardness. Their cleaning abilities, at constant concentration, were found to be approximately equal in moderately hard and very hard waters.

TABLE 7

Average Percentage Soil Removal of Four Detergent Types in Soft, Moderately Hard, and Very Hard Water

	From cotton			From polyester		
	Wash-water hardness					
Detergent type	0 ppm	150 ppm	300 ppm	0 ppm	150 ppm	300 ppm
Low-phosphate powders	112	83	73	116	82	64
Low-phosphate liquids	98	69	59	104	72	50
Nonphosphate powders	88	73	73	83	62	61
Nonphosphate liquids	71	67	70	52	64	70
AHAM reference	111	100	77	112	100	77

The detergency response of nonphosphate liquid detergents to water hardness appears related to the type of fabric being cleaned. As shown by the data in Table 7, their ability to clean cotton is essentially unaffected by water hardness while their ability to clean polyester, at least for some of the liquids, is enhanced in the presence of hardness ions. The reason for this seemingly unusual behavior is believed to be the formation of a magnesium-anionic surfactant compound with superior detersive powers. Evidence has been obtained previously that the magnesium salt of LAS provides better polyester detergency than the sodium form [32].

Based on the detergency studies discussed here it might appear that in high hardness waters nonphosphate detergents could equal or perhaps even outperform phosphate-built products. Generally, however, this is not the case, the reason being that nonphosphate detergent builders contribute little to detergency beyond water softening, while phosphate builders have several additional detersive functions. Because of these added functions, phosphate-built detergents show a correspondingly greater preformance improvement when dosage is increased than do nonphosphate products. Thus under proper usage conditions, i.e., sufficient concentration in relation to water hardness, phosphate detergents usually remain the superior product type regardless of water hardness level.

The effect of wash-water temperature on the performance of phosphate-restricted detergents is illustrated in Table 8, which lists the average soil removal values of phosphate and nonphosphate powdered and liquid detergents obtained in warm (38°C) and cold (10°C) moderately hard water. From these data it is evident that the cleaning ability of all four detergent types is significantly reduced by a reduction in wash-water temperature. Except in

TABLE 8

Average Percentage Soil Removal of Four Detergent Types in Warm and Cold Water

	From cotton		From polyester	
	Wash-water temperature			
Detergent type	38°C	10°C	38°C	10°C
Low-phosphate powders	83	62	82	68
Low-phosphate liquids	69	52	72	49
Nonphosphate powders	73	55	62	51
Nonphosphate liquids	67	58	64	58
AHAM reference	100	82	100	93

the case of phosphate-built liquid detergents, this reduction is greater for cotton detergency than for polyester detergency. The larger negative effect on cotton detergency was expected, since several investigations have shown good cotton detergency to be dependent on adequate thermal energy while good polyester detergency is more dependent on adequate chemical energy.

Low-phosphate and nonphosphate powdered detergents are similarly affected by wash-water temperature reduction. Both powder types lost an average in soil removal of 25% from cotton and 17% from polyester when wash temperature was reduced from 38 to 10°C. Thus, the relative superiority of phosphate powders over nonphosphate powders is constant in warm and cold water.

Detergents least affected by change in wash temperature are the nonphosphate liquids—essentially, the unbuilt liquids. Their average loss in soil removal from both cotton and polyester due to wash temperature reduction is only half that of powdered detergents. This relatively small negative response to wash temperature reduction provides nonphosphate liquids with some advantage in cold-water washing. As previously noted, nonphosphate liquids also have a superior tolerance to water hardness; therefore, they would have a double advantage over nonphosphate powdered detergents under cold, hard-water washing conditions.

D. Problem Areas of Nonphosphate Detergents

When sodium carbonate-built nonphosphate laundry detergents were first introduced in 1970 there were predictions, based on the high alkalinity and precipitation reactions of carbonate, that these products would have serious detrimental effects on fabrics, clothing, and washing machines. To a large extent, these predictions proved accurate. In addition to poor detergency, experience has shown the major problem area of nonphosphate detergents to be buildup of carbonate-water hardness ion precipitates (limestone) on fabrics and washing-machine surfaces. The extent of this problem under actual consumer use conditions is illustrated by a 1977 survey of washing-machine service agents located in phosphate-banned areas of the United States. Survey results showed that 60% of the agents found carbonate precipitate buildup a major problem and the remaining 40% found it at least a nuisance problem [33].

On fabrics, the precipitate buildup problem manifests itself primarily as harshness, apparent fading of colors, and fabric graying. The dewrinkling ability of permanent press finishes and the function of flame-retardant finishes applied to cotton are also adversely affected. In washing machines, carbonate precipitate buildup produces a hard, raspy surface against which fabrics can abrade. In severe cases the buildup, which occurs on all surfaces exposed to the wash water, can be sufficient to cause machine components such as pumps and filters to become inoperable.

Investigative studies have shown that rate of precipitate buildup on both fabric and machine surfaces is dependent on several factors including carbonate content of the detergent, water hardness, water temperature, and the nature of the substrate exposed to the wash solution [34]. All nonphosphate detergents tested in these studies which contained 20% or more sodium carbonate were shown to possess the potential to cause buildup on fabric and machine surfaces under certain conditions. Test results of four detergents containing varying amounts of sodium carbonate, shown in Table 9,

TABLE 9

Carbonate Precipitate Buildup on Fabric and Washer Surfaces[a, b]

Nonphosphate detergent		Ash of test fabric (%)			mg/dm^2
Brand	Na_2CO_3 (%)	Cotton	Polyester/cotton	Acrylic	Porcelain
F	6	0.34	0.35	0.31	0
A	20	1.03	0.91	0.48	2.0
D	36	2.74	1.47	0.87	22.4
O	63	3.25	1.98	1.28	23.0

[a]Fabric tests conducted in 38°C water; porcelain tests in 49°C water.
[b]Buildup in 10 washes in 300-ppm hard water containing 0.20% detergent.

illustrate the buildup that can occur on three fabric types and on the porcelain surface of a washing machine in only 10 washes under the stressful condition of very hard water.

With the increasing consumer trend to cold-water washing, detergent solubility has taken on greater importance. Nonphosphate powdered detergents, generally the low-density type, have been noted to be significantly less soluble at lower wash temperatures than are their phosphate-built counterparts. A screening test procedure developed in our laboratories shows that the average solubility rate of nonphosphate powders in water at 10°C is 15% lower than that of phosphate powders. No significant differences have been found in the cold-water solubility of high-density phosphate and nonphosphate powders.

V. DETERGENTS OF THE FUTURE

Virtually all of the pressures and constraints on the detergent industry which precipitated laundry detergent formulation changes in the 1970s appear likely to continue, some perhaps with escalation, well into the 1980s. Most critical are the interrelated problems of an inflationary economy, an increasingly complex energy situation, and a highly regulatory governmental atmosphere; all of these will undoubtedly continue to play major roles in future detergent technology and in the type of detergents available to U. S. consumers.

From a predictive view of current trends, it appears that aluminosilicate (zeolite), will become a major builder ingredient in many phosphate-restricted detergents. In 1978, Procter & Gamble, the bellwether of the detergent industry, introduced nationally a low-phosphate formulation of their leading brand TIDE containing 20% zeolite. Nonphosphate formulations containing zeolite are also expected to be introduced in the future. These products, however, apparently require a second water softening builder to overcome the inability of zeolite to inactivate magnesium hardness. This second builder could be sodium carbonate, Lever Brothers' CMOS, or a similar hardness precipitant or sequestrant.

Dual-function detergents represent another future direction of detergent technology. These products, in addition to their primary cleaning ingredient, contain materials to impart certain desirable characteristics to a washload. Three such detergents, introduced into consumer test markets in 1977 and 1978, contain ingredients for fabric softening or reduction of fabric wrinkling.

Perhaps the most important current trend in detergent technology is in the continuing development of heavy-duty liquid detergents. Inspired perhaps by an increasing consumer trend to cold-water washing. From a manufacturing standpoint, liquids offer significant advantages over the presently dominant spray-dried powdered products: lower manufacturing cost and energy requirements. From the consumer view, liquids are perceived as being better suited for cold-water washing than are powders; they also provide a reduction in cost and energy. Numerous recently issued patents [35-38] attest to a considerable research effort on the part of detergent manufacturers in the area of heavy-duty liquid detergents.

From a broader and more speculative view, it is possible to visualize changes in detergent technology more radical and dramatic than those already described. For example, a recent patent teaches a new laundering concept based on the separate addition to the washbath of individual detergent components [39]. The advantage is that detergent ingredients—surfactant, builder, and auxiliary materials—can thus be more effectively utilized for specific washing conditions. Another recent patent discloses a laundering method designed for optimum oily-soil removal based on sequential washes using two detergent types [40]. The first wash uses a detergent containing a lipophilic surfactant, and the second uses a hydrophilic surfactant. Savings in water and, in turn, in energy are the objectives of a third recent patent which describes the use of a detergent foam rather than an aqueous solution as the washing medium [41].

The examples of new laundering concepts described above, each requiring the development of new detergent types, illustrate several possible future directions for detergent technology.

REFERENCES

1. S. J. Silvis, Chem. Week, 105(12), 68 (1969).
2. H. Schott, in Detergency: Theory and Test Methods, Part I, pp. 105-148, Surfactant Science Series, Vol. 5, Dekker, New York, 1972.
3. W. W. Morgenthaler, in Detergency: Theory and Test Methods, Part II, pp. 453-502, Surfactant Science Series, Dekker, New York, 1975.
4. W. G. Mizuno, in Detergency: Theory and Test Methods, Part III, pp. 816-896, Surfactant Science Series, Vol. 6, Dekker, New York, 1980.
5. Anonymous, Chem. Eng. News, 45(32), 16 (1967).
6. W. S. Rukeyser, Fortune, 85(1), 71 (1972).
7. Anonymous, Chem. Week, 121(7), 20 (1977).
8. F. J. Fitzgerald, Soap Cosmet. Chem. Specif., 50(2), 51 (1974) (Soap and Detergent Association Annual Meeting, January 16-19, 1974).

9. A. H. Hammond, Science, 172, 362 (1971).

10. P. J. Borchert and J. L. Neff, Soap Cosmet. Chem. Specif., 49(6), 31 (1973).

11. Monsanto Industrial Chemicals Co., Builder M Inf. Bull., pp. 21-23 (1975).

12. Monsanto Industrial Chemicals Co., Builder M Inf. Bull., p. 19 (1975).

13. B. J. Rutkowski, unpublished work, 1976.

14. Anonymous, Chem. Eng., 84(11), 104 (1977).

15. Lever Brothers Co., An Empirical Isomer of Citric Acid for Use as a Detergent Builder, U.S. Department of Health, Education, and Welfare, Washington, D.C., 1977.

16. V. Lamberti and W. J. Urban (to Lever Brothers Co.). U.S. patent 4,015,023. 1977.

17. V. Lamberti, Soap Cosmet. Chem. Specif., 53(7), 34 (1977).

18. V. Lamberti, Soap Cosmet. Chem. Specif., 53(8), 40 (1977).

19. V. Lamberti, M. D. Konort, and I. Weil (to Lever Brothers Co.). U.S. patent 3,692,685. 1972.

20. Anonymous, Chem. Week, 108(9), 15 (1971).

21. Anonymous, Chem. Week, 110(3), 31 (1972).

21a. H. A. Segalas, Hydrocarbon Process., 54(3), 71 (1975).

21b. Anonymous, Chem. Week, 108(17), 10 (1971).

22. National Chemical Laboratory for Industry, Tokyo, Imidobissulfate as Detergent Builder, 1976 (Tech. Bull.).

23. P. H. Shimizu, Soap Cosmet. Chem. Specif., 53(6), 33 (1977).

24. A. C. Savitsky, Soap Cosmet. Chem. Specif., 53(3), 29 (1977).

25. T. P. Matson, J. Am. Oil Chem. Soc., 55, 66 (1978).

26. W. M. Linfield, Soap Cosmet. Chem. Specif., 35(3), 51 (1959).

27. A. J. Stirton, F. D. Smith, and J. K. Weil, J. Am. Oil Chem. Soc., 42, 114 (1965).

28. R. G. Bistline, Jr., W. R. Noble, J. K. Weil, and W. M. Linfield, J. Am. Oil Chem. Soc., 49, 63 (1972).

29. W. R. Noble, R. G. Bistline, Jr., and W. M. Linfield, Soap Cosmet. Chem. Specif., 48(7), 38 (1972).

30. N. Parris, J. K. Weil, and W. M. Linfield, J. Am. Oil Chem. Soc., 50, 509 (1973).

31. W. M. Linfield, J. Am. Oil Chem. Soc., 55, 87 (1978).

32. B. J. Rutkowski, unpublished work, 1971.

33. J. C. Holme and K. C. Spelman, Whirlpool Marketing Services Survey, Oct. 1977.

34. B. J. Rutkowski, unpublished work, 1976, 1977.

35. J. H. Collins (to Procter & Gamble Co.). U.S. patent 4,079,078. 1979.

36. E. J. Maguire, Jr., and E. J. Pancheri (to Procter & Gamble Co.). U.S. patent 4,090,973. 1979.

37. J. Rubinfeld (to Colgate-Palmolive Co.). U.S. patent 4,075,118. 1978.

38. T. W. Gault and E. J. Maguire (to Procter & Gamble Co.). U.S. patent 4,075,118. 1978.

39. R. Graf, L. Brodzina, R. Ströbele, and H. Stache (to Bosch-Siemens Hausgeräte GmbH, Fed. Rep. of Germany). U.S. patent 4,110,075. 1978.

40. D. M. Flower (to Economics Laboratory, Inc.). U.S. patent 4,137,044. 1979.

41. E. Reinwald and M. J. Schwuger (to Henkel Kommanditgesellschaft auf Aktien, Fed. Rep. of Germany). U.S. patent 4,118,189. 1978.

AUTHOR INDEX

Numbers in parentheses are reference numbers and indicate that an author's work is referred to although his name is not cited in the text. Underlined numbers give the page on which the complete reference is listed.

Pages 1-452 comprise Part I
Pages 452-728 comprise Part II
Pages 729-1020 comprise Part III

A

Abe, R., 175(59), 229
Abram, J. C., 166(34), 228
Abruzov, K. N., 628, 629, 659
Adam, N. K., 107(1), 119(1), 121, 123(48, 50), 124(1, 48, 50), 126(48, 50), 126(48, 50), 128(50), 132(1, 48), 143(50), 148, 150, 635, 642, 660, 661
Adams, D. A. W., 732(25), 807
Adams, J. M., 788(154), 812
Adamson, A. W., 107(4), 148, 626 (12), 635, 658, 660
Addison, C. C., 644, 661
Adler, H., 524(12), 608, 825(5), 874, 876, 890
Albin, T. B., 378(327), 411
Albinson, E., 308, 320
Alexander, A. E., 162(41), 172(41), 228
Alexander, B. H., 458(5), 502
Aliman, W. T., 297, 299, 309, 319
Allan, A. J. G., 108(8), 149
Allard, R. P., 538(43), 556(43), 609
Allen, C. H., 330(35), 399
Allen, E., 326(16), 327(16), 390, 398, 409, 424(24), 432(48), 450 451, 732(55), 764, 786(107, 124, 125), 808, 810, 811
Allison, R. C., 430(42), 451
Allmand, J. A., 563(52), 610
Alter, H., 211(175), 234
Amel, R. T., 730(10), 807
Ames, L. L., 961(54), 989
Anders, G., 333, 363, 399
Anderson, D. E., 825(4), 871, 876, 890
Anderson, J. R., 116(29), 117(29), 149
Anderson, R. M., 173(48), 174(48), 229, 381, 407, 827(11), 830, 831 (11), 840(11), 851(11, 64), 890, 892
Anliker, R., 732(51), 757(51), 805 (162), 808, 813
Appel, W. D., 415(3), 449
Arai, H., 365, 386, 405, 408, 476 (30), 491(39), 503, 504
Arkley, T. H., 394(257), 408
Armbruster, E. H., 376(229), 380, 407, 840(47), 841, 842, 892
Armstrong, L. J., 282, 317, 341, 342, 387, 401
Arnold, H. W., 617(8), 623
Arnold, R. S., 668(12), 678
Arthur, J. C., Jr., 340(316), 410
Ashcraft, E. B., 351, 376(138), 380, 389, 390(138), 395, 403

Pages 1-452 comprise Part I
Pages 453-728 comprise Part II
Pages 729-1020 comprise Part III

Aspland, J. R., 754(89), 766, 810
Azad, H. S., 935, 987
Azorlosa, J. L., 285(36), 286(36), 318

B

Babulak, S. W., 671(16), 678
Bach, H., 754(85), 809
Bacon, L. R., 223(191), 234, 247 (38), 248(38), 253(38), 254(38), 266, 357(166), 404, 771(101), 773 (101), 810, 879(125), 895
Bacon, O. C., 191(118), 200(118), 222, 223(192), 224(118), 225, 232, 234, 253(48), 266, 341, 342, 354 (114,140), 355, 386(140), 387, 389 (140), 390, 395, 401, 402, 403, 409, 419, 449
Baer, R. L., 700(13), 705
Bailey, A. E., 309, 320
Bailey, G. L. J., 108(6), 149
Bailey, J. C., 379, 408
Bain, G. W., 376(231), 407
Baker, C. L., 866, 894
Baker, H. R., 136(63), 151
Baker, R. W. R., 833, 835(33), 891
Balog, J. A., 341(104), 402
Banerjee, H. K., 732(40), 808
Barail, L. C., 697, 705
Barbee, R. B., 730(19), 807
Barel, A. O., 667(10), 678
Barker, G. E., 341, 357, 402, 404
Barkin, S. M., 671(16), 678
Bartell, F. E., 12(11), 29, 87(51), 102, 108(7), 116(34), 149, 150
Barth, E. F., 953(42), 988
Bartholomé, E., 283, 290, 317
Barton, L. J., 520(2), 608
Bascom, W. D., 136(62), 151
Bastin, E. L., 190(116), 219, 232, 305, 320, 376(219), 378(219), 383 (219), 386(219), 406
Batdorf, J. B., 286, 287(39), 291, 318
Bavika, L. I., 652(112,113), 662
Baxter, S., 109, 110, 113(17), 149
Bayley, C. H., 69(13,14,18), 101, 173(43), 179(76), 180(76,79,82), 182(76,79), 183(76), 191(122), 194 (122,136), 197(136,147), 202, 204 (122), 229, 230, 232, 233, 273(6), 274(6), 276, 278, 279(16), 280(16), 283, 292(6), 301(6), 311, 312, 313, 317, 318, 337(79), 343, 389, 390, 394, 401, 402, 408
Becher, P., 246, 266, 766, 810
Beck, E. C., 395(300), 410
Beckhorn, E. J., 666, 678
Beckley, J. H., 688(22,23), 693
Bell, R. N., 849, 852, 855(66), 858, 868(104), 874, 892, 894
Beninate, J. V., 286, 295, 318, 353, 390(276), 403, 409
Benischeck, J. J., 325(10), 329(10), 333, 398
Benjamin, L., 181(93), 183, 231
Bennett, M. C., 166(34), 228
Bensing, P. L., 333(46), 347(46,123, 124), 357(46,124,174), 358(124), 361(46), 399, 402, 404, 589(87), 604(94), 611, 612
Benton, W. W., 839(42), 841, 843 (42), 891
Benzel, W., 89(55,56), 102, 103, 185(106), 186(106), 187(106), 188 (106), 190(106), 231, 289(50), 318
Berch, J., 12(14), 29, 120(40), 128, 129, 130, 131, 150, 194(135), 199(155), 202, 203(158-160), 205, 215(158-160, 179), 217(158,160), 232, 233, 234, 289(50), 295, 296, 297, 308(92), 319, 320, 324(1), 325 (1), 336(67), 340, 349, 390(277), 398, 400, 403, 409, 626(10), 658

Berg, R. H., 253(55), 266
Berger, A., 356, 404, 732(53), 741 (66), 746(66), 786(115,119,120, 132), 796(120), 801, 808, 809, 811
Berkman, S., 626(7), 658
Bernett, M. K., 652, 662
Bernstein, I. A., 376, 406
Bernstein, R., 180(81), 230, 308, 320, 340, 376(216), 381(216), 403, 406
Bersworth, F. C., 861(74), 893
Bertrich, F., 1(1), 3
Bettley, F. R., 703, 706
Betz, J. M., 954(46), 988
Bey, K., 39, 41, 42, 46, 48(24), 61, 61, 62, 120(45), 150
Biefer, G. J., 184(98), 231
Bierman, A., 195(139), 232
Bikerman, J. J., 220(189), 234, 626 (3), 631, 632, 635, 658, 659, 660
Bille, H. E., 340(314), 410
Billica, H. R., 130(60a), 139(75, 76), 140(75), 142(75), 145(75,76), 147(75,76), 151, 204, 205(162-164), 208(166), 213(166), 215(166), 217 376(208,213), 377(208,236), 382 (208,213,236), 385(260), 406, 407, 408, 411
Billmeyer, F. W., Jr., 326(13), 328, 398, 427(34), 427(43), 436 (57), 450, 451, 579, 611
Bishop, D. F., 958(51), 960(51), 961(51), 989
Bistline, R. G., Jr., 355(153), 403, 1008(28,29), 1019
Black, W., 107(2), 113(2), 123(2), 126(2), 148, 162(21), 228, 626(2), 658
Blakey, R. R., 327(20), 363, 398
Blank, I. H., 696(1), 702(1), 705
Blickensderfer, P. W., 59(44), 60, 63
Bloching, H., 730(13), 732(38), 794 (155), 805(167,168), 807, 808, 812, 813
Blohm, S. G., 704, 706
Blomfield, R. A., 563(56), 610
Bloom, R., Jr., 953(41), 988
Bloomberg, C. S., 833(32), 834(32), 836(32), 837(32), 840(32), 841(32), 891
Blumbergs, J. H., 531(35), 609
Bochard, V. G., 746(77), 809
Boeck, A., 730(15), 807
Böhme, G., 67(6,7,8,9), 71(33,34), 87(33,34), 98(8,9), 100, 102, 212 (176), 234
Bogaty, H., 111(13,14), 112(13,15), 113(15), 118(13), 149, 243(11,12), 265
Bonnar, R. U., 137(72), 151, 356 (163), 378(163,237), 383(163), 404, 407
Bonte, E., 786(117), 811
Booth, G. E., 730(10), 807
Borchardt, J. A., 935, 987
Borchert, P. J., 1000(10), 1019
Borgnan, R. D., 954(46), 988
Borza, P. F., 881, 896
Bosworth, R. C. L., 241(1), 264
Boucher, E. A., 12, 29
Bourne, M. C., 356, 382, 404
Bowers, C. A., 45, 61, 62, 155(5), 220, 221(5), 227
Box, G. E. P., 393, 409
Boyd, T. F., 180(81), 230, 308, 320, 376(216), 381, 406
Brainard, S. W., 254(57), 267, 309, 320, 334, 392, 400
Bramfitt, T. H., 832, 891
Bretland, R. A. C., 173(50), 186 (50), 229, 343, 376(117), 402
Brier, R., 675(17), 678
Briggs, D., 833, 891
Briggs, H. B., 263, 267
Bright, W. M., 15(23), 29
Britt, C. J., 136(65), 137(65), 138 (65), 139(65), 151, 221(190), 226 (190), 227(190), 234, 382(245), 407
Brocklehurst, P., 531(34), 609

Pages 1-452 comprise Part I
Pages 453-728 comprise Part II
Pages 729-1020 comprise Part III

Brodowski, H., 84(48), 102
Brodzina, L., 1018(39), 1020
Brooks, J. H., 226(200), 235
Brouwers, G., 786(145,146), 796 (145,146), 802(145,146), 812
Brown, B. S., 573(65), 610
Brown, C. B., 120(46), 139(74), 146(74), 148(74), 150, 151, 330, 331(33), 332(33), 350(136), 351, 353(136), 366, 367(33), 386, 399, 403
Brown, C. D., 244(18), 265
Brown, E. L., 141(80), 142(80), 151, 335, 336, 354(60), 400, 632 (36), 659
Brown, F. E., 637, 638, 660
Brown, M. B., 607(95), 612
Brown, R. C., 642, 661
Brown, R. M., 926(9), 987
Browne, C. L., 293, 297, 308, 319, 732(54), 808
Brownlee, K. A., 394(298), 410
Brunelle, T. E., 882(140,141), 886 (140), 896
Brungs, C. A., 861(84-87), 893
Bryce, A. J., 383(249), 408
Bubl, J. L., 368, 369, 387, 411, 623(16), 624
Buehler, E. V., 688(24), 693
Buerk, E. J., 420(18), 450
Bunge, K., 746(76), 809
Burcik, E. J., 634, 660
Burckhardt, W., 701(18), 703, 706
Burg, A. W., 805(161), 813
Burnett, C. M., 698, 705
Burtis, J., 520(3), 608
Buschmann, K. F., 283, 290, 317
Bussius, H., 702, 703, 706
Byrne, G. A., 340(316), 365, 373, 375(188), 391, 405, 410

C

Calandra, J. C., 682, 684, 692
Callis, C. F., 457(4), 459(7,8), 460(9), 462(12), 466(7,19,20), 467(4,19), 468(9), 501(47), 502, 503, 504
Calvery, H. O., 680(2), 692
Calvin, M., 466(21), 503
Camp, M., 49(33), 51, 62, 113(16), 118, 149, 190(112), 231, 357, 404, 652, 662
Campanella, D. A., 455(1), 502
Campbell, G. H., 219(187), 234, 253(51), 258(51), 266
Campbell, K. S., 618(9), 623
Cann, H. M., 680(5), 692
Carberry, J. B., 935(26), 940(34), 954(34), 987, 988
Carlin, B., 263, 268
Carrie, C., 702(28), 706
Carter, J. D., 279, 317, 330, 399
Carter, R. O., 682(15), 690(27), 692, 693
Carter, R. P., 493, 504
Carver, E. K., 637, 660
Caryl, C. R., 274, 317
Casey, E. A., 643(79), 661
Cassie, A. B. D., 109, 110, 113 (17), 149
Cates, D. M., 618(9), 623
Cayle, T., 668(11), 670, 671, 674 (11), 675, 678, 883, 897
Cenker, M., 832(30), 873(30), 891
Chaberek, S., 456, 466(3), 499(3), 501, 502
Chandrasekhar, S., 246, 266
Chang, R. C., 634, 635, 660
Chantrey, G., 45, 61, 62, 155(5), 220, 221(5), 227
Chapin, R. M., 521(7), 608
Chapman, D. L., 84, 102
Chase, M. W., 700, 705
Chenicek, A. G., 873, 894
Chessick, J. J., 376(224), 406

Chetaev, P. M., 637, 660
Chrisman, C. H., 640, 661
Christison, H., 415(3), 449
Chung, H. P., 379(252), 408
Ciriacks, J. A., 186(111), 231
Clark, G. L., 627, 659
Clark, J. R., 341, 401
Clark, W. A., 379, 408
Clarkson, R. G., 11, 28, 648(92), 662
Clayton, R. L., 257(65), 267
Cochran, S. D., 222(195), 224(195), 235, 254(59), 267, 356(155), 404
Cochran, W. G., 389(155), 394(296), 395(296), 410
Cohen, B., 871, 894
Cohen, H., 424(23), 450
Cohen, J. M., 957(50), 958(50), 959 (50), 961(50), 989
Cohn, E. J., 878(123), 895
College, R., 620(11), 623
Collie, B., 107(2), 113(2), 123(2), 126(2), 148, 162(21), 228, 626(2), 658
Collins, J. H., 1018(35), 1020
Compton, J., 67(11), 101, 191(123), 196, 197(123,146,148,150), 198(146, 148,150), 199(150), 200(148), 209 (170), 217(170), 218(150,170), 219, 220(188), 224(148,150), 232, 233, 234, 256, 267, 273, 293, 301, 302, 305, 309, 311, 312(7,8,80, 100,101), 317, 319, 320, 354, 388, 389(149), 403
Compton, J. W., 858(71), 893
Condit, H. R., 435(53), 451
Connick, W. J., Jr., 295, 319
Connoell, C. H., 954(43), 988
Conroy, L. E., 145(97), 152
Consdorf, A. P., 937(32a), 988
Cookson, J. T., Jr., 952, 988
Cooper, C. F., 626, 658
Copeland, H. R., 621(13), 624
Copeland, J. L., 861(88), 894
Coppock, W. A., 333, 364, 399, 786(109,133), 788(133), 790(133), 796(133), 797(109), 810, 811
Corkill, J. M., 170(39), 171(39), 172(39), 183(39), 228
Corliss, D. S., 825(6), 873, 890
Coulson, J. M., 241(4,5), 264, 265
Courchene, W. L., 174(54), 229
Coward, T. L., 778(104), 810
Cowie, J. M. G., 771, 810
Cox, G., 394(296), 395(296), 410
Crecelius, S. B., 825, 841(10), 881, 882(141), 885(156), 890, 896, 897
Crisp, D. J., 109(11), 149
Cross, H. D., III, 337(76,86), 338 (76), 344(76), 347(86,129), 348 (86), 378(76), 401, 403, 538(38), 609
Crounse, N. N., 730(11), 807
Crowe, J. B., 338, 401
Crowther, J. P., 852, 892
Crutchfield, M. M., 493(41), 504
Culp, G. L., 977(61), 978(61), 989
Culp, R. L., 977(61), 978(61), 989
Cuming, B. D., 158(16), 165, 227
Cummins, R. W., 501(48), 504
Cunliffe, P. W., 384, 408
Cuny, K. H., 638, 660
Cupples, H. L., 643, 661

D

D'Alessandro, P. L., 619(10), 623
Dallenbach, H. R., 15(22), 29
Danielsson, I., 144(91), 152
Dannenberg, E. M., 301, 302(76), 313, 319
Darbey, A., 390, 409
Darby, D. O., 938(46), 841(46), 843(46), 846(46), 847(46), 848 (46), 892
Davies, J. T., 78(42,43), 102
Davis, B. L., 528(26), 609

Pages 1-452 comprise Part I
Pages 453-728 comprise Part II
Pages 729-1020 comprise Part III

Davis, D. S., 582(76), 611
Davis, E. M., 949, 950, 986
Davis, G. A., 338(95), 341(95), 390 (95), 395(95, 105), 401, 410, 654 (121), 663
Davis, R. C., 2(2), 3, 69(21, 22), 101, 181(88, 89, 90), 230, 286, 292 (57), 305, 306, 308(57), 311, 313, 318, 320, 336, 337(69, 83), 344, 346, 376(222), 378(222), 382(222), 390, 396, 400, 401, 406, 410, 623 (15), 624, 925(8), 986
Davison, S., 589(88), 611
Dean, J. A., 335(58), 400, 427(32), 429(32), 430(32), 450
Dean, R. B., 961(54), 989
Deininger, R. A., 926(9), 987
deJong, J. C., 702, 706
Dekker, H., 116(33), 150
DeLange, R. J., 667(2-4, 7, 9), 678
Derjaguin, B. V., 72(32, 35), 73(35), 83(35), 86, 102, 195(138), 232
Dervichian, D. G., 144(90), 152
Desalme, R., 652, 662
Deville-Boorsma, G., 309, 320
deVries, A. J., 626(5), 658
Diamond, W. J., 253(53), 266, 376 (227, 228), 378(228), 381, 407
Diaz, R. B., 324(8), 350, 352(8), 353(8), 362, 398
Dickerson, R. W., Jr., 829(21), 890
Dickinson, J. C., 365(183), 405
Diehl, F. L., 330(25), 334(25), 338, 348, 356, 362, 397(25), 398, 401, 861(83), 893
Dierkes, H., 629(19, 20), 659
Dobias, G., 644, 661
Domingo, E., 842(51), 892
Doscher, T. M., 174(58), 177(58), 186(58), 187(58), 229
Dostmann, J., 732(33), 807
Douglas, R. W., 880, 896
Dowas, W. L., 680(3), 692
Dowben, R. M., 145(100), 152
Downey, T. A., 501(46), 504, 861 (81), 864(81), 893
Draize, J. H., 689, 693, 697, 699, 705
Drake, G. L., Jr., 120(40), 128 (40), 129(40), 130(40), 131(40), 150, 194(135), 212(179), 232, 234, 286(40), 295(40), 318, 353 (139), 390(276, 277), 403, 409
Draves, C. Z., 11, 28, 648, 651, 662
Duerham, K., 652, 662
Dugan, G. L., 979(64), 989
Dunken, H., 638(59), 660
Dunlap, R. K., 297(70), 299(70), 309(70), 319
Dunning, H. N., 177(70), 182(95), 230, 231
Dupin, S., 651(100), 662
Dupre, J., 882, 896
Durham, K., 72(37), 73, 102, 113 (16), 118, 149, 195(137), 198 (137), 199(137), 209(137), 210 (137), 211(137), 232, 301(78), 310, 319
Duwez, P., 243, 265
Dwyer, F. P., 466(22), 503

E

Earing, M. H., 882(149), 896
Eastlund, B. J., 979, 989
Eberhart, D. R., 730(4), 806
Echelberger, W. F., Jr., 935(22, 26), 955(49), 987, 989
Eckell, A., 340(314), 410
Eckhardt, C., 805(163), 813
Edelstein, S. M., 651, 662
Edgar, R., 614(1), 623
Edsall, J. T., 878(123), 895
Edwards, G. R., 181(94), 231, 376 (226), 378(226), 383(226), 407

Edwards, P. R., 637, 660
Egey, Z., 67(7), 100
Egloff, G., 626(7), 658
Ehrenkranz, F., 376(209-211), 381, 394, 406
Eisenbrand, J., 747(78), 809
Ekwall, P., 144(91), 152
Ellestad, R. B., 524(14), 608
Elliot, J., 839(41), 841(41), 891
Elliott, T. A., 644, 661
Ellzey, S. E., Jr., 295, 319
El-Shamy, T. M. M., 880, 896
Elsworth, F. F., 137(71), 151
Emendorfer, E., 52(35), 62
Engberg, M. J., 561(50), 610
Epstein, M. B., 145(97), 152
Epton, S. R., 180(86), 230
Erban, F., 651, 662
Ester, V. C., 614, 623
Evans, H. C., 180(84), 181(84), 230
Evans, N. A., 732(58), 808
Evans, P. G., 376(218), 382, 406
Evans, R. L., 929, 987
Evans, R. M., 328, 398, 425(27), 450
Evans, W. H., 667(5-7,9), 678
Evans, W. P., 49(33), 51, 62, 190 (112), 231, 357, 376(218), 382 (218), 404, 406
Everson, H. E., 381(242), 407
Eyring, H., 180(82), 230

F

Fair, G. M., 904(2), 986
Fairchild, G. D., 136(64), 137(64), 141(64), 151
Fancher, O. E., 682, 684, 692
Fargues, A., 786(117), 811
Farquhar, G. J., 961(53), 989
Fava, A., 180(82), 230
Feigl, F., 842, 892
Feit, A. J., 282(25), 317
Feldman, H. F., 979(69), 990
Fennell, F. L., 585(80), 611
Fenner, T. W., 334(54), 400
Ferguson, A., 642, 645(72), 661
Ferguson, F. A., 932(20), 936, 987
Fernelius, W. C., 456, 502
Ferris, R. C., 337(75), 344, 394, 400, 409
Feuell, A. J., 311, 320, 354, 394, 403, 410
Findley, W. R., 670(14), 678, 732 (49,50), 735(50), 759(50), 761(50), 808
Fineman, M. M., 832, 891
Fineman, M. N., 839(45), 848(45), 881, 887(45), 892
Finger, B. M., 341, 378(327), 402, 411
Firsching, F. H., 381, 407
Fischer, E. K., 176(63), 219(187), 229, 234, 253(51), 258(51), 266, 648, 662
Fitzgerald, F. J., 999(8), 1018
Fitzhugh, O. G., 680(1), 692
Fitz-William, C. B. R., Jr., 878 (118), 895
Flath, H. J., 391, 409
Flitcraft, R. K., 16(25), 29
Flower, D. M., 1018(40), 1020
Fong, W., 285, 288, 291, 292, 318, 365(182), 405
Ford, O. E., 177(72), 230, 336, 400
Fordyce, D. B., 882, 896
Formaini, R. L., 525(18), 608
Forster, R. B., 651, 662
Forsythe, R. H., 884(153), 897
Fort, T., Jr., 139, 140(75), 142 (75), 145, 147(75,76), 151, 204, 205(162), 233, 258(69), 267, 376 (213), 377(236), 382, 385, 406, 407, 408
Fortess, F., 204(161), 233, 757 (94), 810
Foster, F. G., 288, 290, 293, 318
Foster, W. R., 168(36), 169(36), 228

Pages 1-452 comprise Part I
Pages 453-728 comprise Part II
Pages 729-1020 comprise Part III

Foulk, C. W., 631, 659
Fourt, L., 243(13), 265
Fowkes, F. M., 107(3), 108(5), 114 (3, 21), 116(3), 117(3), 118(3), 119 (3, 39), 148, 149, 150, 626(8), 658
Fox, H. W., 640, 661
Francis, C. V., 881(137), 887(137), 896
Frankel, S., 626(6), 658
Frantz, A. J., 395(299), 410, 416 (7), 449
Freeman, G. S., 341, 357, 402
Frenkel, J., 247, 266
Friedman, S. D., 671(16), 678
Friele, L. C. F., 356(161), 404
Friele, L. R. C., 786(118), 811
Fries, B. A., 180(80), 230, 308, 320, 376(215), 380(215), 406
Frishman, D., 243(13), 265, 390, 409
Fruhner, H., 653, 662
Fu, Y., 87(51), 102
Fuchs, R. J., 878(119), 895
Fuerstenau, D. W., 195(140), 232
Fuller, R. K., 380(241), 407
Fulton, G. P., 324(4), 357, 398
Furry, M. S., 333, 334(54), 347, 357, 358(124), 361(46), 399, 400, 402, 404, 589(87), 604(94), 611, 612

G

Gadberry, H. M., 839(46), 841(46), 843(46), 846(46), 847(46), 848(46), 892
Gaddum, J. H., 643, 661
Gailey, I., 362, 404
Gans, D. M., 648, 662
Ganz, C. R., 805(164, 165), 813
Ganz, E., 796, 801, 812
Garber, W., 954(46), 988
Gard, A. J., 497, 504
Gardon, J. L., 115(26), 149
Garrett, W. D., 641(67), 661
Gault, T. W., 1018(38), 1020
Gebelein, C. G., 839(45), 848(45), 881(45), 892
Gedge, B. H., III, 16(24), 29
Gelezunas, V. L., 383, 408
Gerard, P. H., 72(29), 101
Gerhardt, W., 341, 402
Gertz, W. J., 243(15), 265
Getchell, N. F., 154(2), 197(2), 201(2), 202, 227, 258, 267
Getmanskii, I. K., 652, 662
Getty, R., 281, 311, 317
Geyer, J. C., 904(2), 986
Ghionis, C. A., 293, 297, 308, 319
Gilbert, A. H., 527(24), 528(24), 557(24), 604(93), 608, 612
Gilcreas, F. W., 841(49), 892
Giles, C. H., 754(87), 810
Gillies, G. A., 181(94), 231, 341 (111), 376(226), 378(226), 383 (226), 402, 407
Ginn, M. E., 69, 101, 141(80, 81), 142, 148(81), 151, 165(31), 173 (48), 174(48), 180(87), 181(87), 228, 229, 230, 324(6), 338, 341, 376(6), 390, 395(95), 396, 398, 401, 490(38), 504, 654(120), 663, 851(64), 892
Glazer, A. N., 667(8, 10), 678
Glazman, Y. M., 174(57), 175(57), 229
Gleason, M. N., 680(6), 692
Gleichert, R. D., 873(108), 894
Gloxhuber, C., 805(167, 168), 813
Gloyna, E. F., 947(35), 949, 950, 986, 988
Gobeil, N. B., 619(10), 623
Goddard, E. D., 145(98), 152
Goebell, K., 218, 234
Götte, E., 67, 89, 101, 103, 192,

[Götte, E.]
197(128), 198(128), 232, 372, 406, 632, 634, 652, 659, 662, 696, 701 (17,19), 705, 706
Goette, E. K., 164(29), 165(29), 228
Goglia, M. J., 244(18), 265
Gold, H., 730(6), 806
Goldstein, S., 241(7), 243(7), 265
Goldwasser, S., 261, 264, 267, 268, 327, 329, 398, 786(135), 796, 811
Gollmick, H. J., 263, 268
Golucke, C. G., 979(64), 989
Goodman, J. F., 170(39), 171(39), 172(39), 183(39), 228
Gordon, B. E., 137, 138(73), 139, 147, 148(73), 151, 181(92,94), 182 (92), 183(92), 190(116), 219, 231, 232, 305, 320, 356, 376(214,219) 377, 378, 379(214), 383, 386, 404, 406, 407, 411, 538(41,42), 609
Goring, D. A. I., 184(100), 231
Gosselin, R. E., 680(6), 692
Gotaas, H. B., 979(63), 989
Gouy, M., 84, 102
*Graf, R., 1018(39), 1020
Grahame, D. C., 84(44,47), 102
Gray, V. R., 115(23), 117(23), 149
Grebenshchikov, B. N., 628, 629, 659
Green, L., 243, 265
Green, R. L., 879, 895
Green, S. R., 671, 678
Greenwald, H. L., 839(45), 848(45), 881(45), 892
Greiner, L., 177(67), 229
Griesinger, W. K., 330(36), 342, 399
Griffith, J. F., 680(7), 682, 690(26), 692, 693, 698, 705, 805(171), 813
Grim, R. E., 50(34), 54(34), 57(34), 62
Grinchuk, T. M., 12(10), 29
Grindstaff, T. H., 130(60a), 139(75, 76), 140(75), 142(75), 145(75,76), 147(75,76), 151, 208, 213, 215
[Grindstaff, T. H.]
(166), 217(166), 220, 233, 376 (208,213), 377(208,236), 382(208, 213,236), 406, 407, 411
Grove, C. S., 634(44), 660
Grove, J. E., 376(227), 381(227), 407
Groves, W. L., 882, 896
Grum, F., 435(53), 451, 786(136, 137,140), 788(150,151), 801, 812
Gruntfest, I. J., 343, 402, 648, 650, 662
Grzeskowiak, U., 563(55), 610
Guastalla, J., 643, 651, 661, 662
Guastalla, L. P., 643(78), 651, 661, 662
Guillaumin, R., 392, 409, 644, 661
Guiteras, A. F., 873, 894
Gustafson, J. H., 338, 339(87), 401
Guter, K. H., 935(26), 987

H

Haahti, E., 42, 45(17), 62
Haak, R. M., 832, 891
Hackford, J. E., 627(14), 658
Hadzekriakides, N. S., 526(19), 608
Haefle, C. W., Jr., 264(85), 268
Haeusermann, H., 730(7), 732(21), 752(81), 766, 806, 807, 809
Hageboeke, E., 186(105), 187(105), 231
Hager, O. B., 648(99), 650(99), 662
Hamaker, H. C., 70, 86, 101
Hamilton, L. R., 730(12), 807
Hamilton, W. C., 827(13), 890
Hammond, A. H., 999(9), 1019
Handy, C. T., 617, 623
Hansen, R. S., 87(51), 102, 644, 661
Harborne, R. S., 272(1), 316
Harder, H., 137(69), 151
Hardy, A. C., 427, 450, 565, 610

Pages 1-452 comprise Part I
Pages 453-728 comprise Part II
Pages 729-1020 comprise Part III

Hardy, E. E., 874, 895
Hardy, W., 632, 659
Harker, R. P., 126(58), 128(58), 137(58), 143(58), 144(58), 151, 396, 410
Harkins, W. D., 12(9), 28, 108(5), 118(5), 148, 635, 637, 638, 640, 660
Harnsberger, H. F., 166(33), 228
Harris, J. C., 2, 3, 12, 29, 69(20, 23), 89(58), 101, 103, 141(80, 81), 142, 148(81), 148(101), 151, 152, 173(48), 174(48), 180(78, 87), 181 (87), 191(119), 193(127, 130), 226 (199), 229, 230, 232, 235, 243(13), 246, 265, 266, 324(3), 335, 336, 338, 354(3, 60), 355, 376, 377, 380, 381(243, 244), 389, 390(272), 395, 398, 400, 407, 409, 415, 418, 419, 423, 449, 490(38), 504, 632(36), 659, 747(80), 809, 827(11, 12), 830 (11), 831(11), 840(11, 12), 841, 851 (11, 12, 64), 890, 892
Harris, M., 111(14), 149, 163(25), 181(25), 186(103, 107, 109, 110), 228, 231, 243(13), 265
Hart, W. J., 67(11), 101, 191(123), 196, 197(123, 146, 148, 150), 198 (146, 148, 150), 199(150), 200(148), 209(170), 217(170), 218(150, 170), 219, 220(188), 224(148, 150), 232, 233, 234, 256, 267, 273, 293, 301, 302, 305, 309, 311, 312(7, 8, 80, 100, 101), 317, 319, 320, 354, 388, 389(149), 403
Hartley, R. S., 137(71), 151
Hartwig, G. M., 181(94), 231, 341 (111), 356(162), 376(226), 378 (228), 383(226), 402, 404, 407, 418, 449
Hatch, G. B., 12(11), 29, 851(65), 892
Hattiangdi, G. S., 632, 659
Haussner, I., 126(57), 143(57), 150
Hawthorne, C. H., 954(45), 988
Haydon, D. A., 72(27), 101
Hayes, F. R., 933, 987
Hayes, W. J., Jr., 682, 692
Healey, T. W., 195(140), 232
Hefti, H., 732(51), 757(51), 808
Heiges, E. O. J., 293(60), 296(60), 297(60), 301(60), 313(60), 319, 339(306), 390(278), 409, 410
Hemmendinger, H., 326, 327, 329 (14), 398, 786(123), 811
Henderson, S. T., 435(54), 451
Hendricks, M. H., 690(27), 693
Henjum, J. E., 856(69), 858, 863, 876, 893
Hensley, J. W., 190(114, 115), 191 (114), 192(114), 194(114), 217 (114), 220, 231, 290, 291, 302, 303, 304, 314, 318, 319, 320, 334(52), 336, 339, 341, 342, 343, 344, 350(52), 376(97, 212, 217, 220), 378(52, 64, 106, 220), 380, 381, 382 (64, 106), 390, 400, 401, 402, 406, 538(40), 609
Herbert, S., 145(99), 152
Herbig, W., 651, 662
Herzberg, J. J., 701(17), 706
Hesse, S. H., 882, 896
Hiemenz, P. C., 177(66), 229
Hierons, S. H., 340(317), 410
Hill, E. F., 275(10), 280(10), 310 (10), 317
Hilton, T. B., 537(36), 609, 643, 661
Hinshaw, J. O., 520(1), 521(1), 608
Hintermaier, J. C., 111(14), 149
Hochreiter, R., 154(3), 201(3), 227
Hock, C. W., 315, 321
Hock, G., 644(90), 661
Hodge, H. C., 680(3, 6), 692

Hodgkiss, E., 435(54), 451
Hoerle, R., 675(17), 678
Hoerner, S. F., 244(19,20), 265
Hoff, G., 344, 365, 402
Hoffman, J. I., 632(34), 659
Hofstetter, H. H., 137(67), 151
Hogan, T. H., 540(44), 609
Hogg, R., 195(140), 232
Holland, V. B., 341, 368, 401, 405
Hollen, N., 614(1), 623
Hollies, N. R. S., 111(13,14), 112, 113(15), 118, 149, 243(11,12), 265
Holme, J. C., 1016(33), 1020
Holmes, F. H., 365(188), 373(188), 375(188), 391(188), 405
Hoogerheide, J. C., 884(154), 897
Hopkins, C. B., 953(41), 988
Höpfer, P. 290, 318
Hopper, G., 680, 692
Hörig, K., 69, 101, 179(77), 182(77), 183(77), 230
Horning, E. S., 873, 894
Howard, P., 654(127), 663
Howarth, E., 205(163), 233
Hsiao, L., 177(70), 230
Hucker, G. J., 839(40), 841(40), 891
Hudson, R. A., 337(80), 344(80), 401
Huffington, J. D., 654, 663
Hughes, V. L., 462(13), 502
Huisman, M. A., 615(2), 623
Humphreys, T. W., 932(19), 987
Hunter, R. S., 328, 335(57,59,62), 350, 351(62), 352(62), 354(62), 356, 362, 363(179), 398, 400, 404, 431, 451, 565, 610, 786(131,138), 788(131), 798(131,158), 801, 811, 812
Hunter, R. T., 313, 320, 327(17), 328, 344, 345, 362(17,120), 398, 402, 570(61), 610
Hurley, H. J., 37, 61
Husson, F., 144(86,87), 152
Hutchinson, E., 179(75), 230
Hutchison, A. W., 173(46), 229

I

Iimori, M., 40, 61, 120(42), 145 (42), 150, 366, 368(191), 405
Ilg, J., 193(131), 232
Illman, J. C., 356, 378(327), 404, 411
Imamura, T., 395(334), 411, 482 (34), 483, 484, 489, 490(37), 503, 504
Inamdar, V. E., 732(60), 808
Inamorato, J., 283(31), 318
Inks, C. G., 190(115), 231, 290, 291, 303, 304, 318, 319, 376(217, 220), 378, 381, 406, 538(40), 609
Irani, R. R., 457(4), 460(9), 462 (12), 466(19,20), 467(4,19), 468 (9,26), 493(41), 501(47), 502, 503, 504
Iwadare, Y., 186(109a), 188(109a), 189(111a), 231

J

Jacobi, O., 696(5), 702, 705
Jager, R., 701, 706
Jakob, C. W., 145(97), 152
Jakob, M., 241(2), 264
Janauer, G. E., 195(141), 232
Jarrell, J. G., 285, 318, 330(37), 399
Jayson, G. G., 180(85), 230, 376 (230,232), 381, 407
Jeanne, M. L., 786(126), 811
Jebe, E. H., 376(210,211), 381 (210,211), 394, 406
Jenkins, D., 954(48), 989
Jennings, W. G., 356, 382(154), 404, 827(13), 890
Jensen, L. B., 89, 102, 173(45), 174(45), 229
Jeschke, D., 140(77), 151
Joeder, H., 747(79), 809
Johns, C. K., 932(19), 987
Johnson, A. E., 615(3), 623

Pages 1-452 comprise Part I
Pages 453-728 comprise Part II
Pages 729-1020 comprise Part III

Johnson, G. A., 161(20), 173(50), 186(50), 195(20), 206, 207(20), 228, 229, 288, 290, 293, 318
Johnson, J. B., 263(79), 267
Johnson, L. D., 350, 411
Johnson, M. L., 347(124), 357(124), 358(124), 402
Johnson, P., 162(41), 172(41), 228
Johnston, G. A., 343, 376(117), 402
Jones, D. V., 682, 692
Jones, G., 637, 660
Jones, T. G., 158(15), 197(154), 198 (15,154), 199(15), 227, 233, 337 (78,80), 344, 401, 476, 503
Jordan, H. F., 12(9), 28, 640, 660
Jordan, J. W., 163(23), 168(23), 228
Jordan, M. E., 301(77), 313(77), 319
Joseph-Petit, A. M., 72(31), 101
Jost, W., 247(36), 266
Judd, D. B., 326(12), 327(12), 356, 387(12), 398, 425(26), 426(26), 435(55), 450, 451, 565, 577, 578 (60), 579, 610, 786(105), 788(105), 810
Jung, D., 730(15), 807
Jungermann, E., 165(31), 226(198), 228, 235, 338(95), 341(95,105), 354(105), 390(95), 395(95,105,300), 401, 402, 410, 654(120), 663, 937, 988
Juva, K., 42(18), 62

K

Kabanov, B. N., 72(32), 102
Kaessinger, M. M., 111(13,14), 112 (13,15), 113(15), 118(13), 149, 243 (11,12), 265
Kaestner, W., 805(167), 813
Kahler, D., 829(20), 839(20,43), 841(20,43), 846(20), 873(43), 890, 891
Kamp, R. E., 380(239,240), 407
Kanamaru, K., 186(108), 231
Kaneko, T. M., 858, 882(148), 893, 896
Kang, M. S., 379(252), 408
Kantor, E. A., 366, 405
Karimova, A. Z., 746(77), 809
Kariyone, T., 365(187), 405, 476 (30), 503
Karrer, P., 89, 102, 186(104), 231
Kashiwa, I., 364(321), 411
Kashiwagi, M., 217(183), 220, 234
Kasper, C. B., 667(9), 678
Kasperl, H., 732(51), 757(51), 808
Kato, H., 732(39), 808
Kazella, I. J., 155(9), 227
Keast, R. R., 598(92), 612, 825(6), 868(100), 873(6,100), 890, 894
Keller, E. R., 730(7), 766, 806
Keller, R., 752(81), 809
Kellum, R. E., 32(2), 33(2), 38(2), 61
Kelly, E. L., 286(40), 295(40), 318, 353(139), 390(276), 403, 409
Kelly, W. R., 827(14), 881, 890, 896
Kendall, H., 596, 611
Kenjo, K., 144(89), 152
Kennedy, J. M., 298, 319, 331, 335, 344, 345, 399
Kensler, C. J., 805(161), 813
Kern, C. R., 357, 404
Kerr, G. P., 435(56), 451
Kerst, A. F., 459(8), 465(17), 502, 503
Keup, L. E., 935(23), 987
Kevan, D. F. B., 372, 406
Kiefer, J. E., 330(31), 399
Kimmel, A. L., 839(46), 841(46), 843, 846, 847, 848, 892
King, A. M., 272(1), 316

King, G., 654(129), 663
King, S. M., 979(65), 990
Kingsbury, S. A., 379(253), 408
Kinney, F. B., 69(20), 101, 180(87), 181(87), 230, 490(38), 504
Kinney, J. A. S., 786(106), 810
Kip, C. E., 204(161), 233
Kipling, J. J., 155(7), 162(22), 227, 228
Kirk, R. E., 832(28,29), 833, 849 (61), 850(61), 867(98), 868(98), 868(102), 872(158), 873(28,107), 875(29), 891, 892, 894, 897
Kirkley, J. L., 333(46), 347(46), 357 (46), 361(46), 399, 604(94), 612
Kirkpatrick, D., 743(74), 809
Kitchen, E. A., 882(147), 896
Kitchener, J. A., 72(26), 101, 161 (19), 173(49), 174(49), 228, 229, 626, 658
Kitchener, J. S., 155(8), 227
Klauck, A., 747(78), 809
Klaus, W., 290(52), 318
Kligman, A. M., 700(14), 705
Kling, A., 137(67), 151, 326(15), 398, 786(134,143), 804, 811, 812
Kling, W., 67(8,9,12), 71(33,34), 87(33,34), 88(52), 89(55,59), 98(8, 9), 100, 101, 102, 103, 121, 124 (49,54,55), 125(55), 126(55,57), 127, 132(49), 135, 143(57), 150, 173(47), 174(47), 185(106), 186 (106), 187(106), 188(106), 190(106), 192, 197(128,129), 198(128), 212 (176), 218, 229, 231, 232, 234, 244(23), 246, 256, 265, 267, 289 (49,50), 318, 396, 410, 701, 706
Knaggs, E. A., 420(18), 450
Knaggs, R., 350(319), 411
Knapp, K. W., 879, 895
Knight, J. R., 379(253), 408
Knowles, D. C., Jr., 308, 320
Koch, O., 746(76), 796(120), 809, 811
Koehler, W. R., 145(100), 152
Kölbel, H., 69, 101, 179(77), 182 (77), 183(77), 230
Kokorudz, M., 832(30), 874(30), 891
Kolat, R. S., 289, 291, 292, 318
Koltoff, I. M., 829(18), 830(22), 890
Kolugin, A. P., 652(112), 662
Komeda, Y., 385, 386, 393(331), 395, 411
Konecny, C. C., 177(68), 229
Koning, J., 163(27,28), 179(27,28), 181(27,28), 228
Konort, M. D., 1001(19), 1019
Koppe, H., 124, 150
Kornreich, E., 311, 320
Kovalsky, S. J., 878(120), 895
Krais, P., 730(2), 806
Kramer, H., 290(52), 318
Kramer, M. G., 190(114), 191(114), 192(114), 194(114), 217(114), 220 (114), 231, 302, 319, 336, 339 (97), 376(97), 380(97), 400, 401, 863, 865(94), 894
Kramer, M. R., 423(20), 450
Kratohvil, S., 195(41), 232
Kretzschmar, G., 223(194), 234, 248(40), 266, 653, 662
Krieger, R. S., 372, 405
Kritchevsky, J., 282(25), 317
Krupp, H., 66(3,4), 67(3-5,7-9), 71(33,34), 82(3), 87(33,34), 98 (8,9), 100, 102, 195, 211, 212 (142,176), 233, 234
Krupp, K., 67(6), 100
Krynitsky, J. A., 641(67), 661
Kubelka, P., 387, 408, 551(49), 577, 610
Kubitschek, H. E., 253(56), 267
Kuentzel, L. W., 935, 987
Kullbom, S. D., 430(41), 451
Kullman, R. M. H., 197(153), 202 (153), 209(153), 216(153), 233, 293(63), 295(63), 319, 390(275), 409

Pages 1-452 comprise Part I
Pages 453-728 comprise Part II
Pages 729-1020 comprise Part III

Kung, H. C., 145(98), 152
Kuno, H., 175(59), 229
Kuno, Y., 35, 36(6), 61
Kurgan, C. R., 313(106), 320, 344 (120), 345(120), 362(120), 402
Kurz, J., 326(15), 398, 741(67,68), 786(134,143,148,149), 809, 811, 812
Kurzendoerfer, C. P., 173(51), 174 (51), 229
Kyes, C., 607(95), 612
Labbee, M. D., 666(1), 678
Labhard, L., 340(313), 410
Lambert, J. M., 55, 56, 62, 154(1), 190(117), 191(121), 227, 232, 252 (43), 266, 283, 292, 293, 300, 302 (33,73), 303(33), 308, 310, 313, 315, 318, 319, 325(11), 326, 327, 329(14), 331, 336, 337(72), 339, 354, 398, 400, 403, 406, 407, 416 (8), 449, 786(123), 811
Lambert, S. M., 462(10,11), 502
Lamberti, V., 1001(16-19), 1019
Landau, L. D., 72, 73(35), 83(35), 102
Landon, M., 667(5-7,9), 678
Lange, H., 67(8,9), 69, 71(33,34), 72(38-40), 84(38,39), 86(38,40), 87(33,34), 88(17,52), 89(40), 95 (60), 97(61), 98(8,9,17), 100, 101, 102, 103, 113(20), 124(55), 125 (55), 126, 127, 135, 149, 150, 173 (47,51), 174(51,53), 175(53), 212 (176), 229, 234, 243, 244(23), 246, 247, 265, 289, 318, 396, 410
Lange, K. R., 863, 866(92), 867, 894
Langer, E., 126(57), 143(57), 150
Langguth, R. P., 670, 678, 883, 897
Lankler, J. G., 334, 400
Lanter, J., 739(64), 743(71), 747 (64), 796(157), 809, 812
Lapitsky, M., 607(95), 612
Larose, P., 393, 409
Larrat, E., 654(119), 663
Larsen, R. I., 244, 265
Lashof, T. W., 786(139), 812
Laskey, R. P., 17(26), 29
Latourette, H. K., 531(35), 609
LaVier, H. W. S., 244(18), 265
Law, P. B., 283, 317
Lawrence, A. S. C., 144, 152
Lawrence, C. A., 932(18), 987
Leach, P. B., 136(63), 151
Leardi, E. B., 355(153), 403
Leaver, I. H., 732(44), 808
Lebensaft, W., 786(148,149), 812
Lederer, E. L., 631, 659
Lee, D. H., 356(163), 378(163), 383(163), 404
Lee, L. H., 115(27), 116(35), 117 (27), 149, 150
Lee, W. W., 273(4), 316
Leenerts, L. O., 337(75), 344, 400, 839(41), 841, 891
Leigh, G. M., 347, 362, 403
Lemberger, A. P., 177(69), 230
Lenker, S., 651, 662
Lennon, W., 732(30), 807
Leonard, E. A., 365(184), 395, 405
Lesser, M. A., 523(10), 527(22), 608
Levans, A. S., 582(75), 611
Levin, G. V., 954, 989
Levin, H., 253(53), 266, 376(228), 378(228), 381(228), 407
Levine, I. E., 59(43), 63
Lewis, E. L., 429(37), 450
Lewis, J. T., 747(80), 809
Lewis, K. E., 161(20), 195(20), 206, 207(20), 228
Lieber, M., 929(15), 932(15), 987
Liebert, C., 670(14), 678, 730(49), 805(164), 808, 813

Liechti, P., 730(8), 806
Lin, J., 732(43), 808
Lindsey, C. H., 342, 347, 357(113), 358(113), 402
Linfield, W. M., 226(198), 235, 341, 354(105), 395(105, 300), 402, 410, 654(121), 663, 1008(26, 29-31), 1019, 1020
Lintner, A. E., 879, 895
Linton, G. E., 424(23), 450, 586(84, 85), 611
Lipson, M., 654(127), 663
Liss, R. L., 16(25), 29, 537(36), 609, 883, 897
Lister, M. W., 521(4-6), 608
Little, A. C., 579, 610
Lloyd, H. B., 177(65), 229
Loder, E. R., 861(84, 87), 893
Loeb, L., 222(195), 224(195), 225, 235, 254(59), 267, 347, 356, 389, 403, 404, 828, 888, 890, 937(32), 988
Löcher, W., 330(26), 331(26), 399
Lohnstein, T., 638(58), 660
Longdon, W. K., 832(30), 874(30), 891
Lord, J., 365(188), 373(188), 375 (188), 391(188, 285), 405, 409
Loughran, E. D., 462(10), 502
Ludeman, H. A., 341, 402
Ludwig, A. C., 422(19), 450
Ludwig, H. E., 979(63), 989
Luechauer, L. F., 333, 399
Lundgren, H. P., 285(35, 37), 291, 318, 365(182), 405
Lunsted, L. G., 886(157), 897
Luthi, C., 730(17), 807
Luzzati, V., 144(86, 87), 152
Lyklema, J., 70, 101
Lyman, F. L., 805(165), 813
Lynch, W. O., 950, 988
Lyness, W. I., 365, 405, 730(10), 807
Lyon, J. B., 702, 706
Lyons, J. W., 459(8), 502

M

Macadam, D. L., 327(19), 398, 435 (55), 451
McAdams, W. H., 241(3), 243(3), 264
McBain, J. W., 141(79), 151, 156, 227, 272, 273(4), 316
McBain, M. E. L., 179(75), 230
MacBeth, N., Jr., 333, 399, 434 (51), 451
McCabe, E. M., 358, 404
McCarthy, R., 56, 62
McCarty, P. L., 931(6), 938(6), 986
McCauley, D., 954(45), 988
McClain, C. P., 527(25), 529(29), 608, 609
McClellan, A. L., 166(33), 228
McClelland, N. I., 926(9), 950, 987, 988
McCoy, L. R., 275(10), 280(10), 310(10), 317
McCullough, N., 369(196), 405
McCune, H. W., 879, 895
McCutcheon, J. W., 417, 449
Macek, K., 805(165, 166), 813
McGilvery, J. D., 850(62), 892
McGregor, R., 754 (90), 810
Machemer, H., 335, 357(61), 400
Machin, J. L., 380(241), 407
Mack, P. B., 330(35), 399, 589 (89, 90), 611
MacKenthun, K. M., 935, 987
MacKinney, G., 579, 610
McLaren, J. R., 961(53), 989
McLaren, K., 333, 399
McLean, G., 621(13), 624
McLendon, V., 120(46a), 150, 308, 320, 347, 384, 402
McLendon, V. I., 589(88), 611
McPhee, J. R., 226(200), 235
McPhilip, J., 805(169), 813
Madden, R. E., 837, 839(36), 869, 885, 891
Maes, E., 176(61), 229, 337(77),

Pages 1-452 comprise Part I
Pages 453-728 comprise Part II
Pages 729-1020 comprise Part III

[Maes, E.]
340, 357, 401
Magill, P. L., 563(54), 610
Magnuson, L. W., 837(34), 891
Maguire, E. J., Jr., 1018(36,38), 1020
Maguire, H. C., 700, 705
Mahl, H., 67(12), 101, 192, 197 (128,129), 198(128), 232, 256, 267
Maitra, N. K., 241(4,5), 264, 265
Makarov, V. V., 746(77), 809
Mallman, W. L., 829, 839(20,43), 841(20,43), 846, 873, 890, 891
Mancy, K. H., 950, 988
Mandell, L., 144(91), 152
Manegold, E., 626(4), 658
Mankowich, A. M., 272, 316
Mann, E. H., 839(39), 841(39), 842, 891
Mann, H. B., 362, 405, 575(66), 610
Mann, J., 688(21), 693
Marboe, E. C., 880, 895
Marder, H. L., 313(106), 320, 344 (120), 345(120), 362(120), 402
Markezich, A. R., 617, 620, 621 (13), 623, 624
Markiewicz, W., 383, 408
Markland, F. S., 667(7,9), 678
Marks, A., 697, 705
Marple, W. L., 46(23), 62, 121(47), 141(47), 150, 301(75), 308, 313, 319, 320, 337, 364(82), 373, 374 (82), 401
Marples, M. J., 32(1), 33, 38(1), 61
Marshall, C. E., 252(45), 266
Martell, A. E., 456, 462(13), 466(3, 21), 472, 499(3), 501, 502, 503
Martin, A. R., 69(21,22), 101, 181 (88,89,90), 230, 281, 286, 292 (57), 305, 306, 308(57), 311, 313, 314(108), 317, 318, 320, 324(4), 336(65,69), 337(65,69), 346, 357, 368, 369, 376(221,222), 378(221, 222), 382(221,222), 389(65), 390, 398, 400, 405, 406, 409, 654 (128), 663
Martin, A. T., 825(9), 839(42), 841(42), 843(42), 881, 883(150), 890, 891, 896
Maruta, I., 365(187), 386, 405, 408, 476(30), 503
Marx, T. I., 644(90), 661
Masland, C. H., 258, 267
Mason, S. G., 184(98,99), 231
Mason, W. P., 263(79), 267
Mast, R. C., 181(93), 183, 231
Matalon, R., 634, 660
Mathai, K. G., 171(40), 228
Mathias, L. D., 523(11), 608, 874, 895
Matijevic, E., 195(41), 232
Matlin, N. A., 372, 406
Matson, T. P., 1007(25), 1019
Matsukawa, T., 295, 319
Mattoni, R. H., 33, 61
Mazzeno, L. W., Jr., 197(153), 202(153), 209(153), 216(153), 233, 293, 295, 319, 390(275), 409
Meader, A. L., Jr., 180(80), 230, 308, 320, 376(215), 380, 406
Mecey, L. W., 670, 678
Meek, D. M., 47(27), 62, 347(127), 358(127), 385, 402, 408
Meeker, D. A., 818(2), 889
Mehendale, S. D., 732(60), 808
Mehltretter, C. L., 458(5,6), 502
Meister, M. M., 704(37), 706
Melikhov, S. A., 653, 663
Mellon, M. G., 426(31), 428(31), 450
Mellor, D. P., 466(22), 503

Meloy, G. K., 462(14), 493, 496 (14), 503
Menar, A. B., 954, 989
Mendenhall, E. E., 837, 843, 869, 891
Mercer, B. W., 961(54), 989
Merian, E., 754(88), 810
Merkel, K., 878(122), 895
Merrill, R. C., 273(4), 281, 311, 316, 317
Merritt, L. L., Jr., 335(58), 400, 427(32), 429(32), 430(32), 450
Meyer, H. R., 730(8), 806
Meyer, L. O., 282(25), 317
Mihalik, W. B., 347, 403
Milbury, W. F., 954(45), 988
Milenkevich, J. A., 856(69), 858, 863, 876, 893
Miles, G. D., 12, 28, 630, 659
Miller, J. N., 631, 659
Millsaps, W. A., 350(135), 403
Milwidsky, B. M., 525(17), 608
Mino, J., 385(328), 386(329), 393 (331), 395(334), 411
Miranda, T. J., 57, 63
Mischke, W., 631, 659
Mita, R., 730(14), 807
Mittlemann, R., 654(128), 663
Mizuno, K., 837(34), 891
Mizuno, W. G., 825(8), 861(88), 868(8,99), 873(8,99), 878(8), 882 (140), 885(156), 886(140), 890, 894, 896, 897, 995(4), 1018
Mizushima, H., 308, 320, 339, 340 (311), 343, 346, 383(248), 386, 401, 407, 410
Moilliet, J. L., 69, 101, 107(2), 113(2), 123(2), 126(2), 148, 162 (21), 228, 626(2), 658
Montagne, J. B., 124(53), 150
Moore, A., 177(69), 230
Moore, A. T., 197(152), 233, 258 (67), 267
Moore, C. E., 56, 62
Moore, H. B., 197(153), 202(153), 209(153), 216(153), 233, 293(63), 295(63), 319, 390(275), 409
Moore, W. P., 878(118,121), 895
Moravec, R. W., 115(28), 149
Morey, E. D., 347(130), 403
Morey, G. W., 880, 896
Morgan, D. M., 279, 317
Morgan, J. J., 928(10), 933(10), 987
Morgan, O. M., 11(5), 28, 334, 399, 648, 661
Morgenthaler, W. W., 464(16), 492(40), 503, 504, 995(3), 1018
Morris, A. R., 343, 402
Morris, M. A., 340(313), 384(257), 408, 410, 615(2), 623(17), 623, 624
Morrish, D. H., 818(3), 890
Morrisroe, J. J., 354(141), 387, 403
Morton, T. H., 362, 405, 732(36), 808
Mousalli, F. S., 732(54), 808
Moyers, C. G., Jr., 975(58), 989
Mueller, F., 786(116), 811
Müller, G., 805(162), 813
Munk, F., 387, 408, 551(49), 557, 610
Muramatsu, M., 379(251), 408
Murdock, R. E., 376, 406
Muskat, I. E., 523(9), 608, 873, 894
Mustacchi, H., 144(86,87), 152
Mysels, K., 159(17), 161(17), 173 (44), 184(97), 227, 229, 231, 626 (6), 658

N

Nageley, L. M., 293(60), 296(60), 297(60), 301(60), 313(60), 319, 390(278), 409
Nancollas, G. H., 466(18), 503

Pages 1-452 comprise Part I
Pages 453-728 comprise Part II
Pages 729-1020 comprise Part III

Nash, T., 144(88), 152
Neale, S. M., 184(99), 185(99), 231
Needles, H. L., 732(59), 808
Neff, J. L., 1000(10), 1019
Nelson, A. A., 680(1), 692
Nettelnstroth, K., 197(149), 233, 386, 408
Netzel, D. A., 384, 390, 408, 644, 661
Neu, R., 357(165), 404
Neudert, W., 653, 663
Neuhaus, H., 702(28), 706
Neumann, A. W., 115(25), 149, 643, 661
Nevison, J. A., 330(36), 342, 399
Nevoline, F. V., 178(74), 230
Newhall, R. G., 354(141), 387, 403
Newman, S., 117(31), 149
Newmann, E. A., 688(24), 693
Nickerson, D., 356, 404
Nicolaides, N., 32(2), 33(2), 38(2), 42, 45(15), 61, 62
Niederhauser, W. D., 882(139), 886 (139), 896
Nieuwenhuis, K. J., 282, 288, 317, 318, 338, 342, 356, 357, 358, 401, 404, 786(142,147), 812
Nikkari, T., 42(18), 62
Nimkar, M. V., 113(18), 149
Nishizawa, H., 364(321), 410
Niven, W. W., Jr., 246, 266
Noad, R. W., 205(165), 209(165), 217(165), 233, 273, 274, 276, 279, 309, 316
Noble, W. R., 1008(28, 29), 1019
Noguchi, S., 379(251), 408
Noll, A., 634, 659
Nowak, W., 383(247), 407
Nuessle, A. C., 293, 296, 297, 301, 313, 319, 339, 390, 393, 409, 410
Nunez-Ponzoa, M. V., 355(153), 403
Nunn, L. G., Jr., 882, 896
Nussbaum, M., 420(18), 450

O

Oba, Y., 305, 321, 364(321), 410
Oberle, T. M., 825(8), 858, 868(8, 70), 873, 878(8), 883, 890, 893
O'Brien, E. M., 357(173), 404
Obrien, J. E., 841(49), 892
O'Brien, S. J., 390, 409
O'Connor, D. E., 365, 405
O'Connor, J. J., 273(4), 316
Ogilvie, R. B., 979(66), 990
Ohkubo, I., 730(14), 807
Ohsawa, K., 730(14), 807
Ohtaki, E., 730(14), 807
Okamura, I., 654, 655, 663
Okun, D. A., 902, 904(2), 950, 986, 988
Olaf, J., 263(83), 268
Olaitan, S. A., 667, 678
Oldenroth, O., 41(11), 46(11), 47, 61, 62, 120(41), 121(41), 137(41, 68), 145(41), 150, 151, 156(11, 12), 216, 227, 234, 253(50), 266, 298, 308, 319, 340(312, 315), 350, 354, 356, 365, 403, 404, 410, 786(141), 812
Olsen, D. A., 115(28), 149
Olson, R. A., 878(119, 120), 895
Opdyke, D. L., 680(7, 8), 692, 698, 705
Osipow, L., 191(124), 232, 253(49), 254(49), 266, 388(268), 389(268), 408, 626(11), 658
Osmundsen, P. E., 805(170), 813
Osteraas, A. J., 115(28), 149
Ostwald, W. O., 631, 659
Oswald, W. J., 979(63, 64), 989
Othmer, D. F., 832(28, 29), 833, 849(61), 850(61), 867(98), 868(98, 102), 872(158), 873(28, 107), 875 (29), 891, 892, 894, 897

Ottewill, R. H., 72(28), 101, 171 (40), 174(56), 175(56), 228, 229
Overbeek, J. Th. G., 72(30, 36), 73(36), 79(36), 83(36), 84(36), 86, 101, 102, 161(18), 211(18), 228
Owens, D. K., 116(32), 117(32), 150

P

Padden, C., 264(85), 268
Pai, K. G., 732(40), 808
Pai, S., 241(10), 265
Paitchel, H., 324(8), 336(66), 350 (8), 352(8), 352(8), 362(8,179), 363(179), 398, 400, 404
Paixao, L. M., 671, 678
Palin, A. T., 958, 959, 989
Palm, A., 878(122), 895
Palmer, R. C., 126(56), 143(56), 150, 211(174), 234, 365(183), 405
Pancheri, E. J., 1018(36), 1020
Pankhurst, K. G. A., 632, 659
Park, R. H., 577, 610
Park, W. J., 548(48), 610
Parke, J. P., 476, 503, 337(78, 80), 344(80), 401
Parker, E. P., 623(14), 624
Parks, G. A., 155(10), 227
Parris, N., 1008(30), 1020
Pascher, G., 696(2), 702, 705
Patapoff, M., 394(295), 409
Patek, J. M., 786(137,139,140), 801(140), 812
Patterson, H. T., 130(60a), 151, 208(166), 213(166), 215(166), 217 (166), 220(166), 233, 376(208), 377(208), 382(208), 406, 411
Patton, J. T., 881(137), 887(137), 896
Paynter, O. E., 680(7), 692
Peacock, C. L., 194(132), 201(132), 202(132), 232, 258(66), 267
Pengilly, P., 531(34), 609
Peper, H., 120(40), 128(40), 129 (40), 130(40), 131(40), 150, 194 (135), 202, 203(158-160), 205, 212(158-160,179), 217(158,160), 232, 233, 234, 295, 296, 297, 319, 336(67), 390(27), 400, 409
Perry, E. M., 297, 319
Perry, G. S., 180(83), 230
Perry, J. W., 9(1), 12(14,15), 28, 29, 199(155), 233, 324(1,2), 325 (1), 394, 398, 626(9,10), 658
Pescatore, J. J., 377(234), 380 (234), 407
Pestemer, M., 741(66), 746(66), 809
Peters, D., 634, 659
Peters, R. H., 184(99), 185(99), 231
Petersen, S., 730(6), 732(20), 806, 807
Peterson, E. C., 617(4), 623, 643, 661
Petrea, A., 368, 405
Petroe, G. A., 523(8), 608
Petterson, R. C., 521(5), 563(55), 608, 610
Petzel, F. E., 393, 409
Pfeil, E., 754(85), 809
Phansalkar, A. K., 176(62), 180 (62), 190(113), 191(62,113), 209 (62), 227(113), 229, 231, 312, 320, 325(9), 357, 376(9), 380, 389, 398, 404
Phelps, E. B., 914, 986
Philippar, W., 754(85), 809
Phillips, J. E., 933, 987
Pickering, Q. H., 930, 987
Picon, M., 643, 661
Pietz, J. F., 394(295), 409, 839 (41), 841(41), 891
Pilpel, N., 144(85), 152
Pingree, R. A., 293, 319
Piper, L. P. S., 205(163), 233
Pisareva, A. G., 652(113), 662
Podas, W. M., 825, 833(32), 834

Pages 1-452 comprise Part I
Pages 453-728 comprise Part II
Pages 729-1020 comprise Part III

[Podas, W. M.]
(32), 836(32), 837(32), 840(32), 841(10, 32), 881, 890, 891
Podlipny, V., 642, 661
Polano, M. K., 696(3), 702, 704 (26), 705, 706
Polcaro, T., 743(73), 809
Polet, H., 145(99), 152
Poliakova, V. A., 178(74), 230
Pollard, R. P., 462(15), 493, 496 (15), 497(15), 503
Pollard, R. R., 341, 402, 861(80), 893
Poole, N. D., 347(123, 131), 348, 402, 403, 589(87), 611
Pope, C. J., 623(14), 624
Porter, A. S., 476(32), 479(33), 480(33), 489, 503
Porter, B. R., 194(132), 197(152), 201, 202(132), 232, 233, 257(65), 258(66, 67), 267
Postman, W., 754(83), 809
Powe, W. C., 33(5), 41(13), 42(13), 43, 44, 45(19, 21), 46(22, 23), 47 (25, 28), 48(13, 28, 30), 49(32), 50 (32), 51(32), 57(41), 58, 59(42), 61(45), 61, 62, 63, 121(47), 141 (47), 143(83), 146(83), 147(83), 150, 152, 154(4), 190(4), 197(4), 227, 252, 254, 266, 304, 305, 308, 313, 319, 320, 337, 364(81, 82), 365, 373, 374(82), 385, 401, 405
Powers, D. H., 732(56), 808
Powney, J., 205(165), 209(165), 217 (165), 233, 273, 274, 276, 279, 309, 311, 316, 320, 354, 403, 648, 662
Prentice, J. B., 861(76), 893
Pressley, T. A., 958(51), 960, 961, 989
Preston, J. M., 113(18), 149, 180 (86), 230
Preston, W. C., 142, 148(82), 151, 627, 659
Pristine, R. G., Jr., 861(82), 893
Prokopenko, A. M., 652(113), 662
Pryor, A. K., 843, 892
Pullinger, B. D., 688(21), 693
Purchase, M. E., 340(318), 378 (325), 411

Q

Quimby, O. T., 849, 851(59), 861, 892, 893

R

Rabenhorst, H., 67(6), 100
Rabinowitch, E. I., 540(45), 609
Rader, C. A., 341, 402
Rajagopal, E. S., 253, 266
Ramakers, H., 632, 659
Ranauto, H. J., 341, 402
Rao, D. S., 732(41), 808
Raschke, R. M., 654(121), 663
Rathburn, D. W., 384(256), 390 (256), 408
Rauchle, A., 333, 364, 399
Ray, B. R., 116(29, 34), 117(29), 149, 150
Ray, L. N., 173(46), 229
Ray, W. A., 637, 660
Rayleigh, L., 644, 661
Raymond, A. J., 771(102), 773, 810
Read, R. B., Jr., 829(21), 890
Rebello, D. J. A., 704(37), 706
Reerink, H., 72(30), 101
Rees, W. H., 191(120), 194(120, 133), 204(133), 232, 336(73), 353, 386(73), 391, 400, 409
Reese, W., 434(51), 451

Reese, W. B., 33, 399
Reeves, W. A., 286(40), 295(40), 318, 353(139), 403
Reich, H. E., 881, 887(137), 896
Reich, I., 176(60), 178(73), 191 (124), 200(60), 229, 230, 232, 252(46), 253(49), 254(49), 258 (72), 266, 267, 272, 300(74), 309(74), 314(74), 316, 319, 387 (265), 388, 389, 408
Reich, J., 15(22), 29
Reich, M., 754(85), 809
Reid, J. D., 197(153), 202(153), 209(153), 216(153), 233, 293(63), 295(63), 319, 334(54), 390(275), 400, 409
Reinhardt, R. M., 197(153), 202 (153), 209(153), 216(153), 233, 293(63), 295(63), 319, 334, 390 (275), 400, 409
Reinwald, E., 1018(41), 1020
Reitz, D. C., 617(8), 623
Resuggan, J. C. L., 847(57), 892
Reumuth, H., 48, 62, 67(10), 101
Reutenauer, G., 651, 662
Reynolds, J. A., 145(99), 152
Rhodes, F. H., 254(57), 267, 309, 320, 334, 392, 400
Rice, O., 851(65), 892
Richards, S., 384, 408
Richards, T. W., 637, 660
Richardson, A. S., 627, 659
Richardson, E. G., 263, 267
Richardson, F., 120(46a), 150, 308, 320, 347, 384, 402
Rideal, E. K., 78(43), 102, 211 (174), 234
Ridenour, G. M., 376(229), 380 (229), 407, 840(47), 841, 842, 892
Ring, R. D., 190(114), 191(114), 192(114), 194(114), 217(114), 220 (114), 231, 302, 319, 339(97), 376(97), 380(97), 401
Ringeisen, M., 309, 320
Rist, C. E., 458(5), 502
Ristenpart, E., 651, 662
Rivett, D. E., 732(43,58), 808
Rivin, D., 155(6), 227
Rizzo, F. J., 333, 399
Roan, S. G., 958(51), 960(51), 961(51), 989
Roberts, R., 108(8), 149
Robinson, R. A., 563(51), 610
Robinson, T., 376(225), 381(225), 407
Robson, H. L., 523(8), 608
Roddewig, J., 137(73), 138(73), 139 (73), 147(73), 148(73), 151, 356 (162), 376(214), 378(214), 379 (214), 383(214), 404, 406, 538 (41), 609
Röder, H. L., 654, 655, 663
Roecker, J. H., 377(234), 380(234), 407
Roga, R. C., 337(86), 347(86), 348 (86), 401
Rohovsky, M. W., 805(161), 813
Rohrer, E., 288, 318
Rollins, M. L., 194(132), 197(152), 201(132), 202(132), 232, 233, 258(66,67), 267
Rosano, H. L., 124(52,53), 150, 643(78), 661
Rose, G., 704, 706
Rose, G. R. F., 69(13,14,18), 101, 173(43), 180(79), 182(79), 229, 230, 273(6), 274(6), 276(6), 279 (16), 280(16), 283(28), 292(6), 301(6), 317
Rose, H. E., 177(65), 229
Rosenberg, L. D., 264, 268
Ross, E. S., 347(123), 348(132), 402, 403, 589(87), 611
Ross, J., 12, 28, 145(97), 152, 194 (135), 210(172), 211(172), 232, 234, 281, 291, 317, 354(145), 403, 475(29), 489(36), 503, 504, 630, 659
Ross, S., 626, 627, 630, 632, 658, 659, 832, 891
Roth, E. S., 598(92), 612, 868(100),

Pages 1-452 comprise Part I
Pages 453-728 comprise Part II
Pages 729-1020 comprise Part III

[Roth, E. S.]
873(100), 894
Rothman, S., 33(3), 36(3), 37(3), 38 (3), 41(3), 42(3), 45(3), 61, 120 (44), 150
Rubenkoenig, H. K., 680(7, 8), 692
Rubin, F. K., 879, 895
Rubin, L. F., 688(23), 693
Rubinfeld, J., 344, 402, 1018(37), 1020
Ruchhoff, C. C., 839(39), 841(39), 842, 891
Ruckelshaus, W. D., 937(31), 988
Rudnick, I., 330(35), 399
Rue, L. M., 882(140, 141), 886(140), 896
Ruff, E. E., 634(42), 660
Rukeyser, W. S., 996(6), 1018
Rushton, J. H., 242, 244(8, 9), 265
Ruspino, J., 244(22), 265
Russel, T. J., 688(23), 693
Rutherford, H. A., 618(9), 623
Rutkowski, B. J., 69(22), 101, 174 (52), 181(88, 89), 186(52), 229, 230, 281, 289, 291, 292(47), 306, 317, 318, 337(84, 85), 344, 346, 355, 376(221), 378(221), 382, 390, 401, 402, 406, 538(39), 609, 866, 880, 894, 1000(13), 1011(32), 1014 (32), 1016(34), 1019, 1020
Ryan, F. W., 116(36), 150
Ryan, M. E., 191, 192(126), 193(126), 201(157), 219(187), 226(126), 232, 233, 234, 253(51, 52), 258(51), 266, 388, 408
Ryder, E. E., Jr., 341(111), 402
Rydzewska, D., 336(74), 386(74), 391, 400
Ryer, F. V., 144(96), 152

S

Sabatelli, P. M., 861(84-87), 893
Sakai, Y., 732(39), 808
Sakuma, S., 732(45), 808
Saleeb, F. Z., 155(8), 173(49), 174 (49), 227, 229
Saltzman, M., 326(13), 398, 425 (25), 450, 579, 611
Sandell, E. B., 829(18), 830(22), 890
Sanders, H. L., 55, 56, 62, 154(1), 191(121), 227, 232, 252(43), 266, 283, 293, 300, 302(33, 73), 303 (33), 308, 310, 313, 315, 318, 319, 325(11), 331(11), 336, 337 (72), 339, 355(11), 357, 376(72), 387(11), 394, 398, 400, 416(8), 449
Sandstede, G., 67(5, 6, 8, 9), 71(33, 34), 87(33, 34), 98(8, 9), 100, 102, 212(176), 234
Sanford, P. B., 222(195), 224(195), 235, 254(59), 267, 356(155), 389 (155), 404
Santner, J. F., 930, 987
Sarge, C. R., 861(84, 87), 893
Sarin, J. L., 392, 409
Satanek, J., 381(243, 244), 407, 827(11, 12), 830(11), 831(11), 840 (11, 12), 841, 842(12), 851(11, 12), 890
Satkowski, W. B., 16(25), 29
Sauerwein, K., 372, 376(202), 381, 406
Saunders, S., 788(151), 801(151), 812
Savitsky, A. C., 1005(24), 1019
Sawyer, C. N., 922(5), 931(6), 938 (6), 950, 986
Sawyer, W. M., 341(111), 402, 626 (8), 658
Sayato, Y., 732(35), 739(35), 743 (35), 808
Scalera, M., 730(4), 806
Scalzo, A. M., 829, 890
Schaafsma, B. R., 337(76), 338(76), 344(76), 378(76), 401, 538(38), 609
Schaefer, K., 113(19), 114(19), 149
Schaeffer, A., 373, 406

Schappel, J. W., 194(134), 232
Schauer, P. J., 16(25), 29
Scheen, S. R., 704(37), 706
Scheidegger, A. E., 243, 265
Schelhammer, C. W., 732(20), 807
Schenach, T. A., 165(31), 228
Schenkel, J. H., 72(26), 101, 161 (19), 228
Schick, M. J., 766, 767(99), 810
Schilling, H. K., 330(35), 399
Schlachter, A., 629(19,20), 659, 732(34), 807
Schlaepfer, H., 730(18), 807
Schleich, H., 668(12), 678
Schmelkes, F. C., 873, 894
Schmid, P., 703(33), 706
Schmid, R., 703(33), 706
Schmidt, G. A., 340(314), 410
Schmolka, I. R., 341, 344, 378(106), 382(106), 390, 402, 771(101,102), 773(101,102), 810, 832, 858(71), 874, 882, 891, 893, 896
Schneider, W., 701(15), 705
Schoen, H. M., 634(44), 660
Schoenberg, K. A., 324(7), 340, 356, 390(7), 398
Schoenol, K., 732(37), 808
Scholz, J. J., 116(29), 117(29), 149
Scholze, A. E., 861(88), 894
Schonhorn, H., 116(36,37), 150
Schott, H., 123(50,51), 137(70), 138 (70), 147(70), 148(70), 150, 151, 155(9), 156(14), 164(30), 165(32), 166(35), 168(37), 169(32,37,38), 171(37), 174(32), 175(38), 177(71), 181(35,91), 182(35), 183(35), 185 (71), 196(143,144), 212(30,143, 178), 214, 217(185), 223(143), 227, 228, 230, 233, 234, 286, 290, 292, 308, 318, 368, 405, 995(2), 1018
Schramm, K. H., 67(5), 100
Schramm, W., 333, 364, 399, 798 (158), 812
Schubert, P., 89, 102, 186(104), 231
Schulman, J. H., 158(16), 165, 227
Schultz, J. R., 884, 897
Schulz, K. H., 704, 706
Schulze, H. K., 642, 661
Schulze, J., 743(73), 805(164,165), 809, 813
Schumann, K., 383(247), 407
Schutz, F., 632, 633, 659
Schwartz, A. M., 9(1), 12, 28, 29, 199(155), 210(172), 211(172), 233, 234, 281(22), 291(22), 308 (92), 317, 320, 324(1,2), 325(1), 336, 340, 341, 349, 354(145), 394, 398, 400, 402, 403, 475 (29), 489(36), 503, 504, 626(9, 10), 658, 827, 890
Schwarz, W. J., 69, 72(40), 86(40), 89(40), 101, 102, 181(88-90), 230, 292, 308, 318, 376(222,231), 378(222), 382, 406, 407
Schwen, G., 642, 643, 661
Schwuger, M. J., 1018(41), 1020
Scott, B. A., 48(29), 62, 136(66), 145, 146, 147(66), 151
Scott, E. W., Jr., 282(25), 317
Seeley, S. B., 52(35), 62
Segalas, H. A., 1002(21a), 1003 (21a), 1019
Segura, G., Jr., 377(234), 380 (234), 407
Seiler, N., 743(72), 809
Sell, P. J., 115(25), 149
Selling, H. J., 356(161), 404
Seltzer, K. P., 301, 302(76), 313, 319
Semionova, A. M., 178(74), 230
Seve, R., 786(121,127-129), 788 (129), 804, 811
Sexsmith, F. H., 163(26), 179(26), 228
Seyferth, H., 11(5), 28, 648, 661
Shanley, E. S., 363, 405
Shapiro, J., 935, 954, 987, 989
Shapiro, L., 12, 28, 651, 662

Pages 1-452 comprise Part I
Pages 453-728 comprise Part II
Pages 729-1020 comprise Part III

Shebs, W. T., 137(72,73), 138(73), 147(73), 148(73), 151, 181(92,94), 182(92), 183(92), 230, 231, 356 (163), 376(214,226), 378(163,214, 226,237,238,326,327), 379(214), 383(163,214,226), 404, 407, 411, 538(41,42), 609
Shelanski, H. A., 703, 706
Shelanski, M. V., 703, 706, 882 (142), 896
Shelberg, W. E., 380, 407
Shelley, W. B., 37, 61
Sheludko, A., 644, 661
Shen, C. Y., 495, 504, 851, 857, 892
Shenoch, F. A., 654(120), 663
Shepard, J. W., 108(7), 149
Sher, B. C., 282(25), 317
Sherrill, J. C., 226(198), 235, 330 (35), 341(104,105), 354(105), 395 (105), 399, 402, 654(121), 663
Sherrill, J. S., 258(74), 267
Shestakow, P. I. E., 330, 399
Shiflett, C. H., 839(42), 841(42), 843(42), 891
Shimauchi, S., 308, 320, 339, 340 (311), 343, 346, 383(248), 386, 401, 407, 410
Shimizu, P. H., 1005(23), 1019
Shinoda, K., 626(6), 658
Shmeleva, T. A., 653, 663
Short, B. A., 200(156), 233, 243 (15), 249(41), 250, 251(41,42), 252 (41), 253(41,54), 254(54,60), 255 (54), 256(60), 257(60), 260 (75), 265, 266, 267, 310, 320, 370, 371(197), 405, 419, 449
Shuck, R. O., 225, 235, 347(130), 403, 828, 888, 890
Shukov, A. A., 330, 399
Shumway, D. K., 671(16), 678
Shuttleworth, R., 108(6), 149
Shuttleworth, T. H., 197(154), 198 (154), 233
Siegel, S., 573(64), 575(64), 610
Siegrist, A. E., 730(8), 732(28,29, 32,52), 738(52), 806, 807, 808
Siehr, A., 631, 659
Silberman, H. C., 937, 988
Sillen, L. G., 472, 503
Silvis, S. J., 994(1), 1018
Simmons, J. K., 882, 896
Simon, F. T., 436(58), 451
Singer, J. J., Jr., 496(44), 504, 861(75), 893
Singer, P. C., 955(49), 989
Singleterry, C. R., 136(62,63), 151
Sinsheimer, J. G., 632, 659
Sirianni, A. F., 771, 810
Sironval, L., 632, 659
Sisley, J. P., 334(56), 365, 393 (56), 400, 651(100), 662
Sivadjian, J., 393, 411
Skewis, J. D., 173(42), 228, 376 (224), 406
Skinkle, J. H., 584(78,79), 586 (83), 611
Skinner, H. A., 376(212), 380(212), 406
Skoulios, A., 144(84, 86), 152
Sloan, C. K., 204(162), 205(162), 233, 258(69), 267, 385(260), 408
Smeenk, G., 702, 704(26), 706
Smialkowski, E. J., 882(139), 886 (139), 896
Smiljanic, A. M., 120(44), 150
Smith, B. F., 340(318), 378(325), 411
Smith, C. E., 275(10), 280(10), 310(10), 317, 357(167,168), 404
Smith, E. L., 667(2-7,9), 678
Smith, F. D., 1008(27), 1019
Smith, J. E., 191(118), 200(118), 222, 223(192), 224(118), 225, 232, 234, 253(48), 266, 354(140),

[Smith, J. E.]
386(140), 387, 389(140), 390, 403, 409, 419, 449, 651, 662
Smith, M. M., 617, 623
Smith, N. R., 778(104), 810
Smith, P. E., 589(90), 611
Smith, R. L., 466(23), 503
Smith, W. C., 415(3), 449
Smith, W. H., 314, 320, 336, 337 (65), 368, 369, 389, 400, 405
Smith, W. R., 52(36), 53(36), 62
Smyth, H. F., Jr., 680(4), 692
Snell, C. T., 258(72), 267, 300(74), 309(74), 314(74), 319, 387(265), 408
Snell, F. D., 191(124), 232, 253(49), 254(49), 258, 266, 267, 272, 279, 300, 309(74), 314, 316, 317, 319, 355, 387(265), 388(268), 389(268), 403, 408
Snugden, S., 638, 660
Snyder, F. H., 680(7, 8), 692
Snyder, H. E., 884(153), 897
Sobocinski, E. C., 878(117), 895
Soderdahl, P., 282(25), 317
Soljacic, I., 732(61), 809
Soller, W., 211(175), 234
Somers, J. A., 295, 319
Sommer, K., 702, 706
Sommerville, J. L., 376(233), 407
Sonntag, H., 135(61), 151, 393, 409
Sontag, M. S., 340(318), 378, 411
Sookne, A. M., 163(25), 181(25), 186(103, 107, 109, 110), 228, 231, 243(13), 265
Sosson, H., 340, 403
Spangler, W. G., 337, 338(76), 344, 347, 348, 378, 401, 538(38), 609
Spelman, K. C., 1016(33), 1020
Spencer, R. S., 241(6), 265
Sperling, G., 66(2-4), 67(2-4), 82 (2, 3), 100
Spier, H. W., 696(2), 702, 705
Spratt, E. C., 384, 408
Spring, W., 330, 399
Spurny, J., 644, 661
Srivastava, S. N., 72(27), 101
Stache, H., 1018(39), 1020
Stamm, J. K., 876, 895
Stanley, C. W., 384(256), 390(256), 408
Stanley, J. S., 186(102), 187(102), 190(102), 231
Staubly, J. L., 350(135), 403
Stavarakas, E. J., 623(18), 624
Stawitz, J., 289, 290, 292, 318
Stayner, R. D., 10(2), 28
Stearns, E. I., 577, 610
Steinfeld, J. L., 937(31), 983(60), 988, 989
Steinhardt, J., 145(99), 152
Steinhauer, A. F., 563(53), 610
Stenius, A., 788(153), 812
Stensby, P. S., 14, 29, 356, 404, 580(73), 611, 670, 678, 730(3), 732(3, 46-50, 62), 733, 735(50), 736(47), 739(3), 740(3), 742(3), 743(73), 751(62), 752(46), 754(3, 62, 92), 757(46, 47), 759(50), 761 (50, 92), 762(62), 763(62), 765(3), 766(3), 766(92), 766(46), 767, 768(46), 769(46), 770(46), 772 (42), 773(46), 774(46), 775(46), 776(46), 777(46), 778(46), 779 (48), 781(62), 782(62), 783(62), 784(62), 785(3, 47), 788(3, 150), 792(3), 793(3), 795(3), 799(3), 801(150), 802(3), 803(3), 804, 805(164, 165), 806, 808, 809, 810, 812, 813
Stephens, D. W., 350, 351, 353 (136), 403
Stericker, W., 279, 317, 330, 399
Stevenson, D. G., 123(50), 126(50), 128(50), 128(59, 60), 141(78), 143(50), 144(78), 145(78), 150, 151, 384(259), 408
Stewart, G., 139(74), 146(74), 148(74), 151, 330(33), 331(33), 365(33), 367(33), 386(33), 399

Pages 1-452 comprise Part I
Pages 453-728 comprise Part II
Pages 729-1020 comprise Part III

Stewart, J. C., 117(30), 126, 127, 131, 132(30), 133(30), 134, 135, 149, 393(330), 411
Stewart, R. K., 935(23), 987
Stiepel, C., 627, 659
Stigman, S., 377(234), 380(234), 407
Stigter, D., 186(101), 231
Stiles, W. S., 425(28), 450
Stillo, H. S., 289, 291, 292, 318
Stirton, A. J., 355(153), 403, 861 (82), 893, 1008, 1019
Stokes, C. A., 301(77), 313(77), 319
Stout, E. E., 298, 319, 331, 335, 344, 345, 399
Strauss, W., 191(125), 208, 209 (167), 210(167), 215, 216, 217 (167), 218(167), 226(167), 232, 233, 277, 317, 388, 408
Streeter, H. W., 914, 986
Strehlow, H., 84(48), 102
Strenge, K., 135(61), 151, 393, 409
Strobele, R., 1018(39), 1020
Stuart, H. A., 140(77), 151
Stüpel, H., 210(171), 217(171), 234, 288, 318, 324(5), 335, 338, 369, 398, 401
Stumm, W., 928(10), 933(10), 987
Sturm, R. N., 805(166), 813
Styers, F., 979(68), 990
Suchowolskaja, S. A., 632, 659
Suda, M., 730(16), 807
Sullivan, G. H., 33, 61
Sullivan, M. R., 191(119), 232, 389 (272), 390(272), 409
Sullivan, W. T., 929, 987
Sulzburger, M. B., 38(8), 61
Sundstrom, F. O., 786(130), 811
Suskind, R. P., 704(37), 706
Suter, H. R., 190(114), 191(114), 192(114), 194(114), 217(114), 220 (114), 231, 302, 311, 319, 320, 336, 339(97), 343, 357(68), 376 (97,212), 380(97,212), 400, 401, 406, 423(20), 450
Suzawa, T., 186(109a), 188(109a), 231
Swanson, J. W., 196(145), 233
Swisher, R. D., 682, 692, 929, 987
Symes, W. F., 526(19-21), 608, 825(7), 874, 878(7,117), 890, 895
Szakall, A., 696(4), 705

T

Tachibana, T., 366, 405
Takada, K., 379, 408
Talboys, A. P., 384, 408
Talmud, D., 632, 659
Tamaki, K., 69, 101
Tamamushi, B., 69, 101
Tamura, T., 217(183), 220, 234
Tanaka, M., 732(45), 808
Tanner, W., 643, 661
Tate, J. R., 170(39), 171(39), 172 (39), 183(39), 228
Tate, T., 638, 660
Taube, R. K., 347(123,131), 348 (132), 402, 403, 589(87), 611
Taylor, A. H., 435(56), 451
Taylor, E. C., 730(19), 807
Temple, N. S., 825(9), 881, 890
Templeton, G. L., 653(119), 663
Tenney, M. W., 935(22, 26), 940 (34), 954(34), 955(49), 987, 988, 989
Ter Minassian-Saraga, L., 162 (24), 228
Tewksbury, F. L., 754(86), 809
Thatcher, T. O., 930, 987
Theidel, H., 739 (65), 743(65), 747 (65), 809

Theis, T. L., 955(49), 989
Thomas, L. B., 589(89), 611
Thompson, J. S., 525(16), 608, 825 (6), 868(100,101), 873(6,100,101), 890, 894, 895
Thompson, S. H., 139(74), 146(74), 148(74), 151, 330(33), 331(33), 365(33), 367(33), 386(33), 399
Thompson, T. L., 979(67), 990
Thompson, W. E., 354, 403
Thornton, J. L., Jr., 634(42), 660
Tidridge, W. A., 336(66), 400
Tipissova, T. G., 178(74), 230
Tityevskaya, A. S., 72(32), 102
Tobin, W., 732(30), 807
Todd, D. K., 906(3), 907(3), 986
Tokiwa, F., 395(334), 411, 482(34), 483, 484, 489, 490(37), 503, 504
Tolgyessy, J., 376(225), 381, 407
Tomiyama, S., 40, 61, 120(42), 145 (52), 150, 366, 368(191), 405
Touhill, C. J., 961(54), 989
Tozer, R. G., 926(9), 987
Trimmer, T. H., 335(63), 358(63), 400
Tripp, V. W., 194(132), 197(152), 201(132), 202(132), 232, 233, 257 (65), 258(66,67), 267
Troesken, O., 730(5), 806
Trommer, C. R., 648, 662
Tronnier, H., 701(15,16), 702, 703, 705, 706
Trost, H. B., 280(20), 285, 286, 287(20), 292, 317, 318, 330(37), 390, 399, 409
Trotman, S. R., 627(14), 658
Trowbridge, J. R., 344, 395, 402, 411
Tschakert, H. E., 632, 659
Tsubomura, M., 366(189), 405
Tsunoda, T., 305, 364(321), 319, 321
Tsuzuki, R., 258(68), 267, 309, 321, 331, 340(311), 399, 410
Tusing, T. W., 680(7), 692
Tutunov, M. A., 653, 663
Tuvell, M. E., 474, 503, 643(79), 661, 830, 831(23), 891
Tuzson, J., 200(156), 233, 249(41), 250, 251(41,42), 252(41), 253(41, 54), 254(54), 255(54), 260(75), 266, 267, 310, 320, 370, 371 (197), 405, 419, 449
Tyman, J. H. P., 732(26), 807

U

Uehlein, E., 732(27), 807
Uhlich, H., 383, 407
Underkofler, L. A., 666(1), 678
Unterbirker, H., 786(132), 811
Uppal, I. S., 651(102), 662
Uppal, M. Y., 392, 409
Urazovskii, S. S., 637, 660
Urbain, W. M., 89, 102, 173(45), 174(45), 229
Urban, W. J., 1001(16), 1019
Utermohlen, W. P., Jr., 191, 192 (126), 193(126), 201, 219, 223 (193), 226(126), 232, 233, 234, 247(39), 248, 253(39), 253(51,52), 258, 266, 354(142), 372, 387, 388, 403, 408

V

Vacker, D., 954(43), 988
Vaeck, S. V., 176(61), 229, 309, 320, 337(77), 340, 357, 358, 401, 404, 623(19), 624, 786(111-113, 144-146), 796, 801, 802, 804, 811, 812
Vail, J. G., 863, 894
Valenta, J. C., 563(53), 610
Vallee, J., 644, 661
Vallee, J. P., 392, 409
Van Lierde, F., 786(111,112,144), 796(111,112,144), 802(111,112, 144), 804, 811, 812
Van Ruette, R., 746(75), 809

Pages 1-452 comprise Part I
Pages 453-728 comprise Part II
Pages 729-1020 comprise Part III

Van Scott, E. J., 702, 706
van Senden, K. G., 163(27,28), 179 (27,28), 181(27,28), 228
Van Slyke, W. J., 961(54), 989
Van Uitert, L. G., 456, 502
van Voorst Vader, F., 116(33), 150
Van Wazer, J. R., 455(1), 459(7), 466(7), 474, 496, 502, 503, 504, 830, 831(23), 848, 849, 852, 855 (67), 858, 859(67), 860(67), 891, 892, 893
Vance, R., 880, 895
Vance, R. F., 838, 891
Vandergrift, A. E., 344, 402
Vaughn, R. H., 879(125), 895
Vaughn, T. H., 223(191), 234, 247 (38), 248, 253(38), 254(38), 266, 275, 279, 280, 310, 317, 320, 336, 343, 357(68,166-168), 400, 404, 423, 450
Velichkova, V., 644, 661
Venkataraman, K., 651, 662
Verhulst, H. L., 680(5), 692
Vermeer, D. J. H., 702, 706
Verwey, E. J. W., 72, 73(36), 79 (36), 83(36), 84(36), 86, 102
Veselovski, V. S., 301, 319
Viertel, O., 330, 340(312), 399, 410
Villaume, F. G., 14, 29, 732(23), 733, 756(93), 766(23), 807, 810
Visser, J., 212, 234
Vitale, P. T., 210(172), 211(172, 173), 234, 276, 281(22), 291(22), 313, 317, 354(144,145), 395(299), 403, 410, 416(7), 449, 475, 488 (35), 489(36), 503, 504
Vitalis, E. A., 292(55), 318
Vittone, A., Jr., 223(191), 234, 247 (38), 248(38), 253(38), 254(38), 266, 279, 317
Vold, M. J., 174(55), 229
Vold, R. D., 176(60,62), 177(66-68), 180(62), 190(113), 191(62, 113), 200(60), 209(62), 227(113), 229, 231, 252(46), 266, 312, 320, 325(9), 357, 376(9), 380(9), 389, 398, 404
Vollbrecht, I., 703(30), 706
Von Rutte, R., 732(42), 805(163), 808, 813
von Stackelberg, M., 89, 102, 185 (106), 186(106), 187(106), 188 (106), 190(106), 231, 289, 318
Voropayeva, T. N., 72(32), 102

W

Waag, A., 184(96,98), 231
Wadachi, Y., 379(251), 408
Wagener, H. H., 702, 706
Wagg, R. E., 47, 62, 136(64,65), 137, 138, 139, 141(64), 151, 217 (184), 211(190), 226(190), 227 (190), 227(190), 234, 347, 358, 372, 382, 389, 394, 402, 406, 407, 408, 410
Wagner, A., 732(20), 739(63), 741(66), 746(66), 747(63), 807, 809
Wagner, E. F., 197(151), 208(151, 168), 209(151,168,169), 216, 218, 224(181), 225(168,181,196, 197), 233, 234, 235
Wagner, H., 701(15), 705
Wagner, H. G., 216(182), 234
Waite, J. M., 168(36), 169(36), 238
Walker, H. B., 648(99), 650(99), 662
Wallace, E. L., 223(193), 234, 247 (39), 248, 253(39), 254(39), 266, 354(142), 372, 387, 403
Walling, C., 634, 660
Walsh, M., 334(54), 400

Walter E., 41(12), 46(12), 61, 357 (164), 404
Walter, G., 56, 63, 67(9), 71(33, 34), 87(33,34), 98(8,9), 100, 102, 212(176), 234
Walther, G., 730(13), 807
Walton, W. W., 632(34), 659
Ward, W. H., 285(37), 288, 291, 292, 318
Warren, A., 338, 341, 342, 401
Washburn, E. W., 111(12), 149
Waters, P. J., 732(58), 808
Watillon, A., 72(29,31), 101
Watson, B. S., 112(15), 113(15), 149, 243(12), 265
Watson, P. R., 458(6), 502
Watters, J. I., 462(10,11), 502
Weatherburn, A. S., 69, 101, 173 (43), 179(76), 180(76,79,82), 182 (76,79), 183(76), 191(122), 194 (122,136), 197(136,147), 202, 204 (122), 229, 230, 232, 233, 273, 274, 276, 278, 279, 280, 283(28, 22), 292, 301, 311, 312, 313, 317, 318, 337(79), 343, 389, 390, 394, 401, 402, 408
Weaver, J. E., 682(16), 690(26), 693
Weaver, P. J., 936(30), 988
Weber, K., 730(8,9), 732(61), 806, 809
Weber, W. J., Jr., 953, 988
Weder, G., 338, 401, 538(37), 539 (37), 609
Weeks, L. E., 191(119), 232, 350 (135), 389(272), 390(272), 403, 409, 632, 659, 747(80), 809
Wegmann, J., 754(82), 755, 809
Wegst, W. F., 825(4), 871, 876, 879, 890, 895
Weil, I., 1001(19), 1019
Weil, J. K., 1008(27,28,30), 1019, 1020
Weill, M., 124(52), 150
Weisberg, M., 861(75), 893
Weissman, S., 786(106), 810
Weisz, P. B., 754(91), 757, 810
Weitkamp, A. W., 120(44), 150
Weldes, H., 866(97), 894
Wells, W. N., 954(43), 988
Wendt, B., 730(1), 806
Wentz, M., 314(108), 320, 336(65), 337(65), 369(195,196), 389(65), 400, 405
Wenzel, R. N., 108(9), 149
Werdelmann, B., 703, 706
Werner, G., 743(72), 809
Werner, H. U., 878(122), 895
Westfall, R. H., 878(116), 895
Westman, A. E. R., 852, 892
Weyl, W. A., 880, 896
Wheatley, V. R., 120(43), 150
Whewell, C. S., 117(30), 126, 127, 131, 132(30), 133(20), 134, 135, 149, 393(330), 411
Whitaker, S., 827(13), 890
White, H. J., 163(26), 179(26), 228
White, W. C., 258(74), 267
Whitehouse, H. S., 689(26), 693
Whitney, D. R., 575(66), 610
Wiechmann, M., 743(72), 809
Wightman, T., 786(136), 788(151), 801(151), 812
Wiley, R. M., 241(6), 265
Wilhelmy, L., 641, 660
Wilke, F., 89(55), 102, 185(106), 186(106), 187(106), 188(106), 190 (106), 231, 289(50), 318
Wilkins, D. J., 72(28), 101
Wilkinson, P. R., 617(8), 623
Willan, A. L. D., 47(27), 62, 347 (127), 358(127), 402
Willard, H. H., 335(58), 400, 427 (32), 429(32), 430(32), 450
Williams, D. G., 186(111), 196(145), 231, 233
Williams, K. E., 805(166), 813
Williamson, R., 732(57), 808
Willis, I., 700(14), 705

Pages 1-452 comprise Part I
Pages 453-728 comprise Part II
Pages 729-1020 comprise Part III

Wilmsmann, H., 635, 660, 697, 705
Wilshire, J. F. K., 732(43), 808
Wilson, A., 145(97), 152
Wilson, D., 390, 409
Wilson, J. H., 14(21), 29
Wilson, J. L., 837, 839(42), 841 (42), 843(35,42), 869, 883(150), 885(156), 891, 896, 897
Wilson, K. B., 393, 409
Winicov, M. W., 882(142), 896
Witheron, J. L., 954(44), 988
Witzel, R. F., 788(150), 801(150), 812
Wixon, H. E., 357, 404
Wolf, K. L., 638, 660
Wolfram, R. E., 12, 29, 115(22), 117(22), 118(5,38), 119(22,38), 149, 150, 393, 409
Wollner, H. J., 341, 357, 402
Wood, P. J., 334(56), 365, 393(56), 400
Woodhead, J. A., 324(8), 350(8), 352(8), 353(8), 362(8), 395, 398, 410
Woodward, G., 680(2), 692
Work, C. E., 682, 692
Worne, H. E., 884(155), 897
Wuest, W., 730(15), 807
Wyszecki, G., 425(26,28), 426(26), 435(55), 450, 451, 565, 577, 578 (60), 579(60), 610, 786(105), 788 (105), 788(152), 810, 812

Y

Yabe, A., 366(189), 405
Yabuuchi, N., 258(68), 267, 309, 321, 331, 399
Yamakawa, Y., 730(16), 807
Yamaoka, Y., 379(251), 408
Yanko, W. H., 380(239,240), 407
Yeager, J. A., 420(18), 450
Yeiser, A. S., 365(182), 405
Yokoyama, S., 383(248), 407
Yonden, W. J., 395(301), 410
Yoshizaki, K., 386(329), 411, 491 (39), 504
Young, D. O., 201(157), 233
Young, E. M., 343, 402
Young, G. J., 832, 891
Young, K. W., 563(52), 610

Z

Zaikowski, L., 829(20), 839(20), 841(20), 846(20), 890
Zeidler, H., 382, 407
Zettlemoyer, A. C., 12(10), 29, 173(42), 228, 376(224), 406
Ziegler, G., 263, 267
Zimmerman, W., 786(122), 804, 811
Zisman, W. A., 136(63), 151, 652, 662
Zollinger, H., 754(84,91), 757, 809, 810
Zussman, H. W., 14, 29, 732(22, 24,30), 766(24), 797(24), 807
Zweidler, R., 730(7), 732(21,31), 806, 807
Zybko, W. C., 297(70), 299(70), 309(70), 319

SUBJECT INDEX

Pages 1-452 comprise Part I
Pages 453-728 comprise Part II
Pages 729-1020 comprise Part III

A

Abrasion, 614-620
  direct, 617
  frosting, 618-619
  resistance, 614, 618
Absorption spectra of whiteners, 738-741
Acid pickling baths, 721-724
Activated sludge, 948-949
Activators, 527-528
Active chlorine compounds, 872-878
Active chlorine in dishwashing, 832-833
Active content of whitener, 747-749
Adherence of particulate soil, 65-100
Adhesion, 211-213
  of soil particles, 66-67, 70-73
  use of centrifugation to measure, 211-213
Adsorption isotherms, 169-171, 278
Adsorption of surfactants, 68-70
  by fibers, 178-184
Affinities of fibers for surfactants, 182-184
Agitation, high-speed, 258-264
Alkaline builders in dishwasher detergents, 867-872
Alkaline cleaning of metals, 713-720
Alkaline metal descalers, 724
Alkalinity of builders, 482-487
Aminopolycarboxylic acids, 498-500
Analysis of laundry soils from the skin surface, 38-45
Ancillary tests, 625-658
Animal tests, 697-701
Anion
  specific precipitating, 460-461
  specific sequestering, 459-460
Anionic surfactants, 766
Anticorrosion agent, 11
Antiredeposition agent, 10, 282-292
Apocrine sweat, 36-37
Autotrophic microbial activity, 910-913, 919

B

Bacillus subtilis, 667
Barrel finishing of metals, 710-711
Basic detergent components, 994-995
Basis metal metallurgy, 708-709
Bikerman dynamic foam meter, 631
Biodegradable surfactants, 929-930
Biological waste treatment, 944-950
Bleaching, 519-607
Bleaching compounds, 520-532
Bleaching evaluation, 543-544
Bleaching evaluation methods, 546-551
  multiple-bleach method, 547-548
  single-bleach method, 546-547
  weighted-soil method, 548-551

Pages 1-452 comprise Part I
Pages 453-728 comprise Part II
Pages 729-1020 comprise Part III

Bleaching, influenced by
builders, 560-564
concentration, 555-559
pH, 560-564
substrates, 559
surfactants, 560-564
temperature, 555-559
time, 555-559
Bleaching methods
fabric, 545-546
solid surfaces, 551-555
Bleach-produced fabric damage, 588-607
Bleach-stable whiteners, 733-735
Borax, 871-872
Bubble pressure, 637-638
Buffing of metals, 710
Builder, 10, 279
Buildup of whiteners, 764-765
Built detergent, 9-10
Burst strength, 616

C

Calcium hypochlorite, 523
Canvas disc wetting apparatus, 649
Capillary rise, 636-637
Carbon black, 206-211
Carbon cycle, 926-930
Carbonaceous soils, 51-53
Carbonate precipitate buildup, 1016-1017
Carboxymethyl cellulose, 282-292
Cation, specific metal, 458-459
Caustic soda, 867-868
Cavitation, 261-264
Chelates, 861-863
Chemistry of fluorescent whitening agents, 730-748
Chloramines, 875
Chlorinated isocyanurates, 525-526
Chlorinated trisodium phosphate, 523-524
Chloroisocyanurates, 876-878
Chromaticity systems, 794-795
Cleaning processes, 238-246
large-scale effects, 239-242
molecular-scale effects, 244-246
small-scale effects, 242-244
Cleanliness of metals, 724-727
CMC (see Carboxymethyl cellulose)
Cold water detergency, 145-147
Color Eye, 798, 800
Colorimeters, 430-432, 797-798
Color-matching functions, 793
Color measurement, 424-433, 565-571
instruments for, 566-568
Color physics concepts, 326-329
Commercial dishwashing machine, 824
Commercial laundry equipment, 423-423-424
Concentration of whiteners, 765
Consumerism, 996
Contact angle, 7, 114-119
effect of surface roughness and porosity on, 108-110
hysteresis of, 108-114
measurement, 12
of water and bezene on fiber-forming polymers, 117
of water on fibers and yarns, 112
Corrosion inhibitors, 879-880
Corrosiveness of dishwasher detergent solutions, 844-845
Corrosives, 686-687
Cost data for wastewater treatment, 942
Cotton fluorescence, 767-778
Cotton whiteners, 732-733
Coulter counter, 253, 370-372
Critical micelle concentration, 8, 12, 141-143, 490-491
in relation to detergent activity, 141-143

D

Deflocculation, 487-489
Defoamers, 881-883
Dermal toxicity tests, 689-690
Dermatological test methods, 696-704
Dermatology tests, 695-705
Desizing, 533-534
Detergency characteristics of phosphate-restricted detergents, 1012-1025
Detergent
  additives, 749-785
  components, 923-925
  evaluation for dishwashing, 838-847
  formulations, 14-27
  process, definition of, 1
  removal by rinsing, 506
  types of the 1970s, 1009-1011
Detergents of the future, 1017-1018
Deter-Meter, 420
Diffusion process, 246-247
Dipersophthalic acid, 530
Dishwasher detergent
  development for, 823, 825
  evaluation of, 838-847
Dishwasher development, 817-822
Dishwashing, 816-889
  detergents for, 838-847
  theory of, 825-837
Dispersion, 487-489
Draves wetting test apparatus, 649-650
Drop volume, 638-639
Drop weight, 638-639
duNouy tensiometer, 640-641
Dynamic foam meter, 633

E

Eating utensils, 816
Eccrine sweat, 36
Electrical double layer, 73-83, 159, 245-246
Electrodialysis, 963-964
Electrokinetic properties of fibers, 184-190
Electrolyte concentration, 94-98
  influence on adsorption of surfacants, 94-95
  influence on cleaning action, 95-98
Emission spectra of whiteners, 741-743
Emulsification, 9
Energetics of removal of soil particles, 73-83
Environmental aspects of detergents, 903-923
Environmental movement, 995-996
Enzyme characteristics, 666-667
Enzymes, 531-532, 539-540, 670-677
  assay methods, 670-674
  effectiveness, 674-676
  evaluation methods, 674-676
  optimum levels, 675-676
  quick in-plant test, 677
Enzymes in dishwasher detergents, 883-885
Epidermal appendages, 34-38
Equiwhiteness diagram, 327, 803
Esophageal corrosivity test, 690-691
Ethylenediaminetetraacetic acid, 497-498
European home laundry equipment, 436-448
Eutrophication, 917-923
Evaluation of whiteness, 786-804
Excitation spectra of whiteners, 741-743
Experimental design, 393-395

F

Fabric damage due to bleach, 588-607
Fabric damage test methods, 582-588

Pages 1-452 comprise Part I
Pages 453-728 comprise Part II
Pages 729-1020 comprise Part III

Fabric fluorescence, 748-749
Fabric hand, 653-657
Fatty acid composition of sebum, 42-43
Fatty soil (see Organic soil)
Federal Household Substances Labeling Act, 680-681
Fiber characteristics, 213-217
  effect on soil redeposition, 215-217
  effect on soil removal, 213-215
Fiber interaction with surfactants, 178-189
Film and spot measurements in dishwashing, 842-844
Filming in dishwashing, 833-838
Fluorescent soils, 334
Fluorescent whitener evaluation, 14
Fluorescent whitening agent, 10-11, 730-806
Fluorescent whitening agent consumption, 731
Fluorometers, 432-433, 797-798
Foam, 626-635
  dynamic measurement methods, 630-632
  measurement methods, 626-632
  single-bubble measurement methods, 632
  static measurement methods, 627-630
Functions of detergent components, 994-995
Functions of dishwasher detergent components, 848-884

G

General properties of whiteners, 737
Guastalla torsion surface tensiometer, 651

H

Hazardous materials, 680
Heterotrophic microbial activity, 909-913, 919
Home dishwasher shipments, 821
Home laundry equipment, 420-422
Human skin investigations, 701-704
Hydrophilic dispersions, 157
Hydrophobic dispersions, 157-161
Hydrophobic fibers, 757
Hydroxycarboxylic acids, 500-501
Hypochlorite in dishwashing, 832-833
Hypochlorite stability of whiteners, 782-784

I

Identification of fluorescent whiteners, 746-747
"Igepal" surfactants, 768-778
Ingestion of detergents, 682-684
Inorganic builders, 1002-1005
Inorganic chlorine bleaches, 520-524
Inorganic nutritional pollutants, 917-923
Instrumental evaluation of whiteness, 797-805
Instrumental measurements, 334-356, 424-435
Instrument manufacturers, 448-449
Interfacial tension, 7
In vitro methods, 696-697
Ion exchange, 962-963
Ion-pair theory, 755
Irritants, 686

K

Kinetic energy in dishwashing, 827-828
Kinetics of soil removal, 221-227
Kubelka-Munk equation, 191-192, 387-390, 577-579

L

Laboratory laundry test equipment, 415-420, 424-436
Lagoons, 946
Lambert and Beer's Law, 842
Launder-Ometer, 415-417
Laundry detergents, recent changes in, 944-1018
Laundry products containing whiteners, 752
Laundry soils, 31-61
Lethal oral dose, 683
Ligand, 454-456
Light sources, 433-436, 787-790
Linear alkyl benzene sulfonate, 1006-1007
Liquid detergent formulation, 14-15
Liquid detergents, 1010
Lithium hypochlorite, 524
London-van der Waals attraction, 160-161
Low-foaming surfactants, 881-883

M

Mann-Whitney U test, 575-577
Mechanical action
  role in soil redeposition, 309-310
  role in soil removal, 237-264
Mechanical finishing of metals, 709-711
Mechanisms of organic soil removal, 121-145
Mechanisms of soil removal, 205-221
Metal cleaning, 707-727
Methods of soiling, 193-194, 201-203
  dry soiling, 201-202, 204
  wet soiling, 202-203, 204-205
Model soils, 190, 276, 300-309, 336-340
Moisture retention as a function of centrifugal acceleration, 445
Multiple bath systems, 254-256, 313

N

Nephelometers, 432-433
Nephelometric titrations, 467-468
Nitrogen cycle, 926, 930-932
Nitrogen removal, 956-961
Nonionic surfactants, 766
Nonphosphate detergent builders, 998-1005
Nonphosphate detergent formulations, 1003-1004
Nylon fluorescence, 711-778

O

Optical brightener (see Fluorescent whitening agent)
Optical properties of whiteners, 738-743
Organic chlorine bleaches, 524-526
Organic soil, 120-148, 218-221
  effect on removal of particulate soil, 218-221
  nature of, 120-121
  removal of molten, 121-145
  removal of soild, 145-147
Oxygen bleaches, 526-531
Oxygen depletion, 913-916

P

Paired comparisons, 572-574
Particle size, 98-100
Particulate soil, 48-61, 154-227, 205-218
  adsorption of surfactants by, 161-173
  effect of characteristics on soil removal, 213-215
  elements present in, 60
  fat free, 205-218
  interaction with fabric, 190-204
  interaction with surfactants, 157-158
  kinetics of removal, 221-227
  nature of, 154-156

Pages 1-452 comprise Part I
Pages 453-728 comprise Part II
Pages 729-1020 comprise Part III

[Particulate soil]
plus fatty soil, 218-221
properties of, 58
Particulate soil adherence theories, 65-100
Penetration of liquid into fabrics, 110-114
Peptization, 487-489
Peracids, 531
Percarbonates, 530
Performance evaluation test methods, 1011-1012
Performance of phosphate-restricted detergents, 1009-1017
Performance properties of whiteners, 751
Perspiration (see Sweat)
pH, effect on sequestration, 462-466
Phosphate as limiting nutrient, 935-936
Phosphate legislation, 995-996
Phosphates, 495-496
Phosphate substitutes, 936-937
Phosphorus cycle, 926, 932-937
Phosphorus removal, 953-956
Photography, 330-331
as extension of visual assessment, 330-331
pH titrations, 468-470
Physical-chemical waste treatment, 950-953
Physical damage to textile substrate, 613-623
Physical form of whiteners, 738
Physicochemical methods for evaluation of soil removal, 392-393
Physiological/physical factors in whiteness evaluation, 792-796
"Pluronic" surfactants, 771-778
Poison control centers, 681
Polishing of metals, 709
Pollutionary characteristics of synthetic detergents, 923-940
Pollution of aqueous environment, 907-923
Polyamide whiteners, 735-736
Polycarboxylates, 1000-1002
Polyester whiteners, 736
Polyphosphates in dishwasher detergents, 848-861
Potassium monopersulfate, 528-529
Potential energy between particle and substrate, 83-86, 207
effect of surfactants on, 87-98
Powder detergent formulation, 16-17
Primary wastewater treatment, 943-944
Properties of CC/DAS whiteners, 734
Properties of dishwasher detergent components, 848-884
Properties of polyphosphates, 851-861
Proteases, 667-670
activity as a function of pH, 668-669
compatibility with sodium tripolyphosphate, 668-670
stability as a function of pH, 670

R

Rabbit-eye test, 687-689
Radioactive soils, 538-539
Radioactive tracers, 136-141, 281 376-384
in organic soil removal studies, 136-141
in redeposition tests, 281
Radiance of soiled and unsoiled cotton, 749-750
Ranking techniques, 574-575
Redeposition of soil, 269-316
Reflectance measurements, 335-356

Reflectance values, multistimulus, 579-581
Removal
of inorganic films from metals, 720-724
of organic soil from fibrous substrates, 105-148
of organic soils from metals, 711-720
of particulate soil, 153-227
of soils originating from the skin, 45-48
Residual soil, 46-48
Resin-treated cellulosic fibers, 756
Reverse osmosis, 963-965
Ring method, 640-641
Rinse additives in dishwashing, 885-887
Rinsing
effectiveness evaluation of, 511-518
efficiency test for, 507-510
in dishwashing, 885-887
process, 505-511
Rolling-up mechanism, 121-136
Ross and Clark static foam apparatus, 627
Ross-Miles foam tester, 630

S

Schlachter and Dierkes foam tester, 629
Scrub test, 554-555
Sebum, 37-38, 40-45, 120-121
Secondary wastewater treatment, 944-953
Secretions from the skin, 34-38
Sequestrant levels in detergent systems, 480-482
Sequestrants commercially available
phosphates, 495-496
sodium nitrilotriacetate, 496-497
Sequestrants' role in detergent products, 473-495
Sequestration, 454-580, 851-855, 863-864
anionic systems, 474-479
factors affecting, 456-466
measurements, 466-470
nonionic systems, 479-480
values, 470-473
Shrinkage, 614, 621-622
Silicates, 863-867
Siliceous soils, 49-52
electron micrographs of, 50, 52
Skin as a soil, 33-34
Skin cells, 32-35
electron micrographs of, 34-35
Skin sensitivity, 696
Sludge
dewatering, 971-973
disposal, 975-977
handling, 965-979
treatment, 967-975
utilization, 977-979
Slurry test, 553-554
Soap with dispersing agents, 1007-1009
Soda ash, 869-871
Sodium carbonate, 1002-1004
Sodium citrate, 999-1000
Sodium hypochlorite, 520-523
commercial production of, 520-521
Sodium nitrilotriacetate, 496-497, 999
Sodium perborate, 527-528
Sodium silicate, 1002-1004
Softener, effect with and without whitener, 779
Soil content of fabrics, 364-386
by ashing of fabric, 368-369
by chemical analysis, 372-373
by Coulter Counter, 370-372
for dishwasher evaluation, 840-841
gravimetric evaluation, 364-365
by microscopic examination, 384-386
by Millipore filtration, 369

Pages 1-452 comprise Part II
Pages 453-728 comprise Part II
Pages 729-1020 comprise Part III

[Soil content of fabrics]
by radiotracers, 373-384
by solvent extraction, 365-368
by weighing the soil, 368
Soiling of hydrophobic fibers, 298
Soiling methods for dishwasher evaluation, 840-841
Soiling procedure, 536-637
Soil level on fabrics, 190-192
measurement of, 190-192
Soil redeposition, 269-316, 488-489
Soil redeposition evaluation methods, 323-398
Soil redeposition tests, 273-316
for evaluation of antiredeposition agents, 282-292
for evaluation of builders, 278-282
for mechanical action evaluation, 309-310
for soil evaluation, 300-309
for substrate evaluation, 292-300
for surfactant evaluation, 274-278
Soil removal, 45-46, 73-83, 248-264, 475-479
effect of agitation rate, 258-260
energetics of, 73-83
as function of water hardness, 475-479
as mass transfer process, 248-254
mechanical aspects of, 248-264
Soil removal evaluation, 335-364
correlation with subjective evaluation, 357-364
by light transmission, 357
by reflectance, 335-356
Soil removal from metals, 711-724
inorganic soils, 720-724
organic soils, 711-720
Soil removal of major brand detergents, 1013-1015
Soil residue measurements in dishwashing, 841-842
Soil retention, 195-205
effect of fatty soil on retention of particulate soil, 203-205
effect of fibers and finishes on, 201
effect of particle size, 197-200
effect of rate of agitation, 200
mechanical retention, 196-197
mechanisms of, 195-200
by sorptive forces, 195-196
Soil suspendability tests, 272-273
Solubility
of polyphosphates, 855-858
of whiteners, 760-762
Solubilization, 8, 141-143
Solvation, 830
Solvent cleaning of metals, 712-713
Specialty cleaning formulations, 17-27
Spectrofluorometers, 741
Spectrophotometers, 426-430
Spectroreflectometers, 797-798, 800
Spot test, 551-552
Spotting in dishwashing, 833-838
Stability constants, 470-473
Stability of sodium hypochlorite solutions, 521-522
Stabilization of dispersions, 173-178
by anionic surfactants, 173-174
by nonionic surfactants, 174-175
Staining of solid surfaces, 540-542
Stain removal, 519-608
Stains
acceptability of stained test fabric, 534-535
coffee, 535-536
dry soil, 536
tea, 535-536

Static foam meter, 628
Stiepel foam meter, 627-628
Storage of test substrates, 542-543
Street dirt, composition of, 55-56
Strength retention, 615-616
Sudsing tests, 12
Sulfur cycle, 926, 937-938
Surface active agent (see Surfactant)
Surface and interfacial tension, 635-645, 830-831
Surface tension, 7
  values of, 645-647
Surface tension measurements, 12, 636-645
  dynamic methods, 644-645
  static methods, 636-644
Surface wetting, 106-107
Surfactant adsorption, 162-173, 178-190
  by fibers, 178-190
  configuration of adsorbed molecules, 166-169
  isotherms, 169-171, 180-182
  mechanisms of, 162-166, 178-180
  thermodynamics of, 171-173
Surfactant changes in laundry detergents, 1006-1008
Surfactants, 9, 87-98, 205-217, 683-684, 923-924, 929-930
  acute oral values, 683-684
  effect on soil redeposition, 215-217
  in dishwashing, 831-832
  influence on particle adhesion, 87-89
  influence on potential energy, 89-98
  use in prevention of redeposition, 205-213
Suspendability of solids, 175-177
Suspending power of soaps, 274-275
Sweat, as a soil source, 34-37, 38-40
Synthetic fibers, 60-61
Synthetic soils (see Model soils)
Synthetic surfactants, 1006-1007

T

Tea- and coffee-stained cotton, 543-544
Tea stain removal, 545
Tear strength, 616
Temperature, 217-218, 466, 780-781
  effect on cotton affinity of whiteness, 780-781
  effect on sequestration, 466
  effect on soil redeposition, 217-218
  effect on soil removal, 217-218
Tensile strength, 614, 616
Terg-O-Tometer, 417-420
Tertiary wastewater treatment, 953-965
Test equipment, 413-449
Test methods, definition of terms, 11-14
Test methods for dishwasher evaluation, 845-848
Test pieces for dishwasher evaluation, 840
Tests for metal cleanliness, 724-727
Thermal energy in dishwashing, 828-829
Thin-layer chromatography, 743-745
Toxicity of detergents, 685-687
Toxicity of surfactants, 679-680
Toxicological methods, 685
Toxicology
  of whiteners, 805-806
  tests, 679-692
Toxic substances, 685-686
Trickling filters, 946-948
Trisodium phosphate, 868-869
Tritimulus whiteness measurements, 799

U

Ultrasonic cleaning, 260-261

Pages 1-452 comprise Part I
Pages 453-728 comprise Part II
Pages 729-1020 comprise Part III

V

Vacuum filtration, 971-972
van der Waals forces, 67, 70-73, 74-83, 245-246
Viewing conditions, 333
Visual evaluation
  of wash quality, 329-330
  of whiteness, 796-797

W

Wash action in dishwashing, 839
Washload composition, 757-760
Wastewater treatment, 940-979
Water cycle, 904-905
Water distribution in United States, 907
Water quality
  common measurements of, 908
Wet soiling, 202-204, 294, 298
Wettability of surfaces, 114-119
  effect of chemical nature on, 114-119
Wetting, 645-653
  agent, 7
[Wetting]
  of fabrics, 105-120
  tests, 11-12
Whitenening process, 753-783
Whitener-fiber interaction, 754-757
Whiteners
  as detergent additive, 749-785
  properties in solution, 760-765
Whitener/surfactant interaction, 766
Whiteness
  effect of substrate, 791-792
  evaluation, 786-804
  functions, 799-805
  of fabric, 784-786, 791-792
Wilhemy slide method, 641-642
World water supply, 906
Wrinkling, 621

Z

Zeolite, 1005
Zeta potential, 8, 184-190, 206, 245-246
  of cotton and carbon black in water, 206
  effect of surfactants on, 185-190
  of fibers, 185-190
  measurements of, 184